Unsettling Cities

John Allen is Senior Lecturer in Economic Geography at The Open University. His recent publications include *A Shrinking World?* (1995, co-edited with Chris Hamnett) and *Rethinking the Region* (1998, with Doreen Massey and Allan Cochrane).

Doreen Massey is Professor of Geography at The Open University. Her recent books include *Spatial Divisions of Labour* (2nd edition, 1995) and *Space, Place and Gender* (1994). She is co-founder and joint editor of *Soundings: a journal of politics and culture*.

Michael Pryke is Lecturer in the Faculty of Social Sciences at The Open University. He has published a number of articles on issues relating to finance and the City. In particular, his work examines the geographies of money and finance and the production of space.

UNDERSTANDING CITIES

This book is part of a series produced in association with The Open University. The complete list of books in the series is as follows:

City Worlds, edited by Doreen Massey, John Allen and Steve Pile

Unsettling Cities: Movement/Settlement, edited by John Allen, Doreen Massey and Michael Pryke

Unruly Cities? Order/Disorder, edited by Steve Pile, Christopher Brook and Gerry Mooney

The books form part of the Open University course DD304 *Understanding Cities*. Details of this and any other Open University course can be obtained from the Courses Reservations Centre, PO Box 724, The Open University, Milton Keynes, MK7 6ZS, United Kingdom: tel. (00 44) (0)1908 653231.

For availability of other course components, contact Open University Worldwide Ltd, The Berrill Building, Walton Hall, Milton Keynes, MK7 6AA, United Kingdom: tel. (00 44) (0)1908 858585, e-mail ouwenq@open.ac.uk.

Alternatively, much useful information can be obtained from the Open University's website http://www.open.ac.uk.

Unsettling Cities
Movement/Settlement

edited by
John Allen, Doreen Massey and Michael Pryke

London and New York

in association with

The Open University

First published 1999 by Routledge; written and produced by The Open University

11 New Fetter Lane, London EC4P 4EE

Simultaneously published in the USA and Canada
by Routledge
29 West 35th Street, New York, NY 10001

The opinions expressed are not necessarily those of the Course Team or of The Open University.

Edited, designed and typeset by The Open University

Index compiled by Isobel McLean

Printed in Great Britain by T.J. International, Padstow, Cornwall

British Library Cataloguing in Publication Data

A catalogue record for this book is available from The British Library

Library of Congress Cataloging in Publication Data

A catalogue record for this book has been requested

ISBN 0-415-20071-7 (hbk)
ISBN 0-415-20072-5 (pbk)

1.1

CONTENTS

THE OPEN UNIVERSITY COURSE TEAM

John Allen *Senior Lecturer in Economic Geography*

Sally Baker *Education and Social Sciences Librarian*

Melanie Bayley *Editor*

Andrew Blowers *Professor of Social Sciences (Planning)*

Christopher Brook *Lecturer in Geography*

Deborah Bywater *Project Controller*

David Calderwood *Project Controller*

Margaret Charters *Course Secretary*

Allan Cochrane *Professor of Public Policy*

Lene Connolly *Print Buying Controller*

Michael Dawson *Course Manager*

Margaret Dickens *Print Buying Co-ordinator*

Nigel Draper *Editor*

Janis Gilbert *Graphic Artist*

Celia Hart *Picture Research Assistant*

Caitlin Harvey *Course Manager*

Steve Hinchliffe *Lecturer in Geography*

Teresa Kennard *Co-publishing Advisor*

Siân Lewis *Graphic Designer*

Michèle Marsh *Secretary*

Doreen Massey *Professor of Geography*

Eugene McLaughlin *Senior Lecturer in Criminology and Social Policy*

Gerry Mooney *Staff Tutor in Social Policy*

Eleanor Morris *Series Producer, BBC/OUPC*

John Muncie *Senior Lecturer in Criminology and Social Policy*

Ray Munns *Cartographer*

Kathy Pain *Staff Tutor in Geography*

Steve Pile *Lecturer in Geography and Course Team Chair*

Michael Pryke *Lecturer in Geography*

Jenny Robinson *Lecturer in Geography*

Kathy Wilson *Production Assistant, BBC/OUPC*

External Assessor

John Solomos *Professor of Sociology, University of Southampton*

External Contributors

Ash Amin *Author, Professor of Geography, University of Durham*

Stephen Graham *Author, Reader in the Centre for Urban Technology, University of Newcastle upon Tyne*

Kerry Hamilton *Author, Professor of Transport, University of East London*

Mark Hart *Tutor Panel, Reader in Industrial and Regional Policy, University of Ulster*

Susan Hoyle *Author, Research Associate in the Transport Studies Unit, University of East London*

Linda McDowell *Author, Director for the Graduate School of Geography and Fellow at Newnham College, University of Cambridge*

Ian Munt *Tutor Panel, Researcher, London Rivers Association*

Phil Pinch *Tutor Panel, Senior Lecturer, Geography and Housing Division, South Bank University*

Jenny Seavers *Tutor Panel, Research Fellow, Centre for Housing Policy, University of York*

Nigel Thrift *Author, Professor of Geography, University of Bristol*

Sophie Watson *Author, Professor of Urban Cultures, University of East London*

Preface

Unsettling Cities is one of a series of books, entitled *Understanding Cities*, that takes a new look at cities. The standard approach to thinking about the future of cities is to consider them as free-standing and geographically discrete places that rise and fall as a result of their strategic location, their economic viability or their political power. Typically, the history of cities is charted *from* the rise and fall of ancient cities (such as Athens and Rome), *through* the rise of mediaeval cities (such as Antwerp and Naples), *onto* the spectacular growth of cities during the industrial revolution and the age of empire (such as London and Paris), and finally *to* the sprawling 'post-modern' cities of today (normally exemplified by Los Angeles). The future of cities is then extrapolated on the basis of this historical vision – usually that urban areas will continue to sprawl across the surface of the Earth, eventually joining up to form '100-mile' or 'mega-'cities. The analysis developed in this series is, however, radically different.

In order to understand cities, we argue that it is necessary to rethink their geography. This is more than simply extending the range of cities considered beyond the London–New York–Los Angeles axis, whether from São Paulo to Sydney, from Manchester to Moscow. It also involves using a geographical imagination to understand how cities are produced, on the one hand, in a context of social relations that stretch *beyond* the city and, on the other, by the intersection of social relations *within* the city. This argument has widespread implications for our understanding of cities. These implications are comprehensively investigated in the three books that comprise this series, *City Worlds, Unsettling Cities* and *Unruly Cities?*. Through the series, we tease out, for example, the ways in which cities bring people from different backgrounds into close proximity; how the juxtaposition of different people and activities in cities can change and alter social interactions; how these juxtapositions can result in, or result from, urban conflicts and tensions; how different parts of cities are connected to, or disconnected from, other cities; how people network within and between cities.

Developing these arguments shows that the city cannot be thought of as having one geography and one history (and therefore one future). Instead, cities are characterized by their openness: to new possibilities, and to new interactions between people. This book series gradually reveals the difficulties and paradoxes that the unavoidable openness of cities presents, as different histories and geographies intersect and overlap. Significantly, then, it is these issues that must be understood, if people are to learn to live in our increasingly urbanized world.

A few words about the books themselves. Each book is self-contained: each chapter provides all relevant supporting readings and materials to enable the reader to come to grips with a new understanding of cities. The three textbooks are, also, a significant component of the Open University third level course, DD304 *Understanding Cities*. The book series is joined by a television series, *City Stories,* and other materials, including audio-cassettes and course guides.

The television series picks up the themes of the three books through case-study material from across the globe – visiting Sydney, Singapore, Kuala Lumpur, Moscow and Mexico City. In particular, the television series develops a line of argument about the different ways in which groups within cities locate themselves within specific networks of social relationships. Local contributors comment on and analyse their own cities, highlighting the issues of importance for their urban futures. The television series is integral to the Open University course and provides a wealth of examples to illustrate the themes and the experiences of living in cities around the world. (Details of how to obtain copies of the TV programmes are available from Open University Worldwide Ltd.: see p.ii.)

Open University courses are produced through extensive and intensive discussions amongst academic authors, a panel of experienced tutors, an external assessor, editors, designers, BBC producers, a course manager and, not least, secretarial staff (listed on p.vi). Every component of the course has been subjected to wide-ranging discussion and critical debate. At every stage, this has led to improvements in the materials, which have benefited from the academic and distance-teaching expertise accumulated at The Open University.

The end-product of this process is to produce textbooks that have a specific style. In particular, authors have sought to make these texts *interactive*. Readers are asked to become involved in the problem of understanding cities, rather than simply to digest the material. Readers are provided, too, with a wide range of readings and extracts, which they can use to enhance their own appreciation of the problems that cities confront. Certain ideas become important to this understanding of cities – and these are registered in two basic ways. First, key points are emphasized using bullet points; and, second, core ideas, that are discussed in detail, are indexed in **bold**. Meanwhile, each book's last chapter looks back over the book as a whole and draws out the key themes that have been discussed. For readers who wish to use the series as a whole to further their understanding of cities, there are a number of references backwards and forwards to chapters in other books in the series; these are easily identifiable because they are printed in **bold** type. This is particularly important since chapters are intended to build on previous discussions. By integrating the material in this way, it is hoped that readers will be able to pull out the themes that run across the series for themselves. Through this interactive and integrated approach, it is hoped that readers will develop their own understanding of cities and ultimately be able to transfer that understanding to other cities, even to other situations.

It only remains to acknowledge those who have worked hard and with such enthusiasm to produce this book series. Most obviously, these texts – and the course of which they are a part – would not be possible without the academic consultants and the tutor panel. All have helped to shape the chapters of others, while also responding constructively to suggestions and advice. The external assessor, John Solomos, provided invaluable intellectual guidance, insights, support and encouragement at every stage of the development of the books and the related course materials. The tutor panel, comprising Mark Hart, Ian Munt and Jenny Seavers, have been rigorous in their comments on the drafts as the chapters developed. Finally, Phil Pinch, who wrote the Course Guide, generously advised on the development of the course as a whole.

Producing an excellent textbook series does not end with the writing. The production process has been co-ordinated and kept on track by Deborah Bywater and David Calderwood. We have been shrewdly supported by our editors, Melanie Bayley and Nigel Draper, who scrupulously scrutinised the chapters before publication. Meanwhile, the design of the books and chapters has been thoughtfully overseen by Siân Lewis. Our cartographer and graphic artist, Ray Munns and Janis Gilbert, developed and drew the maps and diagrams. Together, they have turned the drafts into the excellent books they undoubtedly are. Finding illustrations for this course has proved, at times, a frustrating task, but Sally Baker and Celia Hart have been good humoured throughout. The production of textbooks at The Open University means a never-ending stream of requests for word-processing and the distribution of drafts, often at very short notice. With good grace and efficiency, the course team has been supported by Michèle Marsh and, in particular, the course secretary, Margaret Charters. Within the Faculty of Social Sciences, Peggotty Graham has helped to ensure that the books were produced to high standards, while – at Routledge – Sarah Lloyd has been a constant source of good advice, flexibility and encouragement.

At the centre of this process are the course managers who have worked on this series of books. We have been lucky enough to have gained from the expertise of Christina Janoszka and Mike Dawson. However, it is Caitlin Harvey who has seen the books through the most important stages. Caitlin has never failed to keep her head, even while all about her were losing theirs. Keeping everything together has meant that course team meetings have been intellectually lively and thoroughly thought-provoking. We hope that the vitality and diversity of these debates shine through in these textbooks, which offer a new way of *Understanding Cities*.

Steve Pile
on behalf of the Open University Course Team

Introduction

In one sense, we are over-informed about cities. There is an abundance of studies which dwell on the historical span and fortunes of cites, and numerous sociological monographs which capture the detail of particular cities. Likewise, there is an array of engaging theoretical approaches to city life, as there are informative statistical packages on this or that economic or political dimension to cities. But if we know a lot about cities, there is always the possibility of a different understanding, a different approach, to what may at first sight appear relatively familiar or well-known. In this book, it is the stress upon the openness of cities, their place within wider networks of connections, that forms the basis of our approach. In particular, we develop the view that cities are essentially open, mobile, mixed places. If cities now are where most people happen to be, then that is above all because cities are open and global in character. They represent what many societies have become and what others have long been: sites at which a multitude of social relationships and ties intersect, giving both a sense of their worldly nature, the different times and mixes they embody, and a sense of the resultant intensity and diversity.

To think about cities as open in this way is perhaps to unsettle any preconceptions that we may have about cities as, first and foremost, bounded, self-contained locations fixed in their geography. Opening up the geography of cities, we would argue, avoids reducing cities to the solid boundary lines often drawn around them and draws attention to their permeable nature. Among other things, such thinking draws attention to the mobile, fluid, interconnected character of city life. Indeed, the title of this book, *Unsettling Cities,* is aimed at precisely this recognition. It is intended to convey something of the material fact that cities are shaped as much by events and processes that happen elsewhere as they are by what happens within them. Thus, the very make-up of cities and their development can only be understood in a global, historical context. Yet if this has long been the case, the consequences of this open geography have yet to be fully absorbed.

There are a number of aspects to this observation which, in a variety of ways, run through the chapters in this collection and indeed help to shape their basic stance and the questions they pose.

Principally, there is a concern across the chapters to show how the extensive networks of relations, the cross-cutting connections between cities, lie behind much of what we understand as the rich diversity and heterogeneity of urban life. Such a mixture of relations, often held in tension, has largely come about through the myriad flows and connections – of people, cultures, ideas and practices – combining and settling in very specific ways. To take but one example, what marks out a city like São Paulo today, with its distinctive mix of defensive communities and difference, of dangerous spaces and carnivalesque spaces, is the peculiar combination of what has largely come from elsewhere: from the different times and spaces of its European past, its American connections, its pivotal role in Brazil's coffee economy, its attraction as a magnet for the country's northern immigrants, and so forth. As with any city, its singular history is transformed into many histories when read through the terms of its equally diverse geographies.

A further consequence of the openness of cities is that they are places where something is always going on. The dynamism of cities, the economic and cultural vitality that is often associated with the dense clustering of peoples, cultures and activities, is in many ways a testament to what comes together and blends within particular city spaces. In that sense, cities are the points of intersection, where the flows of money, information, ideas and communications meet up, often to produce something entirely new in form and fashion. Of course, not all parts of a city experience this intensity, nor does the experience of cultural contact necessarily lend itself to an atmosphere of mutual respect and tolerance for others. In the big cities of the world – London, Los Angeles, Shanghai, Tokyo, Moscow, Manila, Mexico City and so on – the experience of 'cityness' may manifest itself through isolation and anxiety as much as it does through creativity and liveliness. Yet – and this is a point that reveals itself throughout the book – such a juxtaposition of feelings may itself arise through the combination of relationships and practices drawn to the city from beyond its permeable limits.

In addition to this, thinking through the open, interconnected nature of cities also alerts us to their co-dependent quality. Development and change within cities in one part of the globe may have a direct impact on the economic and social fortunes of cities elsewhere. Historically, this is perhaps best known through the relationships between imperial cities and their colonial counterparts. But equally, in contemporary relationships of trade, commerce, finance and power, or through more subtle forms of coercion and constraint, the dominant cities in their field of influence may check the progress of others, undermine their living standards, and effectively constrain their livelihoods. In a globalized world not all interconnections produce outcomes of this kind, but they do nonetheless draw attention to the shifting dependencies and inequalities between cities, many of which arise specifically from the openness of their networks and connections.

◆　　◆　　◆　　◆

If one of the aims of this book is to open up the geography of cities through their interconnections, another of equal importance is to stress the fluid, mobile character of such relationships and their fixed, settled nature. The two sides – movement and settlement – have been a familiar theme in urban literature, especially through the influential work of Lewis Mumford, and in particular his book *The City in History*, published in the early 1960s. However, where Mumford tended to view city life as the result of swings between the two poles, here we consider it best to view them as in tension. Rather than contrast the two as separate elements, it is perhaps more useful to think of the mobility of city life and its fixed quality as something that is part and parcel of the experience of urban living. For many a city dweller, it is something that is negotiated on a regular, ongoing basis, not something that is there to be resolved or smoothed out. As another urban theorist, George Simmel, voiced it, the tension between movement and fixity represents one way of reading the city and a means through which its complexity may be comprehended.

This formulation is a little cryptic, so it is perhaps useful to elaborate on what is really at issue here. Cities are often considered to be rather physical, settled places. After all, they do mostly comprise a dense concentration of structures, buildings, roads and infrastructures of varying pattern and durability. In addition to bricks and mortar and strips of tarmac, many cities also display a built form of steel and glass made over in corporate style. More than that, cities are places where people put down roots, make a living on a regular basis, and establish themselves loosely or otherwise in a neighbourhood complex. Clearly, in all of this there is a strong physicality to the many forms of settlement and stability.

Equally – and without the slightest hint of contradiction – cities are often expressed, as we have seen, in terms of movement, fluidity and motion. The constant flows of traffic and people personify the city, even caricature it. Rapid transport systems and fast communications intensify the impression. What is more, information, goods, travellers and migrants appear to move endlessly to and from cities, with the latter, according to Manuel Castells in his recent writings on the information age, acting as nodes or hubs in some kind of 'space of flows'. There is a real sense here in which the qualities of pace and motion hold the key to grasping the nature of city life.

On the one hand, then, as Mumford might have said, we have settlement and stability and, on the other, there is movement and mobility. But if these are considered to be two poles of urban existence, they are more apparent than real. In truth, as we have intimated, both are part of the same rich texture of city life, woven together in inextricable ways. To repeat the point: much of the world and its peoples flow through cities, and as they do so they mix and combine in ways which, more or less, settle in some shape or form. And so the process goes on, with cities embodying motion and fixity as part of their pulse and rhythm, although not always in harmony and not necessarily without friction. In other words, they are also a source of urban tension and ambiguity.

This tension between movement and settlement manifests itself in any number of ways within cities, with varying consequences for those involved. In the chapters that follow we consider some of the more familiar ways in which the mobility of city life and its fixed spaces come into conflict and exacerbate urban tension. There are at least three that deserve recognition.

The first, which appears in the earlier part of the book, concerns the movement of people to the city and their settlement in particular locations. The juxtaposition of different cultures, races, ages and classes in urban settings, often in close contact and proximity, prompts the question of how difference is negotiated in the city. The physical fact of settled groups and communities placed close together in cities may conjure up a meeting of differences that can lead just as easily to conflict and intolerance as it does to respect and mutual recognition. The same can be said for the comings and goings of different groups within shared public spaces, where the challenge of difference may lead to the exclusion of minorities for fear of violence, crime or racial intolerance. As we shall see, such questions of urban justice can extend to other forms of exclusion when the differences of nature – animals, birds or plants deemed to be 'out of place' – are encountered in the urban spaces of the city.

A second way in which the motion of the city and its durable elements may generate friction is simply through the pace and intensity of much city life. On the one hand, there is the rapid movement of people, information, ideas and materials that is made possible by the new transport and communications technologies. On the other hand, there is the attempt to stabilize meaning and to fix social co-ordinates in a world where so much is going on. The tension between face-to-face contacts and relationships mediated by the new communications technologies at a distance is one such illustration. Another is the attempt by the powerful and the not so powerful cities of the globe to fix the flows of knowledge, information and 'creative' resources in order to enhance their relative degree of global control and influence – if not as 'smart cities', then at least as major nodes in the network of flows.

Finally, there is the potential of the different rhythms of the city to clash in ways which may destabilize urban life-chances and life-styles. Motion and movement within and beyond the city beats with a varied pace and tempo. Whether it is the regular comings and goings of different groups at different times of the day or the successive waves of power and influence which reach out across the globe in an attempt to fix the fortunes of distant cities, the varied rhythms hold the potential to align, block or smother the movement and actions of others. Regular and repeatable movements may take on a settled appearance over time, as when the transport flows of individuals appear to converge at peak times within cities. Or the dominance of a regular beat in the world's richest cities, say in relation to the powerful global economic flows of money and private finance, may unsettle existing patterns of everyday life in the more vulnerable cities of the developing world. The economic collapse of certain East Asian cities in the late 1990s is perhaps a case in point, when Bangkok, Kuala Lumpur and Jakarta all found themselves enmeshed on unequal terms in networks of neo-liberal influence controlled through the powerful cities of the West.

When one turns to global interconnections of this magnitude, the significance of movement and settlement as a prism through which cities may be grasped becomes, if anything, more pointed. To be part of the flows of power, money and finance is one thing, even if it is on an unequal basis, but for a major city to be bypassed or, worse, disconnected from the dominant global networks, may prove to be far more eventful. If, by design, cities in the developing world opt for a strategy which creates the possibility for them to stabilize the flows of trade and finance around them, perhaps as an entrepôt or hub in the first instance, their position nonetheless remains vulnerable within the network of connections. If, however, such cities are excluded or actively withdraw from the dominant economic networks, the effect may be to set back economic growth and to further polarize the fortunes between rich and poor cities across the globe. Whatever misgivings governments may have about the current neo-liberal dominance of world markets, if they are on the inside of such networks they may still have the possibility to negotiate them on some basis. Such dilemmas are real for cities in the developing world.

Equally, however, disconnection from such networks is not an entirely negative outcome, for it may lead to the development of other opportunities for cities, other types of connections, other relationships which do not hold the 'western' model in any particular esteem. Disconnection, after all, is not the same thing as closure, and the possibility is always there for other openings to be negotiated or constructed. There is not simply one network of cities 'out there', but rather various networks with different paths of development, many of which are centred outside of the 'West'. The same may be true for excluded groups within the powerful cities of the western world who may find themselves outside of the markets, yet are open to devise alternative networks of support and livelihood. In the latter part of the book, it is to these and other broad, global questions that the analysis turns and which, in a general sense, represents the outlines of a global political economy of cities.

❖ ❖ ❖ ❖

The emphasis in this book on cities as open, fluid, interconnected places develops a number of the insights explored in the first book of this series, *City Worlds*. In particular, one of the objectives of this second book is to reveal the ways in which connections within and between cities are more than simply lines drawn on a map, more than surface relationships which link one place to the next or piece together the different districts of a city. The stress upon the motion, rhythm, stability and tension of city relationships is intentional, as is the emphasis upon the intensity of city living which stems from their proximity and juxtaposition. Considered together, they are intended to bring the city to life – to animate it, so to speak.

Interestingly, then, it is the geography of cities, their very spatiality, which arguably helps to provide this animation; it is the absence over time of a singular logic or trajectory to cities which pushes to the fore the many different stories, the many diverse futures, which meet up and coexist within cities. That they are, in part, the product of the geographical openness of cities, both past and present, is perhaps the most significant message of this book.

John Allen, Doreen Massey and Michael Pryke

CHAPTER 1

Cities of connection and disconnection

by Ash Amin and Stephen Graham

1 *Introduction*

When you think about it, cities have always been places of mixture and diversity. In the nineteenth century, European industrial cities such as Lille, Manchester and Düsseldorf combined working-class squalor with aristocratic splendour and industrial enterprise. Similarly, many cities in the developing world – Bombay, Lima, Mombasa – have always been crossroads of tradition and change, and places of encounter between locals, outsiders and migrants.

As a consequence of the mixed nature of cities, urban life has always been full of tensions. Cities are as much symbols of wealth creation, domination and opportunity as they are of pollution, poverty and struggle for survival. Throughout urban history, such mixture and tension have been closely related with the wide variety of connections which run through cities and link them with the outside world. So, for example, the elite spaces of late-nineteenth-century Manchester and London were partly sustained by the privileged access that their inhabitants, with global economic connections, had to a wide range of then new 'high-tech' services such as electricity, telephones and piped water. In the slum areas of the same cities, however, workers and the poor had to draw on largely local resources and face the disease and low life expectancy that came from an absence of clean water and sanitation. Yet the ambivalence is that these two social worlds were intrinsically locked together within the booming industrial economies of these cities, as exploiter and exploited in the mills and factories, and in their uneasy mingling in the streets and other public spaces.

What, if anything, makes cities of today different? Clearly, cities remain full of social tensions. But perhaps one key difference pointed out elsewhere in the series (**Massey, Allen and Pile, 1999**) is that many contemporary tensions are the product of a greater *intensity* of mixture resulting from the increasing exposure of cities to new global influences.

It is now often argued that the world has in a sense 'shrunk', owing to the rise of two interlinked global processes. The first is the rise of chains of activity that are worldwide in scope, such that *social relations and institutions have become stretched across space*. For instance, transnational firms now have global production networks, there is now a global financial system separated from national money markets and controls, and global cultural values and habits are emerging under the influence of large consumer or media organizations such as Coca Cola, News International and IBM.

Major cities at least have become key dynamic staging points in these global chains of activity (see Borja and Castells, 1997). They are the sites of ceaseless flows of people, money, commodities, ideas, information and cultural influences. They are home to the institutions associated with these flows and, as such, also centres of influence within the chains. This global linkage ≠s increased the heterogeneity of cities by adding a new layer of influences upon older connections between the city and the countryside, the nation, and the rest of the world.

The second related global process is the *intensification of contact between places*, now that day-to-day activities are increasingly influenced by events in far-off places owing to the rise of global transport and communications technologies. Thus a local economic downturn in one region or city of the world may force new migrants and refugees to cities on the other side of the world, while investment decisions in the Far East create or destroy jobs in locations within distant continents.

It could be argued that these two global processes – the stretching of relationships and the intensifying connection between places – are currently involved in shaping virtually all cities of the world. It is clear that traditionally open international cities such as London, Cairo or Bombay (Mumbai) have long been 'gateways' to the world – centres bringing together remarkable mixtures of diverse industries, social groups, and flows of people, goods and information. But now even in more 'provincial' cities, such as Bristol, Bordeaux and Bamako, local traditions are disturbed by cosmopolitan cultures that are introduced by travel, migration, television and global consumerism. Here, too, contrasts and contradictions are an intrinsic part of city life.

Most cities, then, are becoming increasingly permeable – places of multiple and changing connections. Cities juxtapose cultures, people and flows within more or less concentrated material spaces through which they become proximate. Because of this it is not possible – if indeed it ever was – to ascribe to any city a singular purpose or fixed coherence. Dig deep enough and you will find diverse social worlds even in cities which at first sight seem apparently homogeneous – university cities such as Stanford, cities of industry and design such as Bologna or Bangalore, cities of government such as Brasilia or Canberra, and cities of high consumption such as Nice.

Our purpose in this first chapter is to consider the implications of this sense of cities as places which bring together and superimpose diverse connections and disconnections. In section 2 we lay the foundations for a perspective which does this. Following on from, and extending the argument of, Book 1 in this series, we call this a 'relational' perspective because it emphasizes the idea of cities as places of intersection between many webs of social, cultural and technological flow, and the superimposition of these relational webs on the physical spaces of cities.

In section 3 we use this relational perspective to explore, through two examples, some of the *ambivalent tensions* that are produced by this 'bringing together' of many relational worlds within cities. We use this coupling deliberately, to capture a sense of urban life riddled with conflicting demands (tensions), which, however, are not expressed in any straightforward manner (they are ambivalent). The first example focuses on the tensions produced by social and cultural mixture, to contrast situations in which public spaces are shared between different communities with those in which mixture results in segregation and intolerance. The second example examines electronically based communications technologies such as the Internet and satellite television, to explore whether the intensification of electronically mediated distant links between cities represents the destruction of place-based links, or simply their redefinition.

Out of these examples we try, in section 4, to pull out some key aspects of contemporary cities as places. Three strike us as being particularly significant in distinguishing what differentiates cities from other places. These are:

● cities as sites of proximity and co-presence;

● cities as a mix of space/times;

● cities as meeting places.

In the concluding section we speculate on whether contemporary cities, with all their connections and disconnections, 'hang' together in any way as wholes. We ask if there is a fit these days between the physical spaces of cities or between their various social worlds.

2 *Thinking 'relationally' about cities*

Let us start by considering three examples which illustrate the complex ways in which relations of proximity and distance interweave within and between contemporary cities.

2.1 TELEPHONE CALL CENTRES AND BRITISH AIRWAYS' 'VIRTUAL SINGLE OFFICE'

Towards the end of the twentieth century there has been remarkable growth of a whole range of so-called call centres in the richer countries of the North. These deliver services direct to consumers, over the telephone, in whole regions, nations or even continents from one or a few large office complexes. Over 200,000 call centre jobs were created in the UK in the late 1990s in industries such as airline reservations, insurance, banking, information services, computing and a host of other customer services. Call centres effectively remove the traditional geographical ties between the delivery and consumption of these services at the local level within cities. Instead of face-to-face delivery, services are delivered over sophisticated, private telecommunications grids which link together different call centres with a very high degree of capability. So, a person ringing their 'local' electricity company in, say, London will in fact be answered several hundred miles away, in Sunderland in the north east of England.

The case of British Airways' so-called 'virtual single office' (shown schematically in Figure 1.1) demonstrates how call centres bring cities into highly integrated systems of flow. Through an integrated corporate information technology network, which actually straddles the whole world, British Airways has set up a new strategy for dealing with the millions of calls it receives from consumers and travel agents about flight availability and reservations. As the figure shows, these technologies integrate a range of call centre offices that have been

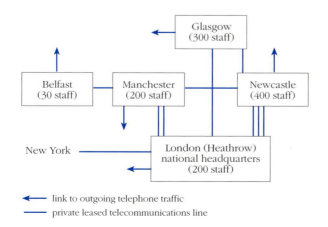

link to outgoing telephone traffic

private leased telecommunications line

FIGURE 1.1
An example of the interconnection of cities by instant systems of electronic communication: British Airways' 'virtual single office'
Source: adapted from Richardson, 1994, p.325

dispersed across the peripheral cities of the UK. Whilst such work could conceivably be located in rural areas, cities dominate call centre work because of their large and diverse labour markets, and property, infrastructure and telecommunications advantages. The system automatically routes calls from all parts of the UK to these various centres based on their opening hours and availability for business. From the perspective of cities, what this example illustrates is how different parts of cities can now be integrated into wider networks of organization and communication with a new degree of intensity and immediacy.

2.2   THE URBAN LANDSCAPE OF NEW YORK: THE LOWER EAST SIDE AND MANHATTAN

Look at the picture of New York in Figure 1.2. In what sense do you think it illustrates the ways in which the physical spaces of a city might or might not relate to each other? To us, it is a good example of how spaces which are physically close within a city can simultaneously be relationally very distant and relatively disconnected.

In the background of the picture are the twin towers of the World Trade Centre and the other skyscrapers of New York's financial centre. Located right at the tip of Manhattan, these buildings house financial services companies which organize global financial transactions of trillions of dollars a day, with a truly global 'reach' using electronic financial trading and, again, highly advanced telecommunications infrastructures. It is said that *one building* on Wall Street

supports global financial transactions on electronic systems of over $1 trillion per day. The top professionals who work in the financial centre are extremely well paid (very often well over $500,000 per year), and tend to live in highly expensive, exclusive housing areas in mid-town Manhattan or upstate New York, beyond the city's built-up area. As well as their very high degree of access to transport and telecommunications infrastructures for work, leisure, tourism and holidays, such groups can also command privileged access to luxury goods, services and commodities from across the planet (food, wine, fashion, furniture, art, etc.). Moreover, their social and friendship networks are likely to include very close ties with similar professionals working in international financial centres all over the world (London, Paris, Hong Kong, Tokyo, etc.).

In the foreground you can see part of the Lower East Side of Manhattan, only a few miles 'as the crow flies' from the

FIGURE 1.2   *Disconnection within the city: Manhattan's financial district rising beyond almost abandoned tenements on New York's Lower East Side*

World Trade Centre. Characterized usually as a 'ghetto', the Lower East Side, which is overwhelmingly populated by African Americans, suffers the characteristic litany of extreme social problems in such places. Levels of poverty, unemployment and poor health are very high, a product of thirty years of collapse in the manual and manufacturing jobs which used to sustain it. As a result, the physical fabric of the area has suffered high rates of abandonment, arson and disinvestment, resulting in the depressing landscape shown in the photograph of burnt-out, derelict tenements separated by fields of rubble. Remaining residents in the area face severe stigma from the rest of the city when trying to find work. Rates of dangerous crime are relatively high, as is incidence of mental illness, drug abuse and alcoholism. Few local shops or banks remain, and insurance and financial services companies tend to exclude local residents from enjoying access to insurance services or bank accounts. In effect, the area has become 'functionally disconnected' from the wider city and its economic opportunities.

Contrasting the marginality of the Lower East Side with the 'super-included' world of workers in Manhattan's financial district and their intense connections to equivalent districts in London, Tokyo or Hong Kong, helps to demonstrate how, as Mulgan (1991, p.3) has put it, 'the centres of two cities are often for practical purposes closer to each other than to their own peripheries'.

2.3 A HIGH-CLASS RESTAURANT

Consider the photograph in Figure 1.3, which shows a high-class restaurant in the central business district of Java, Indonesia. As well as synthesizing local and global cultures, top-class restaurants literally provide a physical interface between the 'relational worlds' of at least two different social groups (Zukin, 1995). First, there are the customers eating expensive food from around the world – often the sorts of affluent executives and business travellers on expense-account lunches that work in places like Manhattan's financial centre.

FIGURE 1.3
A high-class restaurant in Java

As we have seen, their relational worlds frequently extend to cover international business networks, services, travel and tourism, global telecommunications and expensive residential areas. Second, there are the workers – waiters and cooks, usually in low-paid, insecure jobs filled by minority immigrants from across the world or the deep countryside – whose relational worlds might link them into far-flung family and cultural networks.

So even such an apparently simple space as a restaurant reveals the ambivalent tensions surrounding physical proximity between different social, economic or cultural groups in cities. On the one hand, these two groups have fleeting ties of local interdependence. Each is likely to have very different, but probably equally strong, networks connecting them to distant lands, places, cities and cultures. On the other hand, viewed from the perspective of cities as having key roles as 'staging posts' for a highly interconnected global culture and economy, spaces like the restaurant which bring together different social worlds, albeit fleetingly, have become central to the life and rhythms of many large cities as they service international flows of tourists and business people.

ACTIVITY 1.1   Look again at the relational diagram of British Airways' virtual office in Figure 1.1. Note how selective it is: it only shows how the offices are interconnected via telecommunications links and does not reveal how these offices relate to the cities where they are located. Only one aspect of the complex 'relational worlds' which surround the offices is shown. Nothing is shown about how the offices relate to their workers, local property markets, servicing organizations, security firms, and so forth. Try to draw a similar but more comprehensive diagram of the 'relational worlds' that you think might exist in the example of the high-class restaurant.

Start with a square that represents the restaurant. Then draw and label gridlines which represent connections between the workers and the customers and those that each has within and beyond the city. Think also of the flows of money, food, goods, water, energy and services which sustain the restaurant's operation. ◆

2.4   THE MERITS OF A 'RELATIONAL' PERSPECTIVE

The three examples we have introduced above help to demonstrate some of the merits of thinking *relationally* about the connections and disconnections within and between cities. Thinking relationally allows us to explore the interplay between cities as places and cities as nodes within diverse webs of social, economic and cultural relations. In our view, such a perspective on cities is useful for at least three reasons.

First, note how we might have imagined that there is a clear and simple *hierarchical progression* of geographical scales from the first to the third example: from the 'national' level of the UK system of links *between* cities, through the 'urban' level of links *within* the city of New York, to the very 'local' setting of links *within the single building* in the case of the restaurant. It is tempting to see these levels as if they were separate and opposing spheres, each with an independent influence on cities. For example, the British Airways case would be characterized as an example of a set of external forces which impact upon a set of local spaces – i.e. cities – in which the company chooses to locate its call centres. Conversely, the restaurant would be seen as a service for local consumers, and therefore not in any way connected with wider geographical spaces.

Clearly, though, neither of these interpretations begins to capture the complex *relational geographies* that the two cases represent. They fail to support an understanding of the ways in which these geographies necessarily reach across the different geographical scales – the 'local' district in a city, the 'region', the 'nation', and so on. More pointedly,

- a relational perspective allows complex global-national-regional-local interconnections to be placed at the centre of analysis of cities without seeing them as a simple hierarchical progression.

It allows us to hold on to the idea that relational webs, articulated across the different spatial scales, are all involved in the shaping of much contemporary urban life.

Second, a relational perspective allows us to begin to appreciate the *multiple geographies of cities and their various parts.* Going back to our New York example,

- the relational perspective shows the physical proximity but the relational distance between Lower East Side residents and the employees of the World Trade Centre.

This helps us to get away from the simple and still widespread assumption that spatial proximity within cities necessarily produces 'pure' defining relations, whilst distant links that are mediated through telecommunications or transport networks across distance between cities and other places are somehow less 'authentic'.

Third, and perhaps most significantly,

- a relational perspective helps us to appreciate *the fundamentally ambivalent tensions* of cities and the sites within them, precisely because of its understanding of cities as places in which diverse relational webs do or do not connect.

As we saw with the example of the restaurant, these tensions arise from the way in which cities superimpose many relational worlds in more or less concentrated ways within their spaces.

3 *Exploring ambivalent tensions within cities*

What does it mean when we say that contemporary cities are ambivalent places? In this section we provide two illustrative examples which reveal the processes at work and also the conflicting interpretations to which they lend themselves.

The first example focuses on the identities of urban public spaces, which are increasingly assailed by cultures and peoples from around the world. It explores the varied reactions to social and cultural heterogeneity in cities by contrasting strategies designed to *seal off* spaces from a perceived external threat with more open and tolerant, although still ambivalent, public spaces. The second example focuses on the relationships between new information and communications technologies – in other words advanced telecommunications and computer networks like the Internet and satellite television – and the spatial connections of the city, both internally and with the 'outside' world. It illustrates the ever-sharpening tension in urban life between place-based relationships which rely on face-to-face contact and those which are mediated by such technologies over distance.

3.1 PUBLIC SPACES AND PRIVATE ENCLAVES

Our first example illustrates the ambivalences and tensions associated with the use of once 'public' space in cities. Streets, parks, squares, shopping areas, cafés and restaurants are often places of connection where different relational webs meet and overlap – mixtures of migrants, itinerants and locals, as well as different social and ethnic cultures. But the tensions associated with this juxtaposition of difference, perceived or real (such as the fear of crime or violence, racial intolerance, uncertainty and insecurity) often put into question the very definition and usage of the phrase 'urban public space'. The once common understanding of public space as a shared space or arena for social interaction can no longer be taken for granted. For example, urban commentators who see cities with high rates of violent crime such as Los Angeles as paradigmatic examples of the city of the future (for example Davis, 1990; Soja, 1989) argue that public spaces are being re-engineered as places of surveillance from which 'threatening' groups are excluded. However, we can think of many counter examples which show that access to public spaces continues to remain relatively open. These examples help us to sustain the less nightmarish interpretation of public space that we are used to, where some degree of mutual confidence, trust and interdependence helps to balance the tensions between the many groups and identities represented in the space.

Cities thus vary greatly in how they reflect tensions, from the retreat of social groups into enclaves to the more or less tolerant mixing of different social groups in 'public spaces'. In this section we will examine some examples by looking at the enclave and 'fortressing' trends in cities such as Houston and Los Angeles in the USA and São Paulo in Brazil, and the mixing of different social groups in Amsterdam and Calcutta. Through these examples we try to illustrate a relational perspective to bring out how these ambivalent tensions relate to the connections and disconnections that exist in our case-study cities.

3.1.1   The sealed-off gated city

ACTIVITY 1.2   Read Extract 1.1 below by David Dillon, the architecture critic for the *Dallas Morning News*. It typifies the view that the contemporary US city is being transformed into a series of enclave spaces physically and relationally sealed off by gates and walls. Dillon implies that the trend towards living with your own kind in closely guarded walled communities could soon become the norm in the USA (for those who can afford it). Certainly, the latest figures attest to the magnitude of the trend towards gated communities: there were 20,000 gated communities containing three million houses in the USA in 1997. ◆

EXTRACT 1.1
David Dillon: 'Fortress America'

One of the most familiar sounds in the US these days is the clanging gate. Not the garden gate, or the alley gate, but the gate that closes off the street, the block, and increasingly the entire neighbourhood. An estimated one-third of all communities in Southern California are now gated …

Terrified of crime and worried about property values, Americans are flocking to gated enclaves in what experts call a fundamental reorganization of community life …

Walls are only the beginning. Inside may be surveillance cameras, infra-red sensors, motion detectors, and sometimes armed guards. St Andrews, a gated community in Boca Raton, Florida, spends over a $1 million a year on helicopters and canine patrols. Hidden Valley, a private community north of Los Angeles, installed anti-terrorist bollards two years ago to keep non-residents at bay … the bollards rise up to impale vehicles that try to defy them. The tally so far: 25 cars and four trucks.

Ironically, the rush to gated communities coincides with widely reported decreases in violent crime statistics … Developers of gated communities exploit … anxiety by marketing their projects as safer, friendlier, and more economically stable than traditional urban, or suburban, neighbourhoods. Their ads and brochures are sprinkled with words like 'village', 'community', and 'cosy' to suggest a friendliness and manageable scale that is supposedly missing outside …

Gated communities are part of a broader privatization movement, which in turn is linked to a growing scepticism about government's ability to police

FIGURE 1.4   *A gated community in the USA*

streets, stabilize neighbourhoods and property values, and generally look after the public realm … People are responding to taking matters in their own hands … private security guards now outnumber public police three to one …

This desire for control could have dangerous consequences for American cities – and for the world, says urban critic Jane Jacobs … Jacobs sees gated communities as a new brand of urban tribalism that will pit races and ethnic groups against each other. 'It's a gang way of looking at life, the institutionalization of turf. And if it goes on indefinitely, and gets intensified, it practically means the end of civilization.'

Source: Dillon, 1994, pp.8–12

Many other urban commentators are quick to stress that Dillon's portrayal is not confined to US cities, nor to the more affluent North, but in fact captures a phenomenon which is also at work in many cities of the South (see **Allen, 1999**).

ACTIVITY 1.3   Read Extract 1.2 opposite, in which Teresa Caldeira describes a similar process of privatization and social enclosure in the Brazilian 'mega-city' of São Paulo in the 1990s. ◆

EXTRACT 1.2
Teresa Caldeira: 'Proximity and high walls: São Paulo in the 1990s'

São Paulo in the 1990s is a more diverse and fragmented city than it was in the 1970s … there is a tendency among higher-income families to go to the outskirts of the city and/or to create isolated enclaves for themselves … Since the exodus and/or isolation of the rich is not complete, and since various areas to which they move are traditionally low-income neighbourhoods, in the last decade people from different social classes have lived closer to each other than they have since the 1940s. In a context of economic crisis, uncertainty and anxiety about social decay … and fear of crime, this proximity of different social classes … provokes the emergence of new forms of discrimination …

Morumbi is the neighbourhood which best symbolizes the new pattern of urban development taking shape in the city …

FIGURE 1.5

Luxury enclaves, surrounded by walls, gates and electric fences, rising up from the surrounding favela *shanty towns in the Morumbi district of São Paulo*

From 1980 to 1987, 217 buildings were constructed in Morumbi, corresponding to 4,972 units, mostly luxury … Most of them are residential complexes … They offer the amenities of a club, are always walled, have as one of their basic features the use of the most sophisticated security technology, with the continual presence of private guards. Moreover … they have exotic features such as one swimming pool per individual apartment, three maid's bedrooms, waiting rooms for drivers in the basement, special rooms for storing crystals, and so on. All this luxury contrasts with the views from the apartments' windows: the thousands of shacks of the *favelas* on the other side of the high walls which supply the domestic servants for the condominiums nearby.

… for the residents of the new enclosures, the inconvenience seems to be more than outweighed by the feeling of security they gain behind the walls, being exclusively among their equals and far from what they consider to be the city's dangers.

Source: Caldeira, 1996, pp.59–63

Look at the photo in Figure 1.5, which shows the new Morumbi enclaves rising above the roofs of the surrounding shanty town (*favela*) in the foreground. This example shows that while the new patterns of social segregation in the cities of the South are spearheaded by the rich, and are perhaps less of an encroachment on existing public spaces, the motives, instruments and effects of 'gating' are similar to the US example.

Dillon and Caldeira present bleak portrayals of the 'end of civilization', as people retreat from public interchange into the alleged comfort and safety of homogeneous communities, living in fear and mistrust of other 'tribes' beyond the walls. Both suggest that enclave communities are being developed which maximize internal connections within similar socio-economic and cultural groups whilst severing the connections that might force these groups to mix with other social, ethnic or income groups in the city. According to Dillon and Caldeira, this is a development that breaks decisively with the idea of the 'urban' as a shared space, as both US and Brazilian cities physically spread out over huge 'megalopolitan' areas, and as face-to-face interactions in guarded, private enclaves are supported by the growing use of private cars and phone connections from behind the walls of fortressed communities.

But is the story of gated communities really this simple? In reality, they too are ridden with ambivalent tensions. As Dillon implies, much of the retreat is orchestrated by anxieties and fears which are not supported by the crime figures. The irony, then, is that the act of retreat from the 'barbarians' left to roam the few 'public' spaces that remain will only intensify social antagonism by breeding mutual fear, ignorance and tension. What has been called the 'forting up' of US suburbs might become a self-fulfilling prophecy, with a spiral of fear of crime and other social problems prompting an ever-stronger urge by the middle classes to withdraw from the 'city' behind fences, walls and guards. To add to the ambivalence, however, even the most apparently 'homogeneous' of gated communities is full of internal contradictions and impulses which challenge their seclusion (for example, they include sons and daughters who would rather be among the 'barbarians'). Left to develop their own social momentum, gated communities might gradually reveal that their imagined homogeneity and 'tranquillity', as portrayed in estate agent brochures, is a less than accurate description of what goes on in and beyond them.

Again, relational thinking helps us here, in revealing

● how no physically bounded community can ever completely withdraw from the city which surrounds it. No place – even a high-security prison – is ever relationally isolated completely from its surroundings. The relational ties and connections that gated communities have with the rest of the city that surrounds them merely change.

Such places require maids, gardeners, guards and workers from the rest of the city. They are sustained by highway, energy, telecommunications and water grids, and of course television and radio services, which support their rhythms and social worlds and integrate their residents into the wider life of the city. Whilst we would not deny the trends that Dillon and Caldeira emphasize, we must remember that many of the fences that surround fortified communities – even the electric ones in Morumbi – are far from impervious to people determined to cross them.

3.1.2 The public arena as a mixed space

Paralleling the process of gating, some public arenas in cities continue to remain spaces of social mixture and coexistence between diverse groups. We do not wish to suggest that there are no tendencies towards the privatization and surveillance of public space, but that these generally exist alongside other usages which do not necessarily imply the loss of common access. Indeed, the idea of the city as a crossroads suggests that by far the most common usage of public spaces – from streets, squares and parks to shopping areas, markets and popular restaurants – is that of diverse groups and cultures jostling for space to support their own needs, identities and relational worlds. This is not to imply that public spaces are theatres of cordial interchange, civic tolerance and democratic engagement. Rather, it is to suggest that

- they are a setting of contested uses and aspirations, where the contest might result in turf wars, indifference, resigned cohabitation, or some positive social interchange based on mutual trust and respect.

This sense of the mixed ambivalence of public spaces is wonderfully captured in Ulf Hannerz's (1996) description of Amsterdam from his book *Transnational Connections*.

ACTIVITY 1.4   Turn to Reading 1A on Amsterdam by Ulf Hannerz. Focus in particular on the two sections 'City of windows' and 'Making culture from diversity'. Note how the former describes the openness of the city's public spaces, and think about what contributes to this outcome. Note too how the latter illustrates Amsterdam's tolerance for diversity and how in some instances the 'experience of diversity' produces new cultural mixtures. Beyond this primary focus on the character of public spaces, read what Hannerz has to say about the effects of world influences and how these have changed over time in the section 'Commerce, migrants, tourists'. Consider how his analysis relates to our earlier claims in the introduction about the intensification of mixture in cities due to new forms of global exposure. ◆

What conclusions can we draw about the character of public spaces from Hannerz's description of Amsterdam? We are struck by three things. First, and most obvious, the reading shows that not all cities have actively engaged in the privatization and closure of public spaces. In the case of Amsterdam, Hannerz seems to be suggesting (towards the end of 'City of windows') that this is partly the product of tolerance of diverse (economic) activities in the same space, and partly the product of a noticeable awareness by people that the street itself is a 'place for scanning diversity, a source for the life of imagination'. Furthermore, the sections on Amsterdam's long-standing role as a place of global connections and flows seem to imply that the city's openness derives from its very long history of exposure to diversity. It is as though this history has allowed Amsterdam – from its inhabitants and visitors to its planners and decision-makers – to feel less anxious and threatened by the consequences of mixture.

Second, Hannerz makes it clear (at the start of the last section) that diversity does not automatically produce tolerance and the open usage of public spaces. It could produce spatial segregation, or mixture 'at the micro-level of street and neighborhood', and it 'may result in excitement as well as unease', or ambivalent tensions if you prefer the language that we have been using. In the case of Amsterdam, Hannerz is certain that the overarching result has been tolerance of diversity, but what stands out for us is how this tolerance is the product of people being allowed to go about their own affairs, rather than any conscious design or active civic attitude.

Third, Hannerz shows how the intermingling of peoples and cultures – in the context of a mutual tolerance allowed to build over centuries – can result in the formation of new cultural identities and tendencies which, importantly, influence and transform existing identities. This process of cultural 'hybridization' – the construction of new identities out of the mixture of older ones – can have effects across the urban social spectrum, as it comes to affect both the 'mainstream' and its counterpart. Hannerz makes this point forcefully when he describes, for example, how Yiddish words have crept into Amsterdam dialect, and how the cultural identity of young Surinamese Rastas draws ideas and symbols from Kingston in Jamaica, Surinam, London and Amsterdam.

Much of our earlier argument has been that cities, across all their spaces, bring together variety and diversity in close juxtaposition, in their role as nodes of incessant local, national and global flow, movement and connection. The implication of this claim is that tolerance of, or at least resignation to, diversity among urban residents may be much more common than we are inclined to think, especially from alarmist accounts of gating and other forms of urban spatial and social isolation.

ACTIVITY 1.5   Now read Extract 1.3 from urban anthropologist John Hutnyk's engaging book on different representations of Calcutta. It describes the mixed character of Sudder Street, a street where travellers stay, and the tacit codes of cohabitation produced by this mixture. ◆

FIGURE 1.6
Street life in Calcutta

EXTRACT 1.3
John Hutnyk: 'The rumour of Calcutta'

The Sudder Street area … is a fascinating part of town. There are few opportunities to live in such close proximity to people who speak such a wide variety of languages. There is, first of all, the local Bengali, Hindi and Urdu … There are travellers from so many countries that the community is very much a 'polyglot' or 'heteroglot' mix of mostly English, but also French, Dutch, Belgian, and so on … With variations and failures of language due to forced translations among different grammars, with minor nationalisms and hierarchies evident in subgroups, and with comic mistranslations … classics like 'porge' instead of 'porridge' and 'mixed girl salad' instead of 'mixed grill and salad' …

The street … is uneven, the footpath is more often rubble than flat, the holes in the road are large, and dangerous during monsoon flooding, and yet 'the chaos seems uniform' … Everyone seems to find an appropriate path or position, there is a 'code', there are protocols to learn, and patterns into which visitors 'fit' – in spaces or corridors, designated for them … Some hawkers attract tourists to one side of the street, others avoid a particular corner, street children occupy a disused sidestall, the heads of passers-by turn at the more popular café to see who is there, a newspaper-wallah stops customers in the middle of the road – a tracing of the patterns of these trajectories would reveal the 'code' of use of the street in economic as well as cultural and political terms. Another visitor, with unintended aural irony, called the chaos of everyone wandering all over the road the 'staggering urban clutter of Calcutta'.

Source: Hutnyk, 1996, pp.47–8 and 135

We can think of many other cities of the world that could be described in similar terms. What is striking in this example is the role of mixture as a source of new meanings and identities, arising out of the contests and fusions between diverse relational worlds. Such hybridity is as much a feature of contemporary cities as are the attempts at 'purification' highlighted by Dillon and Caldeira. The open public space is not disappearing from urban life.

ACTIVITY 1.6   Think back to the examples of gating in the extracts from Dillon and Caldeira. Think of a city that you know reasonably well. Identify one space that is not open to the general public twenty-four hours a day (for instance a shopping mall or a residential enclave). Who do you think is being excluded, by whom and why? List all the connections that you think the space continues to have with the rest of the city. Is the space internally 'pure'? If not, list some of the mixtures, conflicts and contests that might be at work within it. ◆

You might have selected a shopping mall that is open during the day and early evening but closed at night. Whilst clearly such malls are carefully controlled and monitored to be 'pure' consumer spaces, just listing a few of the connections, mixtures and contests that can characterize these malls shows that purity can never be attained in practice. We would highlight the many physical flows and connections that sustain malls: for example, the daily flows of shoppers, workers, cleaners and security guards, and the connections to transport, water and energy networks. We would also point out that many groups who routinely face close scrutiny in malls, such as youths or ethnic minorities, *actively contest* this scrutiny by exploiting weaknesses in security and policing strategies.

3.2 NEW COMMUNICATIONS TECHNOLOGIES AND CITIES

Our second example considers how new information and communications technologies such as the Internet, satellite television and electronic trading systems might relate to the connections and disconnections and ambivalent tensions within and between cities. Such technologies support contact between people and places, based on exchanges of information, images, media services and money. As they have become increasingly computerized and reliant on 'digital' streams of information (i.e. zeros and ones), the technologies of electronic exchange are now much more capable than during the past century and a half. They offer some of the most striking examples of geographically stretched relational webs operating at national, international and even global scales. At first sight, the distant connections they support seem to contrast starkly with 'traditional' face-to-face interactions in the streets, public spaces and work places in cities, such as the street life in Calcutta and Amsterdam discussed above. This is because advanced telecommunications can effectively make distant interactions *instantaneous*, apparently allowing meaningful social, economic or cultural relations to be maintained over long distances between people who are not physically in the same place.

As we saw with the examples of the British Airways 'virtual single office' and the New York financial centre, certain parts of cities are now linked together into many complex, wider geographies sustained by advanced telecommunications networks. Global financial centres act as hubs on global fibre optic and satellite networks which connect them intimately via hundreds of thousands of daily electronic interactions (phone calls, computerized financial dealing systems, etc.). Networks of cities are now woven together through many electronic grids supporting call centres, the private information technology (IT) systems of corporations, social telephone traffic, and growing flows of terrestrial and satellite television. For those who have them, telephones and wireless phones support mobile lifestyles, access to all manner of personal and information services, and instant contact with friends and family. Furthermore, the Internet system of computer networks now unevenly weaves the sites and places of most cities in the developed world, and many in the developing world, into a widening universe of 'virtual communities' and systems of service and information provision (Mitchell, 1996; Graham and Marvin, 1996).

But how does the growing use of telecommunications and information to mediate such a range of connections relate to the ambivalent tensions of cities and urban life? Two main opposing positions can be identified. Thinking about these not only helps us to imagine the technological futures of cities; it also helps us to think about the sorts of relations that might continue to sustain their social and economic fabric.

Position 1: Electronically mediated relations will destroy the place-based meaning of cities

The first position suggests that relations mediated by new telecommunications will grow rapidly in importance and simply *replace* the more familiar place-based interactions within and between cities. Instead of people physically travelling to work, or meeting socially in cities, new advanced computer networks and telecommunications will provide electronic substitutes that are accessible from the home.

FIGURE 1.7   *The death of distance, according to the front cover of* The Economist, *6 October 1995*

Predictions that distance and space constraints are effectively ceasing to matter are implicit in almost any advertisement for mobile phones, computers or the Internet. The cover of *The Economist* magazine shown in Figure 1.7 captures the idea well. Recent social commentary on cities, especially in North America and Europe, has also featured many predictions of simple substitution and transcendence from urban place, as more and more of the transactions and exchanges traditionally done in cities seem to become possible 'online' from virtually any location (and at any time). For example, the UK architecture critic Martin Pawley (1995) suggests that 'once time has become instantaneous, space becomes unnecessary. In a "spaceless city" the whole population might require no more than the 30 atom diameter light beam of an optical computer system'. Nicholas Negroponte, Director of the world-famous MIT Media Lab, believes that new so-called 'virtual reality' technologies, accessible remotely via home, office and even mobile terminals, will do nothing less than provide multi-sensory electronic 'environments' within which humans become psychologically immersed. The substitution for the experience of cities and places more generally will be complete. He predicts that:

> digital living will include less and less dependence upon being in a specific place at a specific time, and the *transmission of place* itself will start to become possible. If I could really look out the electronic window of my living room in Boston and see the Alps, hear the cowbells, and smell the (digital) manure in summer, in a way I am very much in Switzerland.

> (Negroponte, 1995, p.165, emphasis added)

If one carries through the logic of these ideas, the implications seem very troubling for cities. Surely, distant relations mediated by new technologies are about to replace the very fabric of face-to-face social, economic and cultural exchange that sustain cities and urban life? Economic activities and city residents will use new technologies to allow them to decentralize from cities, as everything they require will be 'one click away' and accessible via 'online' television and computer screens. Affluent groups most troubled by the ambivalent tensions of urban life – perhaps the same groups who are currently moving into gated enclaves in São Paulo or Dallas – might no longer need even to live near cities, as they might sustain all their relational worlds from attractive rural villages.

Extract 1.4 below by the urbanist Anthony Pascal is a good example of such a 'death of the city' idea. He suggests that the whole fabric of urban economic life, even in world cities like New York, will inevitably be doomed, as new technologies will reduce their advantages and expose their relatively high costs and problems as places in which to invest and live.

ACTIVITY 1.7   Now read Extract 1.4 from Pascal's article 'The vanishing city'. Why does he feel that better media technologies will lead to cities 'vanishing'? Make a list of all the traditional face-to-face activities in city economies that he thinks will be replaced by systems of telecommunications, and note down the use of the technology that he predicts for each. ◆

EXTRACT 1.4
Anthony Pascal: 'The vanishing city'

The era of the computer and the communication satellite is inhospitable to the high density city. Clerical and record keeping functions have already begun to deurbanize … The distant suburbs and small towns of the US are dotted with highly computerized complexes performing bookkeeping, billing and archival tasks for banks and insurance companies. The newly emerging technologies will soon begin to provide excellent substitutes for face-to-face contact, the chief remaining *raison d'être* of the traditional city …

Where it was once necessary to concentrate large staffs of managers who could combine to generate quick solutions for nonstandardized problems, teleconferencing will soon provide an alternative. Teleconferencing even has some advantages over face-to-face meetings: participation need not be sequential and can be anonymous, printout and videotape records are available, subcommittees and caucuses can be accommodated simultaneously. People who grow up in an environment where such things are possible will more easily adapt to and more readily use the innovations. The broad implication? A pension fund with a long time horizon would be ill advised to invest in any fourth tower at New York's World Trade Center.

The ease of face-to-face communication and the concentration of specialized support services also made the city an incubator for business … The city acted as the urban hatchery for new companies, many of which subsequently departed for the suburbs or the countryside. As the cost of disseminating knowledge and information declines and as specialized inputs are more easily assembled in remote locations, incubation too can take place almost anywhere …

Although there will always remain functions that require the physical congregation of people and people who prefer tangible proximity to others, substantial new flexibilities will enter the system. Estimates for 1995 suggest that perhaps 15 per cent of the US labor force will telecommute by then. Development of the electronic briefcase and the refinements of the cellular telephone promise even more disengagement between workers and fixed, centralized facilities. In the future then, many will be able to work virtually anywhere. Given a preference for low density residence in amenity-rich areas, which a large fraction of the population expresses, further depopulation of cities seems inevitable …

Through [new technological] innovations, improved communication substitutes for proximity. Because the city specializes in the advantages of proximity, its attractiveness as a focus for human interactions will continue to decrease. The proper late twentieth century reply to Gertrude Stein's complaint about Oakland, 'There is no *there* there', is, 'There is less and less *there* anywhere, anymore. Increasingly, *there* is everywhere'.

Source: Pascal, 1987, pp.590–600

Regardless of whether you agree or disagree with this view, did you notice the number of face-to-face activities destined to disappear?

We would note Pascal's prediction that suburbs and rural areas will start to use teleconferencing to draw in administrative and management functions from cities. We also noted his belief that small firm innovation no longer needs to be in high-density urban 'incubators' and that all sorts of workers will soon be able to 'telecommute' to work rather than physically commute by transport. You may have listed others, but before you consider them, read the counter position.

Position 2: Electronic and the place-based relations interact and complement one another

Counter to the idea of total substitution, there is a second perspective which suggests that new communications technologies are being *subtly integrated* into older urban structures and ways of doing things, in a process of urban *evolution* rather than *revolution*. According to this perspective, new media technologies will not somehow evaporate the meaning of urban places in some gigantic stampede to move 'online'. This is seen as a naive and oversimplified view. Rather, such technologies are seen to complement, and interact with, the place-based relational webs of connections and disconnections within and between cities.

Writers from this perspective stress that the growing use of new communications technologies of all sorts is closely related to the growing urbanization of our planet (Castells, 1996). The implication is that we must explore how technologically mediated relations within and between cities relate to each other and are constructed to combine together. Cities are being remodelled rather than disinvented as their landscapes are unevenly woven into the whole range of emerging 'high-tech' industries surrounding telecommunications and computing. Moreover, emerging so-called 'informational cities', whose economic fabric is heavily sustained by new media industries and applications, display their own tensions and ambiguities, and highly uneven connections and disconnections to the new technological infrastructures (Castells, 1996).

Access to the new technologies of communication remains starkly uneven both socially and geographically within the city, whether it be the global financial districts of London and Tokyo, cities with many call centres like Glasgow, or the cities of research and development like Toulouse or Silicon Valley. Whilst powerful international industries and socio-economic elites maintain their relational power to link to distant places through access to state-of-the-art technologies, even the most 'high-tech' of cities, as we saw with the New York example above, demonstrates that many social groups and geographical areas remain disconnected from the 'liberating' promise of new technologies for lack of funds, infrastructure, skills, equipment, even electricity. In these social worlds, face-to-face physically based relations still predominate.

The associated reading from Madon describes how the internal disparities are emerging in one of the world's most rapidly growing cities, Bangalore in Southern India, a city that has grown to dominate Indian computing and telecommunications industries.

ACTIVITY 1.8   Turn to Reading 1B by Shirin Madon. Note down the kinds of disparities discussed by Madon. Think about how high-quality computing and telecommunications networks, and the electricity that drives them, might extend the relational worlds of the middle-class group of software engineers in Bangalore's high-tech enclaves. Where do you think these networks connect to? Then consider the relational worlds of the rural migrants living in the shanty towns on the edge of Bangalore, drawn to the city by the possibility of work, and with little or no access to electricity or telecommunications networks. What does the article imply about the relational worlds of these social groups? ◆

The Bangalore example demonstrates forcefully to us that new communications and information technologies are being woven very unevenly into the fabric of cities. They tend to support the relational connection of powerful groups and spatial zones within cities, both to each other and, via geographically stretched social and corporate networks, to equivalent far-off cities and spaces. But this very power held by the elite and other groups works in stark contrast to the relational disconnections that often characterize physically close social groups in 'network ghettos' (Thrift, 1995) such as shanty towns and similar places (deprived of decent water, energy and communications infrastructures and often excluded from mainstream political power). In Bangalore we have a good example of how parts of cities can be physically close but relationally distant. In this case, in-migrants from the countryside colonize shanty towns and struggle to find poor-quality work, whilst facing the constant threat of harassment from city agencies and property developers who are keen to 'modernize' Bangalore's fabric by developing shopping malls and corporate work spaces in the place of informal settlements.

It is clear, then, in keeping with our theme, that

● cities are sites which combine and superimpose diverse relational worlds, and that

● IT-mediated relations become subtly combined with the relational worlds within the spaces and places of cities.

ACTIVITY 1.9   You should now read Extract 1.5 by the British geographer Dennis Cosgrove. Cosgrove uses two simultaneous events that occurred in New York in 1995 – the launch of the *Windows 95* software package and a photographic exhibition about the lives of financial traders on Wall Street – to argue against the idea that cities are somehow dissolving into a pure, electronically mediated world. Cosgrove suggests rather that the world of electronic communications is essentially combined with the intensely local and 'messily physical' worlds and relational webs of city life. ◆

EXTRACT 1.5
Dennis Cosgrove: 'Windows on the city'

Two unconnected social events which occurred during 1995 serve as emblems … [of] urban culture and communication in this century's final decade. Stretching to its limits the advertising industry's creative capacity to generate cleverly recomposed images, Bill Gates' Microsoft computer corporation launched its new software package *Windows 95* , through a series of spectacular staged events in cities across the globe. Their common message was the achievement of another giant step towards networking individuals into universal cyberspace. Meanwhile, in a gallery on New York's Upper East Side, a photographic exhibition displayed compelling images taken by Merry Alpern in the heart of Wall Street's financial district. But the images seen through *Dirty Windows* do not reflect those glass walls of corporate office towers, nor the moving headlights of the downtown expressway that [Nigel Thrift has remarked are] the clichés of the post-modern city. They reveal a commerce of human bodies, in various stages and modes of sexual encounter, young women and besuited financial traders engaged in after-hours prostitution and cocaine consumption, photographed through the misted pane of a 'men's club' backroom window.

These events are emblematic, not so much in the banal sense that they reveal 'two sides of the city', as that they represent the coexistence in the same physical space – in this case, the very core of the globalised city of finance – of quite distinct but equally insistent spatialities: the one connected to the very latest mode of disembodied electronic technology, allowing clean, sharp-edged and instantaneous information exchange between people, and the other tied to the most enduring, intimate, messily physical and emotional exchanges between individual bodies. The urban world networked by Gates' technology, 'strung out on the wire', is not disconnected, abstract, inhuman; it is bound in the places and times of actual lives, into human existences that are as connected, sensuous and personal as ever they have been.

Source: Cosgrove, 1996, p.1495

Cosgrove demonstrates forcefully that we will only understand how the new media are developing within and between cities by exploring the rich relational connections between the social worlds of urban places and their subtle links with electronically mediated experience. For him, it is not a question of either city or electronic communications but rather both together.

4 *Key aspects of cities*

What, then, can we generalize about cities, if so much of their contemporary life is about variety and ambivalent outcomes? We believe there are three crucial aspects of cities as *places* which can be teased out from our earlier discussion of relational webs and from the above examples of ambivalent tensions (see Amin and Graham, 1997, for a fuller account). One is quite simply the density of place that co-presence produces. The second is the character of cities as sites of multiple experiences of time and space, resulting from their overlapping relational webs of varying time rhythms and geographical reach. The third, and perhaps most enduring, aspect is that of cities as a place of social encounter. These three aspects are considered in turn below.

4.1   ... AS SITES OF PROXIMITY AND CO-PRESENCE

As we have seen, cities bring together place-based connections (such as local ties of work or sociability) and connections 'stretched' over increasingly distant links, through the operation of transport and telecommunications networks. It is tempting to conclude, as we saw in the case of one of the interpretations in section 2.1 of the implications of global information and communications technologies, that the place-based connections are being displaced by the wider connections, and that, as a consequence, the city itself is being 'stretched' out of existence.

Clearly, though, this is not the case. The smog that hangs over cities, their audible noise, the sky they light up at night, their starkly experienced hustle and bustle, all reveal that the density of cities as places is not disappearing. We therefore need to see the link within cities between the local and the global in a different way. The urban sociologists Dierdre Boden and Harvey Molotch (1994, p.259) claim that the spatial essence of cities in these stretched spaces of relational proximity lies in what they call the 'thickness of co-present interaction'. The meaning of this term is not immediately obvious, but to us it suggests that

● intense and enduring face-to-face interactions within urban space coexist with flows of communication and contact to the broader city and beyond.

So what are the implications and meaning of coexistence? We have already suggested that coexistence does not automatically suggest either complementarity or friction between relational worlds. Instead, coexistence highlights the importance of *place density*, in a world of stretched connections and incessant global flow. Cities, or more accurately parts of cities, are centres of agglomeration and proximity, because of their unique feature as the nodes of diverse relational webs. These properties of density – agglomeration and proximity – give cities a distinctive role in social, economic and cultural life.

The positive effects of agglomeration and proximity can be considerable. For example, in the economy, the pooling of specialized skills, services and know-how, together with the trust that is built through face-to-face contact and interpersonal familiarity, has notable advantages. These advantages of urban agglomeration are potentially sources of vitality, economic competitiveness and innovation within the relevant global business networks, offering collective advantages to firms that are denied to competitors in more isolated settings. Similarly, in the cultural arena the critical mass of creative and talented people that cities might offer – often concentrated in particular neighbourhoods – can be a unique source of cultural innovation and renewal through the daily exchange of ideas, information and knowledge.

These examples should not leave us with the impression that place density has only positive implications for cities. It can also work to their disadvantage. For example, clusters of unemployment, social deprivation, criminality and ghettoization in general can lock cities or parts of them into spirals of decline for exactly the same reasons associated with spatial proximity. Here, face-to-face contact and informal exchange might serve to reinforce alternative strategies of survival beyond the margins of the mainstream (for example petty crime, racketeering, prostitution) in addition to giving legitimacy to life on the margins.

Either way, it is clear that the 'stretching' of the city does not in any way represent the end of the effects of place density – be they positive or negative.

4.2 ... AS A MIX OF SPACE/TIMES

The second key aspect of contemporary cities that strikes us is that they are a place of what we might call *multiple time-spaces*. We started the chapter by claiming that an ever-increasing number of relational webs of connection, each with its own geography and time-span/space, cut across cities and their individual spaces. Take as an example the diverse and changing time patterns of any site. A railway terminal in a neighbourhood at the start of the working day is a place of transit for commuters in a rush, a place of work for its employees during the day, a stop on the leisurely travels of a tourist, a gathering place for teenagers after dusk, or the temporary shelter of the homeless person at night. Each activity has its own rhythm and tempo. Across the city as a whole, a variety of other so-called 'time geographies' are at work. British geographer Nigel Thrift (1996), who has studiously drawn our attention to the varied temporality of human life, notes that contemporary cities display a similar kind of variegated sense of time as the eighteenth-century city – from the intense and global instantaneity of the financial markets and global media flows, to the new urban mythology of new age religions.

Whatever your perspective on whether the time geographies of contemporary cities are more complex than in the past, the important point is that we cannot assume that cities have a single or dominant time-space dimension. Thus it would be wrong to see contemporary cities only as places of very fast and fleeting interchange, associated with their rapid access to virtually all parts of the world. It would ignore the spaces and the people, such as the poor areas or the homeless, that are by-passed by these global flows. It would also ignore those people who integrate these flows (for example the experience of overseas travel or surfing on the Internet) into a wider set of everyday time-space experiences such as local shopping, travel to work, domestic activity, or arranging to meet friends.

It seems to us that a more accurate interpretation is that

● cities, and their individual spaces, are places where multiple time-spaces become intensively superimposed. In terms of the experience of urban life, the negotiation of this multiplicity is perhaps what really matters.

For example, a key challenge for city dwellers may be how to *cope with* the varied experiences of time and space that urban life involves, rather than with the isolated effects of any one relational link to the rest of the world (such as surfing on the Internet).

4.3   ... AS MEETING PLACES

The third striking aspect of cities seen as places of connection and disconnection is their extraordinary dynamic of social mixture and cultural variety, as we saw in our discussion of the culture of public spaces. Cities are woven fully into all the social ambiguities of modern life, with their tensions between shared purposes and conflicting individualism, their contested desires and cultural demands, and their mixtures of exclusion, inclusion, fear and trust.

As we have seen, despite widespread efforts at attempted social enclosure, heterogeneity continues to thrive within contemporary cities. We could go so far as to suggest that cities are in fact *constituted by* such fluid interactions, and separations and connections, across the fabric of urban life. Just think of the sheer variety of the social and cultural activities that cross over one another to make up urban life: dance, performance and play, media activity, socializing and drinking, sexuality, tourism and travel, formal and informal political action, street markets, youth cultures, voluntary and non-governmental organizing, technological innovation, learning, education, caring and health provision, social movements and protests, religion, spiritualism, folklore, shopping and consumption, recreation, trading, street-watching, surveillance, social control and oppression, gentrification, sport, eating, spectacle, festivals, fashion, and so on. In cities, importantly, these practices are *spatially* expressed, and hence they stamp the particular identity of a city. They become expressed in physical forms through ethnic and social clustering, specialist centres, and the continual construction and reconstruction of new social and cultural spaces.

This is not to imply, however, that the physical identity or boundary of the contemporary city is fixed. Quite the contrary:

● the multiple webs of relations which weave to constitute urban life are characterized by complex encounters where different lives meet and intersect. Through these, individuals and communities seek to enact their lives within the multiple choices, demands, constraints and meanings of the 'urban'. Thus, diversity breeds more diversity, as well as endless fluidity, constantly challenging the inherited physical 'settlement' of cities at any one point in time.

'The city', in short, 'is always being stretched' (Demos and Comedia, 1996, p.15) by the continually changing practices of human action embodied in the city's diverse activities.

It is extremely important to note that the 'stretching' of cities, to forge new identities through encounter and mixture (as in Sudder Street) does not necessarily represent a process of genuine cultural interchange or social justice in urban life. Change intermeshes with the highly unequal relations of social and economic power which tie together, as well as disconnect, the diverse groups living in cities. For, as we saw in the examples of gated cities and of New York, Amsterdam and Bangalore, cities across the world continue to display concentrated extremes of affluence and poverty.

Contemporary urban landscapes, as with all the cities of history, are thus *etched with highly uneven social and power relations.* Elites might enjoy rising incomes and the ability to use transport and telecommunications networks to extend their relational reach across time, space and people (for instance employing domestic servants, using the Internet, and enjoying high levels of consumption and access to state power). Their 'super inclusion' into the globalized world, however, is overlaid upon the immiserization and social exclusion of less powerful groups who lack the ability to extend their actions beyond their local neighbourhood. Every aspect of the social fluidity and cultural dynamism of cities leads to complex asymmetries of power and enablement, and constraint and domination.

5 *Conclusion*

Our purpose in this chapter has been to show that the increasing permeability of contemporary cities to influences from around the world is making urban life ever more mixed and full of ambivalent tensions and outcomes. The overall effect, compared to the past, is the intensification of urban mixture and change. This makes it almost impossible – which is not to imply that it was possible before – to interpret cities as singular or coherent entities. Cities are caught in so many relational webs of one sort or another, and are subject to so many changing flows and influences, that any attempt to capture them in uni-dimensional terms is likely to be flawed from the start. This is why we have argued for a *relational* perspective that, first, explicitly recognizes the mixed nature of cities and, second, helps us to understand the ambivalent connections and disconnections which come together in them. Through our two examples, we have explored some of these ambivalences, such as those surrounding spatial fragmentation and integration, social exclusion and inclusion, consensus and conflict, and cultural retreat and mixture.

An important question raised by our interpretation is whether contemporary cities 'hang' together as a whole. Does the intense mixture, change and ambivalence of urban life justify generalization or a search for any 'wholeness' to cities? The answer rather depends on what we mean by wholeness. For instance, it should have become clear from our analysis that there is no wholeness that derives from any single overall purpose – for example, the idea that contemporary cities are the forcing houses, the nodes of economic and cultural vitality and innovation, of the world economy. In addition, it is clear that there is no internal coherence to be found that can be traced to the ways in which social and economic life in the city interrelate, since the multiple spaces and social worlds of cities are bound into wider relational webs. Thus cities are not 'actors' breeding internal coherence.

On the basis of such observations, it is tempting to conclude that cities are now simply mixtures of fragments, and therefore not worth understanding as wholes. This could almost certainly be claimed for contemporary *urban life*, varied as it is, and accessible as it has become through travel and communications to people in non-urban settings (for example through holidays in cities or via television). But what of cities as physical *places*? Is there something more than a series of disconnected fragments to be found in this sense? Cities remain settled places which house an increasing proportion of the world's population; they possess a distinctive physical landscape gathered around past and present infrastructures, buildings and architectural forms; and, in general, they are a more or less concentrated site of human and non-human activity. Is there not a wholeness here?

In our opinion, there is. This is why we attempted, in section 4, to tease out some of the main dimensions of cities as *places* – as a material and physical setting of connections and disconnections. Viewed as such, we would claim that

● there is a certain wholeness, or overarching identity, to cities, which derives from their character as places of *juxtaposed spaces and superimposed relational webs*. It is this intensely material juxtaposition of diversity that seems to mark the city from both other physical agglomerations of social activity such as households, towns and villages, as well as geographically stretched social worlds such as extended family networks, global networks of transnational corporations, and so on.

Ultimately, does it really matter whether cities are more than the sum of their parts? What difference does it make analytically to stress that the wholeness of cities is to be found in their roles as places of juxtaposed diversity? These are not easy questions to answer, but we believe that such a sense of the whole does matter, at least in two ways.

First, it allows us to see cities *as nodes in various time-space webs* (from historical connections to current global geographies of communication). Cities stabilize and fix the flows in these webs within particular urban institutions, communities, buildings and locations. They give material expression to these flows and connections and are constituted by them. Even the most apparently familiar sites of the urban landscape can then be understood through the ways in which they support the intersection and interconnection between multiple relations and time-space webs at varying geographical scales. From small cities to global mega-cities, we can find transport interchanges, shopping and leisure zones, corporate and manufacturing districts, and residential areas which, in these days of deepening and stretching global relations, can only really be understood as sites which hold the multiple relational webs of the city in a state of subtle articulation with those linking it into extended networks of power, interaction and flow to the rest of the world.

Second, stressing the powers of juxtaposition allows us to recognize *the independent effects of proximity* and the physical clustering of people, institutions and artefacts. We saw with the new technologies example that there is more holding cities physically together than will ever be simply pushed 'online' over distance. We also saw in our discussion of the culture of public places that co-presence can lead to social antagonism, separation and indifferent tolerance or intermingling and the rise of new cultures. What is striking is that all these effects are in some way the result of diversity placed side-by-side, resulting in outcomes that would not otherwise occur. Through such outcomes, cities *actively generate* new meanings. The quality of 'city-ness' does have effects. In other words, juxtaposition generates new social dynamics, and therefore also cities in evolution.

References

Allen, J. (1999) 'Worlds within cities', in Massey, D. *et al.* (eds).

Amin, A. and Graham, S. (1997) 'The ordinary city', *Transactions of the Institute of British Geographers,* vol.22, pp.411–29.

Boden, D. and Molotch, H. (1994) 'The compulsion of proximity', in Friedland, R. and Boden, D. (eds) *Now/Here: Space, Time and Modernity*, Berkeley, University of California Press.

Borja, J. and Castells, M. (1997) *Local and Global: Managing Cities in the Information Age,* London, Earthscan.

Caldeira, T. (1996) 'Building up walls: the new pattern of spatial segregation in São Paulo', *International Social Science Journal*, no.147, pp.55–65.

Castells, M. (1996) *The Rise of the Network Society*, Oxford, Blackwell.

Cosgrove, D. (1996) 'Windows on the city', *Urban Studies,* vol.33, no.8, pp.1495–8.

Davis, M. (1990) *City of Quartz*, New York, Verso.

Demos and Comedia (1996) *The Richness of Cities: Rethinking Urban Policy,* London, Demos.

Dillon, D. (1994) 'Fortress America', *Planning*, June, pp.8–12.

Graham, S. and Marvin, S. (1996) *Telecommunications and the City: Electronic Spaces, Urban Places*, London, Routledge.

Hannerz, U. (1996) *Transnational Connections: Culture, People, Places*, London, Routledge.

Hutnyk, J. (1996) *The Rumour of Calcutta: Tourism, Charity and the Poverty of Representation*, London, Zed Books.

Madon, S. (1997) 'Information-based global economy and socio-economic development: the case of Bangalore', *The Information Society*, vol.13, no.3, pp.227–43.

Massey, D., Allen, J. and Pile, S. (eds) (1999) *City Worlds*, London, Routledge/The Open University (Book 1 in this series).

Mitchell, W. (1996) *City of Bits: Place, Space and the Infobahn*, Cambridge MA, MIT Press.

Mulgan, G. (1991) *Communications and Control: Networks and the New Economics of Communication,* Cambridge, Polity.

Negroponte, N. (1995) *Being Digital*, London, Hodder and Stoughton.

Pascal, A. (1987) 'The vanishing city', *Urban Studies*, vol.24, pp.597–603.

Pawley, M. (1995) 'Architecture, urbanism and the new media', mimeo.

Richardson, R. (1994) 'Back officing front office functions – organisational and locational implications of new telemediated services', in Mansell, R. (ed.) *Management of Information and Communications Technologies*, London, ASLIB.

Soja, E. (1989) *Postmodern Geographies*, London, Verso.

Thrift, N. (1995) 'A hyperactive world', in Johnston, R., Taylor, P. and Watts, M. (eds) *Geographies of Global Change*, Oxford, Blackwell.

Thrift, N. (1996) 'New urban eras and old technological fears: reconfiguring the goodwill of electronic things', *Urban Studies*, vol.33, no.8, pp.1463–93.

Zukin, S. (1995) *The Culture of Cities*, Oxford, Blackwell.

READING 1A
Ulf Hannerz: 'Amsterdam: windows on the world'

In Bijlmermeer

Take a seat by the window in a café on one of the squares in the Amsterdamse Poort shopping center on a Saturday afternoon. Enjoy the pancake of your choice; and as you watch the ever-changing scene outside, you may marvel at the way the public spaces of larger cities in western Europe and North America have become transnational commons, where white, black, brown, and yellow people mingle.

And this mingling continues in the Shopperhal block next door. Much of its merchandise is of the kind you can get anywhere in Amsterdam. But at the more distant, I believe northern, end of the Shopperhal, the scene begins to become a little different. One of the stalls sells beauty aids for non-European women, and across the aisle a food stall carries tropical specialities; there are *Surinaamse bollen* as well as *Berliner bollen*. You continue past the Razaq Islamic butchery and leave the Shopperhal through a nearby exit.

Then, in the square outside, and in the housing area beyond it, you see almost only black people. This is Bijlmermeer, sometimes described as the second largest city of Surinam – after Paramaribo, the capital – nestling inside the largest city of the Netherlands. In the 1970s, when the Netherlands gave independence to its sparsely populated colony on the northern coast of South America, the inhabitants (mostly of African descent, but some Asians as well) were offered a choice between becoming Surinamese or Dutch. Something like a third of them opted for the latter alternative, and newly built Bijlmermeer was where many of them ended up.

In the landscaped grounds between the rather rundown ten-storey apartment buildings, some boys are throwing fire crackers, which echo loudly between the walls; from an open window high up in one building you hear an old Bob Marley tune. A little further up in the air, a large jet plane is just coming in from the east to land at Schiphol airport.

On your way back, you come upon a smaller, modest shopping area where the most prominent establishment is the Hi-Lo Supermarket. Here you can get frozen fish straight from Surinam, frozen cassava, also dried fish, fresh okra from Kenya, plantains, and printed-in-Bombay greeting cards with a picture of Ganesh, the elephant god, on the cover and the text 'We wish you that festive Deepawali bring you and your family all round prosperity' inside. A small stall next to the Hi-Lo sells the same kinds of printed cloths which have always tempted me in the market-places of West Africa, and exhibits a male mannequin in a flowing, richly embroidered gown. The record and cassette store across the hall has Indian and Caribbean music, next to it a stall sells roti, and a few steps away the Pacific Travel Centre announces that it is the agent for Surinam Airways, while the Chez Polly hairdressing salon proclaims its expertise in handling Afro, Euro, and Asiatic hair.

Back in the main area of the Amsterdamse Poort again, you may rest in a reasonably credible likeness of that Dutch institution, the brown café; fringed lamp shades, carpeted tables, pool tables and all. Here again you see almost only white people, and only men. But a little dark-skinned girl in bright clothes dashes in through the door, looks among those standing at the bar, and leaves with her burly, blond father.

City of windows

One of the first things to strike any visitor to Amsterdam – any foreign visitor, I should perhaps say, as an indigenous Netherlander will be less surprised – is the windows. A very large proportion of the façade of just about any building in the city seems to be taken up by windows, from the lowest floor to the highest. There may be good practical reasons for this. Houses are often deep, and want to draw in all the daylight they can get.

Yet they also offer the sensation of great openness: culture flows through these windows, as it were, from private to public spaces and vice versa. It may be a conspicuous claim that 'I have something to trade,' made by a scantily dressed young woman in a window framed by red neon lights; or, simply and piously, that 'we have nothing to hide,' in the instance of the elderly couple glimpsed through the next window, with their backs turned, lace curtains not drawn. Through the window the market displays its goods, and forms of life other than one's own can be inspected, at least surreptitiously, in passing. And this is a two-way flow, for windows also allow you to keep an eye on the street scene.

There has been some interest among anthropologists in the meaning of mirrors – what you see when you see yourself. In the village, your neighbours may be like mirrors – they more or less resemble you. In Amsterdam, in some mirrors, you do not see yourself at all. There seem to be more than in any other big city of those little mirrors which, placed sideways from the façade, serve rather as window amplifiers, and again allow you to look at other people (particularly whoever is at your front door).

It is also obvious that Amsterdam has a great many 'other people,' not only of the kind that looks like you, but of many kinds. It has drawn them from the outside for centuries, and whether people have come from far away or not, they have been there in sufficient numbers to maintain a great many viable subcultures. Diversity within the form-of-life framework, then. But the market also seems to offer a noteworthy variety of merchandise. This probably has something to do with the scale of the built environment. Some cities have been built, at least as far as their central business districts are concerned, to give preferential accommodation to large enterprises, and car traffic. In Amsterdam it seems as if there is a hole in the wall, with a window in front, for just about every business idea, with easy access for anyone who happens to pass by, on foot or bicycle.

Nonetheless, I am surprised by some of these ideas. Amsterdam is known throughout the world for some of its museums, and museums are great devices for importing diversity from the outside, or for maintaining it by preserving the past alongside a different present. What I had not really expected, however, was a cannabis museum, or a museum of torture.

Again, the street scene itself can also be a spectacle, a place for scanning diversity, a source for the life of imagination. In Amsterdam, as in many cities, people seem noticeably aware of this. It can be seen in the variety of entertainers – musicians, dancers, jugglers, or whatever – who seek out places almost any day on the Dam (the central square in front of the royal palace), in Kalverstraat, or in front of the Centraal Station. Also, I have been told, the number of sidewalk cafés is increasing. This is hardly just because of the drink and food they offer, or because of fresh air and sunlight, but not least because they are street observatories. If the view of human traffic usually makes up much of the cultural flow in the form-of-life framework, here we note that in choice locations, it, too, can be commoditized.

The world on view

In Amsterdam, the notion comes readily to one's mind that cities can be seen as windows on the world. In Bijlmermeer, and its Hi-Lo Supermarket, you encounter the Caribbean: sights, sounds, smells, tastes. In Kalverstraat, the major pedestrian artery, Europe goes shopping, while a block or so away, you can browse in newspapers and magazines from almost everywhere in the Athenaeum Nieuws Centrum (and thus remind yourself that media also concentrate in major cities). The Jewish History Museum tells the story of people who came to the Netherlands and to Amsterdam, lived there for generations and had a part in shaping it, and then were taken away. At the Tropenmuseum, formerly a colonial museum, you get a sense of where the Dutch went in earlier years, and what they found there.

If cities are windows on the world, you see different things through different windows. I can only hope to have caught a glimpse through the window that is Amsterdam, but I will try to fit whatever impressionistic understandings I may have gained into … the part played by cities in organizing the global ecumene.

All cities do not operate over the same distances, nor do they do the same things in organizing culture and social relationships. Each of them fits into the various frameworks which organize cultural process in its own way, has its own place in a partly hierarchically ordered, partly competitive network of urban places, and is a product of its own evolving history. Somehow certain of them seem to resonate better than others with the idea of a global ecumene. Where, in such terms, is Amsterdam?

While I am primarily concerned with the present, I think in this case there is no way of ignoring the past. Despite perhaps unpromising beginnings as a fishing village, threatened by the destructive powers of wind and water, Amsterdam in time grew quickly in size as well as importance, and did so on the basis of the long-distance relationships reaching out from its port. And thus, in the terminology of Redfield and Singer (1954), it evolved very obviously into a heterogenetic city.

It seems, moreover, that the ethos it thus established shows remarkable continuity and distinctiveness. Some comments by the urban historian Donald Olsen (1993, p.184) are interesting in this regard. First of all, there is his characterization of Amsterdam in its seventeenth-century golden age – 'capitalistic, individualistic, tolerant, republican, cosmopolitan, pragmatic, rationalistic, and scientific'. Many of these terms seem equally true of the city three hundred years later. Yet Olsen goes on to note that even if Amsterdam in its golden age seems to anticipate the twentieth century, it was really rather 'the last great medieval city', a successor to city states like Venice and Florence. As the nation-states of Europe came into being, such independent cities were generally brought into submission.

And so, something perhaps a little reminiscent of what Redfield and Singer called orthogenetic cities became a dominant urban model. In their pure form, best exemplified by the earliest urban centers of human history, orthogenetic cities are surely not around any more, but the centers to which the nation-states gave shape seem to have borrowed something from them. These states are not really, or only partly, theater states, yet their capital cities may have been conspicuously engaged in defining, refining, and celebrating national culture, and somewhat jealously watching the integrity of its own bounded domain. In that sense, there is a greater affinity between them and the mosaic model of world culture.

Again, Amsterdam seems to show little of this; continuing, it seems, proudly heterogenetic, carrying that heritage of medieval urbanism and at the same time becoming very much a city of the contemporary global ecumene. If this is in any way enigmatic, the explanation for it may be found to some extent in the division of labor between Amsterdam and The Hague, as well as in the somewhat peculiar nature of the Dutch nation-state, historically itself to such an extent created by its cities, for its cities.

And now, the time for something resembling the medieval city may be here once more. The cities which in the late twentieth century we call world cities are beginning to lead lives rather distinct from those of their territorial states again, and entities such as Singapore and Hong Kong may even suggest that city states can at least in some ways be viable social forms.

Commerce, migrants, tourists

But does then Amsterdam at this point count as a world city? This is apparently something one can hold different opinions about. Various lists of such cities are in circulation. New York, London, and Paris are no doubt on all of them, but Amsterdam, alone or as part of a Randstad conurbation, is on some and not on others.

Of course, it is not a very big city. New York is the Big Apple; Amsterdam, arranged by the canals into layer upon layer until you get to the core around the Dam, is more like a medium-sized onion. To get back to the metaphor of cities as windows on the world again, perhaps you expect to see more through big windows. That may be true to a degree, but what matters more is certainly how the window is placed relative to what is to be seen.

In other words, I think it is more important to consider what number and variety of linkages a city has beyond the boundaries of its nation. I will want to say something here summarily about Amsterdam and transnational commerce, about migrants to Amsterdam from other continents, and about Amsterdam and tourists. These, I think, have much to do with giving Amsterdam its place in the global ecumene; a place with some features shared with other more or less central places, and others quite distinctively its own.

To reiterate, what are nowadays thought of as world cities are, above all, nerve centers of the global economy, headquarters of managers and entrepreneurs. I am reminded again of Amsterdam in its golden age. In his widely praised cultural history of that period, *The Embarrassment of Riches*, Simon Schama describes the conduct of business in the city:

> News was of the utmost importance in these floating transactions, and the regular publication of the courants supplied political and military intelligence to help investors make informed decisions. Strategically placed relatives or correspondents in ports across the globe helped pass on relevant information, but professional punters on the bourse regularly used couriers, eavesdroppers and spies in the coffeehouses of the Kalverstraat to glean tidbits about an enterprise's prospects, or indeed to propagate optimism or pessimism as their stance required.
>
> (Schama, 1987, p.347)

In this era, it seems, Amsterdam was indeed a world city with respect to global business. It has not descended so very far from the heights of that business now either. If the port is no longer so important for communications, one can see Schiphol as its heir, not the world's largest airport but often said to be its finest (as I heard again and again some years ago in Singapore, whose Changi airport tended to come in second place and thus had to try harder). Yet in business terms, by now, Amsterdam is not at the level of New York, London, or Tokyo, and also has regional alternatives like Frankfurt, Zürich, and Brussels to compete with.

World cities, we have also been told already, must cater to the interests and tastes of the transnational elites. Let us use this understanding to bridge the gap between global managers and global retail business. As windows on the world, for the rich and for those of modest means, for natives and visitors, cities are not least shop windows.

It is one of the theoretical tenets of the globalization of the market-place that a superior product finds its niche, its segment of the local market, anywhere in the world (cf. Levitt, 1983). To modify this a little, I would suggest rather that the market segments for some commodities are now just about everywhere, some are recurrent in a great many places, and others again are highly local. And as far as consumer goods are concerned, they are present to the extent that they match locally represented forms of life, with their characteristic resource bases, values, and everyday practices.

It is of course often those commodities which are everywhere, and those which are recurrent, that provide much of the support for the notion that globalization equals a global homogenization of culture. As there is a branch of Sotheby's on Rokin, we can assume that Amsterdam's elite is not doing too badly, and has tastes not altogether dissimilar from elites in at least a handful of other cities; yet P.C. Hooftstraat, sophisticated as its shops may be, is not really the last word in opulence. With regard to commodities at various points further downmarket, Amsterdam also has Benetton, Marks & Spencer, IKEA, Kentucky Fried Chicken, and McDonalds, all of which may or may not also be found in Singapore. In a way, of course, all these transnational retail enterprises serve to make Amsterdam a window on the world, not least for the Dutch.

Nonetheless, commerce in Amsterdam also makes it a cultural crossroads by way of things found in a few other places; sometimes perhaps nowhere else. No visitor walking through Amsterdam can fail to notice its unusual mixture of fast foods, junk foods, 'out-of-the-wall foods'; it is not all McDonalds, and not all herring, eels, little pancakoid *proffertjes* or croquettes either, but also *loempia* rolls and *saté*. In other words, Amsterdam's market-place also offers the traditions of immigrants from far away in readily accessible commodity form. And immigrants have for centuries contributed to the heterogenetic qualities of Amsterdam, in the form-of-life framework as well as in the market.

This has been an unusually open city, to Sephardic Jews departing in the seventeenth century from persecution in Spain and Portugal, and to Ashkenazic Jews coming later from the east; to Huguenots from France in the eighteenth century, and to a great many others arriving at one time or other through the port or the Centraal Station, in recent years not least to Turks and Moroccans in search of a livelihood. But the walk through Bijlmermeer, and the *loempia* in the street stand and the *saté* sticks in a brown café, remind us that not least, it has been an openness to a lingering empire.

One can perhaps more readily take the Pakistanis and the Jamaicans in London and the Senegalese in Paris for granted, since London and Paris are undisputedly world cities of the first rank. But the Surinamers and the Javanese in Amsterdam, whether first, second, or third generation, are more like the Angolans and the Goans in Lisbon. They demonstrate that the global ecumene, however its coherence may have increased in later years, is still a creation of a turning and twisting history, and derives some of its enduring polycentricity from this fact. Colonial and postcolonial memories are probably never

unambiguously happy, on one side or the other, but in different ways old metropoles, even in what have become small countries, remain centers to which scholars, tourists, and sometimes exiles from old dependencies continue to be drawn. Sometimes these are merely brief visits. In the Dutch case, however, decolonization has also entailed some noteworthy waves of permanent migration. That of the 1970s, when so many Surinamese chose to move to the Netherlands, is the last but perhaps most striking example.

As far as foreign tourists in Amsterdam are concerned (and the first thing to observe here is that Amsterdam is both energetic and successful in presenting itself as a tourist city), they may not mind the McDonalds restaurants and the Benetton stores so much. In a paradoxical way, in the global ecumene it may be such sights that in an alien environment remind them comfortably of home. What positively attracts them more, presumably, is whatever is not everywhere, for tourism must always feed on difference (if only a managed, balanced difference). These things, in Amsterdam, are partly things typically Dutch, for from the tourist perspective, these also make Amsterdam a window on the world: the houses and the canals, the museums, the brown cafés. Yet the tourists, taking in Amsterdam with all their senses, are also likely to be struck by the particular diversity of origins of people and things, in Kalverstraat, or in the less fashionable Albert Cuypstraat street market, or – if they ever get there – in Amsterdamse Poort. And the official tourist brochures are certain to make a point of recommending the exotic *Rijsttaafel*, even as they are mostly silent about the native pea soup.

Making culture from diversity

What, then, do Amsterdamers do with the diversity of their city, as they manage meaning and meaningful form in their daily lives? Cities can handle diversity in different ways. Some are very heterogeneous as wholes, but make the least of it in their parts, by constructing something rather more like a spatial mosaic of its more or less homogeneous, segregated entities. Amsterdam has some of this as well. Yet partly due to the nature of the built environment, and partly to the way the powers that be have handled this resource, much of Amsterdam also exhibits unusual diversity at the micro-level of street and neighborhood.

This may result in excitement as well as unease, and Amsterdam has some of both. But the overarching metacultural stance which the people of Amsterdam have evolved toward diversity, it is frequently said, is one of tolerance, of cultural *laissez-faire*. I am reminded of a little book on San Francisco, where

Howard Becker and Irving Louis Horowitz (1971) describe its 'culture of civility.' San Francisco, they say, has long had a tolerance of deviance, even a readiness to take difference to be a civic asset. Its history as a seaport, its tradition of radical unions, the fact that it has in large part been a city of single people or couples rather than families has contributed to such a climate. It would seem that very similar things can be said about Amsterdam, and given yet greater historical depth. This, of course, is one of its attractions to many natives; but also, as I will note again, to many strangers.

Perhaps this tolerance came early, as the Dutch passed through, and learned from their own historical crises and clashes over differences in belief and practice; it was already on the list of traits which Donald Olsen enumerated with regard to the golden age. Just how it has been shiftingly influenced over the years by a range of quite different experiences – 'pillarization,' *verzuiling*, and its crumbling; the countercultural upheavals of the late 1960s and early 1970s, with their testing of established assumptions (commemorated at the Nieuwmarkt underground station); the rowdiness normally to be expected in a port city where sailors seek release from the harsh constraints of life on board; the influx of groups of aliens – would be difficult for anybody to work out, and impossibly so for a visitor.

Anyway, one may suspect that as far as newcomer groups are concerned, it has probably helped Amsterdamers take a rather relaxed view of their incorporation into the social fabric that these have in several instances not shown very high-profile differences. The Sephardim did not seem so unlike other prosperous burghers. The people from the East Indies, and the Surinamers, had been through an anticipatory socialization for Dutch life by growing up in Dutch colonies. Possibly there is now a little more worry about the Moroccans and the Turks, who do not necessarily fit in quite so readily.

Diversity and tolerance of diversity, however, do not only pertain to matters of ethnicity and religion. Amsterdam also has its reputation for allowing, in its easy-going manner, various other kinds of conduct, and the commerce catering to them, for which greater obstacles tend to be created in many other cities: of such phenomena the red-light district has obviously been around the longest, depending in history no doubt on the demands of a major port. Such acceptance is necessarily in some part a fact of local life, at the levels of official and unofficial native reactions. At the same time, at present, it also has its part in the attractiveness of Amsterdam to what one may term international subcultural tourism. Jansen

(1991), in his study of the geography of cannabis in Amsterdam, notes that many of the visitors to the coffee houses of the central city are young foreigners who know little of the city except the names of some famous outlets for soft drugs. Perhaps Jansen's English-language monograph itself attracts readers as a guide to such facilities.

Generally, I suppose, in Amsterdam or anywhere else, tolerance is very often more of a passive and distracted rather than an active stance. As people go about their own affairs, others can conduct theirs in their own way, as long as they cause no significant disruption. Tolerance as an accomplished fact may likewise have something to do with the microecology of the Amsterdam built environment. Dense, unpredictable, unsurveillable, uncontrollable, it allows some activities to go on unseen, or at least allows the excuse that they are not seen, whenever one prefers not to see. It is a part of the history of the city that for minorities during times of difficulty or persecution, this environment has provided at least temporary sanctuaries: a Catholic church in an attic in the seventeenth century, a Jewish family behind a hidden door in the early 1940s.

Tolerance, however, is not all that has come out of the Amsterdam experience of diversity. Its handling in cultural process has also included the passage of cultural forms between groups and contexts, along with reinterpretation and innovation. There are the Yiddish loan words characteristic of the Amsterdam dialect, and the somewhat bland *nasi goreng* served for lunch in a Dutch old people's home. The cuisine of the East Indies was no doubt introduced to the Netherlands by immigrants from there, and by returning colonialists; then as it was gaining popularity, I have been told, the resulting niche for small restaurants was taken over and expanded by Chinese seamen on Dutch ships, coming ashore in a period of economic depression. Thus almost every small town in the Netherlands seems with time to have been colonized in its turn by that special culinary institution which is part Chinese, part Southeast Asian, and certainly part Dutch. And as a late twist, but not necessarily the last, the Amsterdam street trade in *loempia* rolls now provides an opening for the Vietnamese.

The cultural careers of young black Surinamese, settled in Bijlmermeer or elsewhere, can be no less interesting. The Surinamese who came to the Netherlands were mostly people of fairly limited education, and not very much cosmopolitan sophistication. As they encountered the faster pace and rather bewildering heterogeneity and strangeness of Amsterdam and other cities, some found it a bit threatening, and turned inward to their kin and friendship networks and to the Moravian church they had brought from home. The younger people, as usual, were quicker to explore the new habitat. Boys, predictably, tended to be in the streets and other public places, and at least some of them made contacts with their Dutch counterparts there. Surinamese and Dutch girls likewise sometimes became friends, but more often they went to each other's homes. Thus it was the Surinamese girls who often became more familiar with the domestic aspects of Dutch life.

When the young Surinamese came to Amsterdam, however, they also found transnational youth culture waiting for them, among their peers and in the market-place. Some became 'disco freaks,' with whatever this implied in terms of clothing fashions and musical taste. In their case, as elsewhere in Europe, it turned out that in one particular context at least, being black was no handicap; among the devotees of popular culture inspired from the United States and the Caribbean, belonging to an ethnic minority was rather a social asset, and Surinamers could find themselves as stylistic leaders on the dance floor. If there was a disadvantage to this, it might rather be that they were tempted to devote the larger part of their energies to the night shift.

On the other hand, there were the Rastas. There had been no Rastafarianism in Paramaribo, or anywhere else in their old homeland. In Amsterdam this movement, or style, was apparently largely unknown until Bob Marley and his Wailers came for a concert in 1976, but then it caught on. Groups of the new Surinamese Rastas from Amsterdam might go over to London, European capital of Rastafarianism, and trade Dutch marijuana for Rasta hats, badges, and records. They could find inspiration, as well, in movies playing in a couple of ethnic-oriented theaters in Amsterdam.

Like Rastas elsewhere, they looked for authenticity in their African roots, only with the special Surinamese twist that to the consternation or even horror of their elders, they began to identify with the Bush Negroes of the Surinamese hinterland, treated back in Paramaribo rather as exemplars of the idiocy of rural life. Yet Amsterdam Rastafarianism was also a cultural mélange of varied ingredients. From elsewhere in the countercultural landscape of Amsterdam, it drew on the romantic environmentalism of 'being in touch with nature.' Meanwhile, if in a manner and to a degree Amsterdam Rastas thus became Greens, when Amsterdam coffee shops could no longer be quite so straightforward about being cannabis outlets, they adopted the red, yellow, and green emblematic colors

of Rastafarianism, and the national flag of Jamaica, to advertise their merchandise in a slightly more roundabout way.

And so cultural mingling and recombination goes on and on, weaving its way between contexts and organizational frameworks. Ideas and symbols make their passage from Kingston, Jamaica, to Amsterdam by way of London, and inspire a wistful look back to the bush beyond Paramaribo; not quite the shortest route from one Caribbean place to another. But then that is the way center-periphery relationships work, and again, it means that the centers are often centers not because they are the origins of all things, but rather because they are places of exchange, the switchboards of culture.

References

Becker, H.S. and Horowitz, I.L. (1971) 'The culture of civility', in Becker, H.S. (ed.) *Culture and Civility in San Francisco,* Chicago, Transaction/Aldine.

Jansen, A.C.M. (1991) *Cannabis in Amsterdam,* Muiderberg, Coutinho.

Levitt, T. (1983) 'The globalization of markets', *Harvard Business Review,* vol.61, no.3, pp.92–102.

Olsen, D. (1993) 'Urbanity, modernity and liberty: Amsterdam in the seventeenth century', in Deben, L., Heinemejer, W. and van der Vaart, D. (eds) *Understanding Amsterdam,* Amsterdam, Het Spinhuis.

Redfield, R. and Singer, M. (1954) 'The cultural role of cities', *Economic Development and Cultural Change,* no.3, pp.53–73.

Schama, S. (1987) *The Embarrassment of Riches,* London, Collins.

Source: Hannerz, 1996, pp.140–9

READING 1B
Shirin Madon: 'Bangalore: internal disparities of a city caught up in the information age'

A distinctive feature of most cities in the developing world is the fragmented character of their spatial organization. During the colonial period, this fragmentation was a direct consequence of the need to separate the European population from the indigenous (Balbo, 1993). The few decades that have elapsed since the end of colonialism have not been sufficient to transform the fragmented city into an integrated one. In fact, it appears that the nature of fragmentation has shifted to one based on participation in the information-intensive global economy by a core elite, and nonparticipation by the masses.

Bangalore presents a case study of this type of fragmentation. It is the fastest growing city in India (ranked fifth in the country) with a population rising from 1.66 million in 1971 to 2.92 million in 1981 (a 76% growth rate), and a projected population of 3.8 million by 2000 (Heitzman, 1992). At the turn of the century, Bangalore was a provincial town in Southern Karnataka that had evolved from a medieval temple/fort center and a colonial military encampment. Industrialization until the 1940s centered around textile manufacturing. The city's emergence as a center for information technology stems from decisions in New Delhi shortly after independence to locate strategically sensitive industries well away from borders and coastlands. Bangalore was therefore an obvious place to base the Indian air-force base and other public-sector institutions, which promoted the establishment of a number of universities, institutions, and colleges providing engineering and scientific training (Holstrom, 1994).

Between 1951 and 1971, there was a marked correlation between industrial development and accelerating population growth, with figures doubling during the 20-year period. By the 1970s, the occupational profile of the city showed the distinctive patterns of an emerging information society with 10.5% of the population in professional positions, 16.5% in clerical jobs, and 45% in production of which around half were in electrical fields (Singhal and Rogers, 1989). By the late 1980s, Bangalore included 375 large- and medium-scale industries and had 3000 companies employing 100,000 people in the electronics industry alone. The city contained up to 10,000 small industries and eight large industrial parks (Premi, 1991).

Bangalore now plays a prominent role in international electronics, telecommunications, and information technology, contributing almost 40% of India's production in high-technology industrial sectors (hardware, software, telecommunications) (*Observer*, 1995). The availability of highly skilled technicians, relative political and social stability within the state, the absence of labor conflicts, and an efficient banking network have meant that almost every big player in the information technology scene has its place there. Among indigenous high-technology companies that have established themselves in Bangalore are Wipro Systems (one of the largest Indian hardware and software vendors), Infosys Technologies, and Sonata Software. These, in turn, have been followed by a galaxy of multinational information technology companies attracted by the temperate climate of the city. These companies have adopted varying strategies to gain competitive advantage through offshore software outsourcing using local programmers and satellite links to design and produce customized packages. Some of these transnational companies (TNCs) like Digital Equipment, IBM, and British Aerospace have formed joint ventures with existing Indian players (NASSCOM, 1995).

Many of the high-tech companies in Bangalore are growing by more than 50% per year and are employing increasing numbers of graduates. For example, Siemens Communications, set up eighteen months ago to build software for the group's digital switching systems, already employs 250 people and plans to grow its workforce of software engineers to 1000 by the end of the decade (*Observer*, 1995). The increasing demand for highly skilled 'knowledge' workers threatens to outstrip the potential of local institutes in Bangalore to produce such skills, and many of the high-tech companies are having to recruit their workforce from other cities. In recent years, an increasing number of specialized small-scale workshops have been established to supply the high-technology industries. These workshops, most of which are characterized as being in the informal sector, recruit an increasing number of semiskilled migrant workers and help to bridge the gap between the demand and supply of labor.

To capitalize on its emergence as a popular location for software research, Bangalore has established a software technology park at Electronics City, just three-quarters of a mile out of the city. Hundreds of acres of research laboratories are occupied in the park by the likes of IBM, 3M, Motorola, Sanyo, and Texas Instruments. Companies that locate within the park are insulated from the world outside by power generators, by the leasing of special telephone lines, and by an international-style work environment. Bangalore is also in the throes of constructing a new upmarket international information technology park in the suburb of the city. The $250 million project, due to open in 1996, is a joint venture between the government of Karnataka, a Singapore consortium, and the Tata Group. The project aims to integrate advanced work facilities with recreational and residential attention, and thousands of enquiries regarding investment opportunities in the park have flooded in from the United Kingdom, North America, Europe, and Hong Kong. In parallel with the launching of this new park, a new international airport is also being funded by the Indian Tata Group and Singapore International Airlines, which is intended to provide a boost to industrial activity in Bangalore, in particular to the electronics and software industries (*India Today*, 1995).

Increasingly, the presence of a sizeable modern industrial sector has brought prosperity to the city and has given its central parts a cosmopolitan outlook. Measured in terms of expensive restaurants and pubs, boutiques, shopping plazas, and other signs of available purchasing power in the context of Western behavior patterns, a middle class is strikingly present in the central parts of the city and in other expensive areas beyond the center. A similar indicator of prosperity is the boom in the construction industry catering for the upper segments of the housing market and for commercial use. Cheap real estate has been one of the reasons why entrepreneurs prefer this city to the choked or very expensive cities like Bombay (*Observer*, 1995).

For all the preceding reasons, today Bangalore is conceived internationally as a prosperous and modern Indian city. But this label is misleading. At least three more characteristics have to be added to portray the reality of the situation. First, there exists gross inequality between groups of different socioeconomic status within the city of Bangalore and the region of Karnataka. Second, extreme poverty prevails among many inhabitants in the city. Third, there is an acute problem of civic deficiency both in Bangalore city and in the state of Karnataka, together with poor access to information outside the capital. These three characteristics are described in the remainder of this section.

First, contrary to the international reputation earned by Bangalore, the advent of the information age has yet to make a dent in the overall economic picture of the state, which remains primarily an agricultural state. Out of a total population of almost 50 million people, around 76% live in rural areas, and there is a high incidence of rural poverty in these areas (GOI, 1990). For example, 95% of the rural poor population in Karnataka have an annual family income of less than $102 (Nicoll, 1995). A sizeable share of the increase in the population of Bangalore city is related to migration. Many households of predominantly landless agricultural laborers and marginal farmers have been pushed out of their native rural villages by lack of means to survive and have been forced to move to Bangalore to find employment in the informal sectors as unskilled laborers. Tension has escalated within the city of Bangalore with the vast influx of migrant laborers from other Indian states, mainly from neighboring Tamil Nadu. These Tamil migrants seek menial jobs in construction and pose a real threat to local poor inhabitants, as witnessed in the violent anti-Tamil riots of 1991 …

Second, although poverty could be claimed to be of nationwide concern, in Bangalore the problem is more acute. In terms of absolute numbers, the poor easily predominate over the middle classes and professionals. While the share of urban inhabitants living in huts without access to infrastructure facilities is relatively small in Bangalore (10%) in contrast to other Indian metropoles (25–30%), there has been an exponential growth in this share over the last decade. The condition is continuously deteriorating both in terms of an increase in new slum areas and in terms of an increase in the population density of existing slums (de Wit, 1992). The discrepancy in habitat conditions between the rich and the poor inhabitants is also more extreme and more dramatically visible in Bangalore than in other cities. Due to societal modernization, the state is pressured to cut down trees to accommodate use of central urban land for nonresidential purposes and for expensive housing. Local middle-class residents and the urban poor are driven out of the city because of the rise of real estate and rent in the city center. They are therefore forced to squat on the urban fringe and to incur transport costs of commuting to the city each day in search for work (*India Today*, 1995).

While it is an open question as to whether the proliferation of shanty towns and environmental degradation is as a result of rural distress and increased population growth, or due to urbanization and globalization, what is certainly noticeable is the increasingly negative attitude of policymakers and planners in Bangalore toward slums. In discussions with the Bangalore state authorities, most policymakers did not officially acknowledge slum dwellers as citizens of the city, even though they clearly constitute the majority of the population. This attitude is reflected in the periodic demotion of any visible manifestations of poverty such as the frequent 'clean-up' programs in which squatter settlements are demolished in order to preserve the quality of life of the so-called 'modern' sector. Contrary to policy in other Indian cities, an increasing number of slums in Bangalore are located on private land that does not belong to the government and therefore precludes the eventual transfer of ownership to hut dwellers. In many cases, private developers build high-rise blocks so that less space is consumed for housing and the remaining space is allocated to commercial units for which private developers can reap hefty revenues.

A growing number of small, recently established nongovernment organizations such as CIVIC and the Bangalore Poverty Alleviation Programme (BUPP) have taken issue with the claim that slum dwellers have no right to live in the city. According to these groups, the recent internationalization of industrial activity in Bangalore has had a negative impact on the poor since less money is allocated to improving public services and providing urban development programs. These organizations are striving to give poor urbanites some say in the functioning of the city and are currently fighting against the state government's decision to force slum dwellers to share their already scanty land with private entrepreneurs. The BUPP is also developing an information system on slum activities using data compiled by slum dwellers themselves on the status of the land they occupy, the number of people living in each hut, access to amenities, slum dwellers' prioritization of their own problems, and on the nature of information and communication channels between economic agents in the informal sector. The information from the system will be circulated to city authorities for planning purposes and to slum dwellers themselves to make them more aware of their role in the life of the city.

Third, civic deficiencies caused partially by Bangalore's industrial success need to be faced. Although software companies are not big power users, they need to install voltage regulators, uninterrupted power supplies, and generators to run their computers, because of shortages of power. Similarly, there is a looming water shortage, with municipal pumped water only available two days a week and boreholes drying up. The Bangalore Development Authority is trying to encourage more private-sector involvement in the development of infrastructure, especially transport. In terms of industrial development, the city has almost reached saturation point. Today, the state of Karnataka is home to some 114 companies – all of which are located in Bangalore, due to the poor infrastructure facilities outside the city.

The development of secondary cities within the state of Karnataka has been envisaged for some time by the Bangalore metropolitan regional development authority, with an ambitious project to develop a mega ring road and a light railway connecting all towns in the state. However, little has been done to date. As a result, thousands of young computer professionals continue to struggle along terrible roads to get to their jobs in software factories.

Information infrastructure in the region remains confined to the business community in the city with poor access to information outside the capital.

References

Balbo, M. (1993) 'Urban planning and the fragmented city of developing countries', *Third World Planning Review*, vol.1, no.7, pp.23–35.

de Wit, M.J. (1992) 'The slums of Bangalore: mapping a crisis in overpopulation', *Geographic Information Systems*, vol.2, no.1.

Heitzman, J. (1992) 'Information systems and urbanisation in South Asia', *Contemporary South Asia*, vol.1, no.3, pp.363–80.

Holstrom, M. (1994) 'Bangalore as an industrial district – flexible specialisation in a labor-surplus economy?', Pondy papers in social sciences, no.14, Pondicherry, French Institute.

GOI (1990) *Concurrent Evaluation of IRDP – the Main Findings of the Survey for January 1989–December 1989*, New Delhi, Government of India.

India Today (1995) 'Real estate survey', 31 July.

NASSCOM (1995) *The Software Industry in India: Strategic Review 1995*, New Delhi, National Association of Software and Service Companies.

Nicoll, A. (1995) 'A finger-lickin' uproar', *Financial Times*, 5 October.

Observer (1995) 'Karnataka – a world leader in technology', 29 October.

Premi, M.K. (1991) *India's Population: Heading Toward a Billion – an Analysis of the 1991 Census*, New Delhi, BR Publishing.

Singhal, A.K. and Rogers, E.M. (1989) *India's Information Revolution*, New Delhi, Sage.

Source: Madon, 1997, pp.232–6

CHAPTER 2

Moving cities: transport connections

by Kerry Hamilton and Susan Hoyle

1 *Introduction*

In this chapter, the notion of cities as something 'made' through networks of connections – through the movement and flows of people in particular – is explored further through the issue of transport. We want to invite you to consider transport as the key to the city: to its historical development, to its late-twentieth-century manifestations, and to our individual engagement with it. Transport does not explain everything about the city – no single perspective will do that – but we think it is a persuasive way of looking at the urban experience: the shape of the city and its growth, its power and inequalities, its connections and disconnections, its myriad networks, and its intensity. Transport mirrors so much of the city – the horror and the romance, as we refer to it in the final section. Indeed, we would argue that transport holds the key to urban sustainability. Although we cannot exhaust that issue here (see **Blowers and Pain, 1999**), it is no accident that the control of transport, and in particular of the car, is increasingly the focus of urban planners worldwide. We hope that by the end of this chapter you will find that transport is an obvious approach to the study of the city. What follows seeks to persuade you that this is indeed the case.

Sections 2 and 3 set out the centrality of movement and transport to city life, and the significance that choices over transport infrastructure have made to the development of cities more generally. Following that, in sections 4 and 5, some everyday assumptions about speed are interrogated, and issues of access and inequality in relation to transport and mobility are explored at some length. Then, from section 6 onwards, we engage with some of the ambivalent tensions, to use the language of the last chapter, that are generated by transport and car travel. In particular, the idea of freedom often associated with movement and car ownership is addressed, together with its contradictory aspects. Finally, in a similar vein to the above, the horror and romance attached to matters of transport and travel are considered for the complex feelings they all too often arouse in us.

2 *Transport makes and remakes the city*

Let's start by asking ourselves a simple question: how is it exactly that transport makes the city?

ACTIVITY 2.1   Shortly we will look at the ways in which transport has shaped some specific cities, but for now we want you to think about the general case. Take a few minutes to make a list of the ways in which transport affects, and is affected by, the city. Try to list at least ten points (for example: 'transport takes people to and from their employment', 'transport brings people together').

You will find our list in the appendix at the end of this chapter. ◆

2.1   MAKING THE CITY

In the course of this chapter we hope to touch upon most of the points you have raised. For the moment, however, we want to consider transport in general and to emphasize the central role that it plays in cities. The city, for better or worse, is made by transport. For example, many cities grew up around rivers, because rivers were the major trading routes, but those rivers are also major barriers to movement to this day: the Left and Right Banks in Paris, north and south London, Buda and Pest. Fords, ferries, bridges and tunnels have long been available to help us over or under the barrier, but the expense and therefore scarcity of these links means that where a city has a river, it remains one of the constraints on movement in that city.

We may not normally think of bridges, roads and pathways as constraints on our movements. Rather we might see them as facilitating travel, but their existence does mean that we have to move along *them* rather than elsewhere, even if they are far from the straightest route to our destination. If we do want to go elsewhere in the city, we are likely to find a solidly built environment blocking our path. We need also to be aware of the way in which roads and railways *divide* communities, making movement and contact between them difficult, often socially and psychologically, as well as physically: people are said to live on the 'wrong side of the tracks'.

Of course, it is not only people who move in cities. In most places the canals and railways were originally built to accommodate freight, not people. Different kinds of goods – letters, bricks, food, clothes, furniture – are carried by different means within the city. Until quite recently you would have seen large flocks of animals trotting to market along the streets of British cities, and they can be seen in cities of less developed countries today.

Another reason why transport shapes the city is that transport infrastructure rarely goes away. There will have been considered reasons for building the road, canal or railway in the first place, but those reasons do not necessarily survive. For example, Crystal Palace in London was developed in the 1850s as a major tourist attraction; the road infrastructure is still there, but the tourists are long gone (not least because the building burned down in the 1930s!). This observation applies also to links between cities: **Pile (1999)**, for example, describes the growth of networks in nineteenth-century Chicago. In Britain, canals link eighteenth-century centres of production; railways link nineteenth-century centres of production; and motorways link twentieth-century centres of production and, more recently, consumption. The relative importance of cities changes over time, but the transport links between them, especially roads, are rarely demolished. They remain not just as part of our landscape, but usually as part of our transport infrastructure.

In the same way, the generations that follow us will have to live with our transport priorities. Or will they? Might they reshape their cities by filling in rivers, closing roads and turning throughways into linear parks?

2.2 REMAKING THE CITY

Transport not only makes the city: it can also unmake and remake it, as the historical example of Liverpool demonstrates. Liverpool historically has been surrounded by sea and bog, and until about two hundred years ago it depended almost entirely upon ships and boats for its links with the rest of the world. Because of its marshy hinterland, there were no roads into the city suitable for wheeled vehicles. Most of what the city needed from outside came by sea and, after 1763, by canal; otherwise it had to be brought on foot, or on the back of a horse or donkey. Manchester came to rely on cotton shipped from Liverpool through the Bridgewater Canal, and the canal-owners increasingly exploited that dependence, with escalating charges and huge back-logs of bales stacking up on the Liverpool quaysides. When the frustrated mill-owners of Manchester threatened to build a railway, the canal-owners ignored them, certain that the bog at Chatmoss was impassable. But Robert Stephenson, as well as building the Rocket steam-engine, designed the technology which remade Liverpool. He borrowed from Ireland's experience with peat-bogs: roads, and then the railway itself, were built on, first, layers of sand and gravel, and then (to quote Stephenson's own words) 'bundles thickly interwoven with twisted heath, which form a platform for the covering' (Francis, 1851/1968, p.132).

3 *Island cities*

The transport choices made in the development of a city determine the city's shape. To illustrate this we shall look at three cities, all built on islands and each responding in crucially different ways to the transport problems and opportunities which their sites offered: Venice, New York and Hong Kong. Islands offer their inhabitants a naturally defensible position, which is usually the first attraction of the chosen site. When cities are on the sea coast (as all our examples are), they also provide a convenient landfall and/or foothold for invading colonists. But once a settlement begins to grow, the transport infrastructure choices will profoundly influence how the city subsequently develops.

Venice

The islands on which Venice was founded (traditionally in AD568, although there were people living on the lagoon islands long before) were alluvial mud-banks, which were, and have remained, unfit for supporting buildings. None of the land is fit for agriculture either; the only naturally occurring fresh water is from rain, and the only local food resource is fish. Structures in Venice, and on the other islands in the lagoon, are supported by wooden piles driven deep into the mud. There are two major canals in the city, the Grand Canal and the Giudecca, both probably old river courses. There are some 175 other canals, with a total length (including the Grand and Giudecca Canals) of 28 miles; in addition there are about 90 miles of alleyways, a very few of which are as wide as a narrow street. All this infrastructure is very dense: the main island is about three miles long, and in its heyday 175,000 people lived in Venice.

There are three bridges over the Grand Canal and none over the Giudecca, while there are numerous little bridges over the smaller canals. To get anywhere in Venice, one either walks or takes to the water. There is a causeway to the mainland at Mestre, over which pass the railway and the road. Neither goes any further than the very edge of Venice. The only wheels in Venice, outside the railway-station, the bus-station and the multi-storey car-park (all on the edge), are on wheelbarrows and children's toys or pushchairs. Nor are beasts of burden allowed: Lord Byron may have kept a menagerie in his Venetian *palazzo*, but he kept his horses on the Lido (the island which divides the lagoon from the sea, two miles away to the south-east).

The *vaporetti* (waterbuses, originally steam-powered, but now diesel-fuelled) are intermittently noisy; the water-taxis and private launches, the barges and garbage boats add to the pollution and the swell; but for the most part, the traveller on foot can neither hear nor see what is happening on the water, just as anyone who views Venice from the water would be almost entirely oblivious of the bewildering warren of paths, alleys and lanes which criss-cross the city, crowded with people, with just the sound of their voices and footsteps.

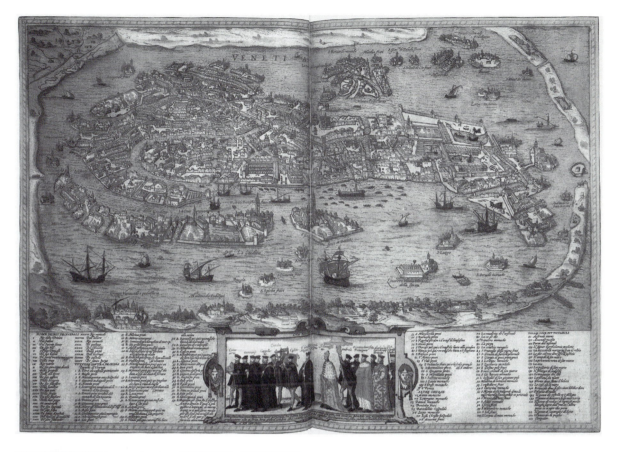

FIGURE 2.1 *Map of Venice, 1572*

FIGURE 2.2 *Goods being transported by barge along a Venetian canal*

New York

New York was founded in 1615 when the New Netherlands Company built a warehouse and a fort on Manhattan Island. Today the city consists of five boroughs – Manhattan, the Bronx, Queens, Brooklyn and Richmond – of which only the Bronx is on the mainland. Manhattan is on Manhattan Island, Queens and Brooklyn are on Long Island, and Richmond is on Staten Island. There are a number of smaller islands (for example Coney and Riker Islands) within the city boundaries. The city's total waterfront is 578 miles, of which Manhattan and Richmond account for only 43 and 57 miles respectively; Brooklyn's waterfront is some 200 miles. New York has a huge natural harbour – in fact made up of several harbours – where the enormous old ocean-going liners, such as the *Queen Mary,* the *Queen Elizabeth* and the *France,* used to tie up alongside the Express Highway on the east side of Manhattan, just across from the New Jersey shore.

Travel between the islands was originally by private boat and by ferry, but these services were unreliable, particularly in winter, when ice formed in the rivers. The construction of the Brooklyn Bridge (which opened in 1883) was the first important link from Manhattan Island. Other bridges followed later, and also road tunnels.

Probably most important of all in the subsequent shaping of New York was the development of the subway system. In 1904, the first of New York's underground railways opened, and the process which has made and kept New York quite different from any other North American city was begun. As Hood (1993, p.12) points out, 'The subway integrated New York, overcoming river barriers, joining the boroughs, and shaping neighbourhoods. It kept New York compact and socially diverse'.

The average annual boardings per head of population for public transport in US metropolitan areas of more than one million population is 53.3; for New York it is 143.4. The New York-Long Island-Northern New Jersey mass transit accounts for over a third of all US usage of public transport, even though usage has for some years been in gentle decline (Wendell Cox, 1998).

FIGURE 2.3
Manhattan, an island city

Hong Kong

ACTIVITY 2.2   Now turn to Reading 2A by Timothy Hau, 'Transport for urban development in Hong Kong'. Despite our best editorial efforts, much of the 'transport economist' flavour of the original article survives. If you have an economics background, this will not be a problem, but for those of you who do not, parts may seem heavy-going. Don't worry about the details of the economic recommendations, but concentrate on the implications for city transport, and the overall strategies that Hau is examining.

Hong Kong is another city built on islands, but this piece does not dwell on either the city's origins or how its transport shaped its growth. Rather it looks at how its future growth will be helped or hindered by its transport systems.

Given what Hau calls 'the fundamental law of traffic congestion', you may find the global concern with the demand for cars in urban areas more understandable: his answer is to manage demand as well as supply, and specifically to look at road-pricing. How would this impact upon the idea of the city as a place of freedom? ◆

The ways in which cities differ from each other are as many as the ways in which they are similar.

● People's transport needs in Venice, London, Denver and Colombo are different. History, topography, population density and distribution, levels of income, and the proportion of day-time city-users who don't live there all influence what can be done.

● Whatever the differences between them, however, cities can and do learn from each other: road-pricing in Hong Kong (see Reading 2A) is influencing thinking throughout the world; attempts in US cities like Atlanta, Los Angeles and Houston to reintroduce public transport systems have been followed in European cities facing similar emergent problems; and the role of transport in allowing the planned development of Curitiba in Brazil, particularly through its very successful bus network, has attracted international attention through conferences and publications (see, for example, Birk and Zegras, 1993).

We have said that transport makes the city. In our eagerness to stress the often overlooked contribution that transport makes to the shape and very experience of the city, perhaps we have overstated the case. We admit it: the city is more than movement. A glance at the rest of this book makes that obvious. But movement is central, and that is what we are going to focus on now.

4 *Beam me up, Scotty!*

Transport is traditionally referred to as a derived demand: that is, it is secondary to the ultimate purpose of the journey. Part of the assumption in treating transport as a derived demand is that if people didn't have to move, they wouldn't, and that if everything needed to support life was within an arm's length, travel would be a very rare taste. These assumptions feed into planners' and politicians' attitudes to transport provision, and mean that cities tend to provide rudimentary means of getting about for both people and goods. The experience of travel is often ignored, and the needs of travellers are rarely investigated, let alone accommodated. The movement of goods is barely tolerated, but little is done to reduce its impact. It is not only planners and politicians who fail to get to grips with this central problem of the modern city; other city-dwellers also fail to think through the implications of transport, and no one is encouraged to do otherwise. It is treated as a necessary evil. The dream, too often, is that people and goods should be transported in an instant.

Although this dream may have its origins in economists' concept of a derived and thus secondary demand, it is probably also related to the nightmare that if anything kills the city, it will be transport. But what about a different dream:

●   That the city is the ideal place for people-transport which is financially and operationally efficient, visually enriching, secure, comfortable and available, and for goods-transport which is timely, unobtrusive and economically viable?

●   That there is great scope for city transport which is minimally polluting, in terms not only of vision and noise, but also, and above all, in terms of air quality?

And what are people saying if they indulge the dream of instantaneous transfer: that they will not have to go anywhere for anything; that 'no transport' is a good thing?

ACTIVITY 2.3   We are clearly sceptical about the dream of a travel-free life. This is not to say that reducing the demand for travel is not a 'Good Thing', but how that would be achieved is highly problematic.

Think of some technological or social ways of reducing the need to travel, and consider how they would affect transport demand. Be as imaginative as you can: the technology doesn't have to exist or have been adapted yet, and the social systems may not be in place. We suggest that the objectives behind the 'fixes' will be:

●   bringing things to people rather than people going to them (e.g. deliveries from supermarkets);

- moving desired destinations closer together (e.g. home and workplace); and

- substituting a quite different activity (e.g. virtual tourism rather than the real thing).

Try to list five examples for each objective, and remember to analyse how the changes would impact on transport demand.

Now turn to Reading 2B by Stephen Graham, which is about information technology (IT) and the city. How do the developments he describes affect the viability of your 'fixes'? ◆

One of the ironies, or just peculiarities, of IT is that it can and does coexist, within the same city, with a level of poverty and deprivation about which personal computers and the information revolution have nothing at all to say. In São Paulo in Brazil, for example, 'São Paulo never stops' is a popular saying. A variant on this is 'especially when looking for a parking space'. Nevertheless, in spite of the high rate of car ownership, two and a half million people who live there walk to work; for many of them even the cheapest buses are too expensive.

FIGURE 2.4
'Sao Paulo never stops'

5

Speed, mobility and access

5.1   THE QUEST FOR SPEED

Speed, even in cities, is a virtue in many societies. It was not always so. In other societies time has had a different, not necessarily lesser, value. Part of the twentieth-century attachment to speed is the sheer fact of its possibility: human beings feel compelled to do certain things just because they find the means to do them – climbing Everest, genetic engineering, splitting the atom, flying to the moon. This urge may well be related to the way humans learn to speak, walk and explore the environment, although you may agree with us that this is not the place to pursue that thought. However, we do want to pursue the question of the desirability of speed in the city.

Let's consider what the quest for speed has done to the city. In New York, London, Paris, Tokyo, road traffic speeds are more or less the same now as they were one hundred years ago. The modern motor-car goes no faster through the metropolis than did the horse and carriage. The whole paraphernalia of today's transport infrastructure – underpasses, flyovers, roundabouts, traffic-lights – has left city travel just as slow, just as fast, as it ever was. This is despite the wholesale destruction of community assets (housing, parks, footpaths, pubs, shops) to make way for the car and lorry, and despite the vast financial cost.

TABLE 2.1:   *The costs of road-building in the UK*

Location	Length (miles)	Cost (£m)	Average cost per mile (£m)
Aberdeen	17.5	80	4.6
London Limehouse link	1.1	448	408.0
Newbury bypass	9.0	17	1.8
Stonehenge tunnel	2.5	200	80.0

Source: House of Commons, 1997; averages calculated by authors

Table 2.1 shows the costs of of four major road-building projects in the UK. Even taking account of the high cost of labour and land in the UK, why has this investment been so self-defeating? Why are other cities queuing up to do the same? All the evidence is that the more roads are built, the more traffic arrives to use them, and any speed-gain disappears within months if not weeks. One paraphrase of this situation (known as the Lewis-Mogridge position) is that 'traffic expands to meet the available road-space' (Mogridge, 1990).

The propositions we are working towards are:

● that speed, if you can get it, is rarely worth the price;

● that, in the city, you are unlikely to be able to get it anyway; and

• that travel does take time, and travellers need to accept that fact – or else make other arrangements.

None of this is simple, and we are able to do little more than raise the issues. What is a reasonable length of time to spend travelling a given distance? What distance is it reasonable to live from work? What is a reasonable price to pay for transport (fares, resources, opportunity costs)? Who is to judge what is 'reasonable'?

16 VALVES PUMP PURE ADRENALIN.

Now and again a new car is launched that sends the blood racing through your veins.

The moment you see it, you have the overwhelming urge to inspect the engine.

Followed by an even greater urge to get behind the wheel.

The new Saab Turbo 16 is one such car.

Its third generation 16 valve double overhead cam turbo engine with new head, LH-Jetronic injection,

5.2   SPEED IS RELATIVE

ACTIVITY 2.4   We present below four extracts which relate to the experience of speed. As you read through them, try to draw out the similarities and differences in the way in which speed is evoked. In particular, you should bear in mind the obvious point that speed can be slow as well as fast, leisurely as well as rapid. You should also consider how the 'thrill' of speed is often related to the passage of time. ◆

FIGURE 2.5
Speed in the city: an advertiser's dream?

The first extract is from George Eliot's introduction to *Felix Holt, the Radical*. This novel was written in 1865–6, during the agitation preceding what became the UK's second Reform Act (1867), and is set during the disturbances that preceded the first Reform Act (1832). The quotation is from a famous passage which describes the Midlands as seen from a stage-coach in the years immediately before the arrival of the railway. George Eliot (real name Marian Evans) would have been a child of 11 or 12 at the time her novel was set, and it is most unlikely that the 'memories' she describes are her own:

> … you have not the best of it in all things, O youngsters! the elderly man has his enviable memories, and not the least of them is the memory of a long journey in mid-spring or autumn on the outside of a stage-coach. Posterity may be shot, like a bullet through a tube, by atmospheric pressure from Winchester to Newcastle: that is a fine result to have among our hopes; but the slow old-fashioned way of getting from one end of our country to the other is the better thing to have in our memory. The tube-journey can never lend much to picture and narrative; it is as barren as an exclamatory O! Whereas the happy outside passenger seated on the box from the dawn to the gloaming gathered enough stories of English life, enough of English labours in town and country, enough aspects of earth and sky, to make episodes for a modern Odyssey.

(Eliot, 1866/1972, pp.75–6)

The second extract is from Fanny Kemble's account of her first railway journey, in 1830. This was the year the first passenger railway (Manchester to Liverpool) opened, when she would have been 20 or 21 and at the height of her fame as an actress. Train travel was a new experience for everyone:

> You can't imagine how strange it seemed to be journeying on thus, without any visible cause of progress other than the magical machine with its flying white breath and rhythmical, unvarying pace, between these rocky walls, which are already clothed in moss and ferns and grasses; and when I reflected that these great masses of stone had been cut asunder to allow our passage thus far below the surface of the earth, I felt no fairy tale was ever half so wonderful as what I saw. Bridges were thrown from side to side across the top of these cliffs and the people looking down upon us from them seemed like pygmies standing in the sky … The engine … was set off at its utmost speed, 35 miles an hour, swifter than a bird flies … You cannot conceive what that sensation of cutting the air was; the motion as smooth as possible too. I could either have read or written; and as it was I stood up, and with my bonnet off drank the air before me. The wind, which was strong, or perhaps the force of our own thrusting against it, absolutely weighed my eyelids down. When I closed my eyes this sensation of flying was quite delightful, and strange beyond description; yet strange as it was, I had a perfect sense of security and not the slightest fear.

(quoted in Hamilton and Potter, 1985, pp.12–13)

The next two quotations are from the 1990s. The first, by Bruce Graham, is from the Internet:

> WOW! On Saturday November 11, 1989, I was a passenger on a British Airways Concorde … from London via Washington to Miami. I will never forget that day. It's not often that one gets treated to a gourmet dinner while being moved through the sky faster than a rifle bullet.
>
> Of course, terrific meals are not all one gets to enjoy in flying on one of these fabulous aircraft … Here are a few additional details about Concorde travel:
>
> - As Concorde breaks the sound barrier, there is no unusual sensation inside the cabin. You experience only an increase in engine thrust.
>
> - No 'sonic booms' are heard inside your plane at any time …
>
> - If your flight lifts off after sunset, as mine did, you may see a sunrise in the west. The sun appears to move backward in the sky, as your conveyance outruns it.
>
> - At 60,000 feet, you can enjoy a view of the curvature of the Earth …
>
> - Tiny dots, far below, appear to be sitting on top of the clouds. These are 747s and other jumbo jets, laboring to cross the Atlantic. The jumbos are about five miles below you …
>
> - Going west, you land several hours before you took off.
>
> In a word … WOW!

(Graham, 1996)

The last extract is from Martin Amis:

> Here … is my auto biography [*sic*].
>
> I hit the road in 1966 at the age of 17, perched on the saddle of a scooter. The very choice of vehicle was a statement of pack aggression, for this was the age of Mods and Rockers. Mods rode – and crashed – scooters; Rockers rode – and crashed – motorbikes. Whenever we encountered one another, we would rev and veer and dice, race, chase, flee, yaw, skid, crash. But I needed no Rockers to help unseat me – to send me skittering like a hockey puck over the bitumen. I did it all by myself, at least once a day. Invariably travelling much too fast I had great difficulty slowing down when things appeared in my path: cars, trucks, pedestrians, brick walls.
>
> One winter night, enjoying my usual speed trail through the side streets of Chelsea, I noticed an obstacle up ahead. It was a main road. My attempt to slow down met with the usual snag: I was travelling much too fast. Rammed sideways by a tomato-red Jaguar MkII, I spent that Christmas in hospital. My Vespa Sportique was a cuboid ruin. I had owned it for four months.
>
> With my apprenticeship served, I graduated to four wheels. I no longer crashed scooters. Now I crashed cars. I crashed my car all the time. My friends crashed their cars all the time. We crashed each other's cars all the time. Occasionally, while practising handbrake-skid U-turns or hot rubber starts we crashed our cars into each other's cars. We drove everywhere as fast as we could, with busted indicators, faulty brakes and flat tyres. We drove drunk and we drove stoned and we drove on LSD.
>
> (Amis, 1998, p.1)

These widely differing accounts of the effect of speed on travel in and between cities also share some attitudes. We note that while Fanny Kemble writes in 1830 of travelling swifter than a bird flies, Bruce Graham's Concorde is faster than a rifle bullet; he sees the jumbo jets far below as 'tiny dots', 'laboring' across the ocean, while Kemble sees 'pygmies standing in the sky' on the bridges high above the cuttings. She tells us that one could read or write on her train; Graham eats a gourmet dinner on his plane. Kemble 'felt no fairy tale was ever half so wonderful as what [she] saw', while Graham is reduced to 'WOW!'. George Eliot and Martin Amis are both accomplished writers of fiction: Amis may be proffering his 'auto biography' as real autobiography, but perhaps it is to be taken with a pinch of salt; *Felix Holt* is certainly fiction, but George Eliot intends us to believe every word of her coachman's slow world-view. Both of them use words with particular care and attention: they are describing what once was ordinary (but no less amazing for all that) to audiences for whom (it is assumed) these experiences are foreign, or at least forgotten. George Eliot and Fanny Kemble were writing about the speed revolution of the 1820s and 1830s, George Eliot to regret what had been lost (even if she accepted the change as necessary), Kemble to embrace the new, bonnet off and eyes shut. Amis and Graham are writing about the speed revolution of the last few decades, but while Amis is apparently ashamed of his boy-racer past, Graham is quite uncritical of his speed.

5.3   MOBILITY AND ACCESS

There is not only the misunderstanding of speed (which concentrates on time); there is also the misplaced value accorded to the related concept of *mobility* (which concentrates on space). It is misplaced because it is too readily confused with the rather different matter of access. For too many years, 'mobility' has been the prime focus in transport provision: the aim has been to keep the traveller moving at all costs. This focus obscures the need for an engagement with the purpose (as opposed to the fact) of movement. Consider a car journey from Birmingham to London (about one hundred miles, mostly along motorways): easily half the journey time can be taken up within London, as you approach your destination through very slow-moving traffic. Once there, half the time again can be spent finding a parking-space for the car. Although the car may be moving for virtually all the time, and therefore the driver and passengers may well be mobile, this emphasis misses the obvious point of the journey, which is to reach the destination. Thus the popular definition of 'mobility', which concentrates on movement and speed to the detriment of access, will always be misguided, especially within an urban context.

Mobility has to be about *access*. Everyone needs access, not only people in wheelchairs and not only to buildings. By 'access' we mean the (relative) ease of physically reaching a person or place: the shops may be in full view across the road from the place where you live, but if there is a three-lane dual carriageway in between, and the nearest footbridge is half a mile away, then the shops are certainly not easily accessible to you. If you have a car and the next roundabout is a mile off, then this requires a two-mile drive to shops which are only a matter of yards away – this again is an example of (relative) inaccessibility. Transport corrupts our concept of space (or at least undermines it) by introducing differential experiences of distance. Equidistant destinations within the city may take radically different lengths of time and effort to reach, while different people, for a variety of reasons, may have quite different problems in reaching the same destination from the same origin. Spatial arrangements are only part of the story, as transport reveals.

It is because of problems of mobility/access (rather than mobility/speed) that there are such stark inequalities in transport provision and use. It is to transport and inequalities that we now turn.

6 *Transport, inequality and power in the city*

6.1   POWER AND CHOICE

Inequality provides a perspective on power, and an investigation of a city's transport draws a picture of the power relationships in that city. Although we concentrate below on the transport inequalities signalled by gender, race and age, there is nothing inherent in femaleness, skin colour or old age that renders people less likely to drive or more likely to use the bus. Having power is about having choices and controlling change; it is also about controlling other people's choices.

Power is exercised when voters refuse to pay towards the cost of transport investment or revenue subsidy. It can be exercised indirectly and (perhaps) unknowingly by driving a car down a street: by taking space from other uses, polluting the air, creating danger. It can be done directly by successfully opposing a bus-route along your street. In London at the turn of the nineteenth century, people living in Kensington and Chelsea kept trams off their streets in order to keep out the labouring classes which were perceived to be the main patrons of trams. Gated streets are common in many cities to bar entry to undesirable elements (see Chapter 1 and also **Allen, 1999**).

Power is not a simple concept; it is everywhere, latent or manifest, in any relationship. It provides a key to the deconstruction of the concepts of speed, mobility and access which we discussed earlier. There is an obvious relationship between speed and power: there is the growth of speed as a status symbol, and there is the way in which even those without access to speed are encouraged to tolerate and admire it. An example of this is the history of Concorde, a joint project between France and the UK to produce a supersonic aircraft which would transport small numbers of people between rich urban centres at very high ticket prices. The refusal of most of the countries of the world to allow supersonic travel over their land further restricted Concorde's market, but it has also enhanced the exclusivity of what was always intended to be an exclusive experience. At the same time, the emphasis on Concorde's speed (recall the quotation from Graham above) has allowed it to meet its objective of becoming a source of national pride for people, the vast majority of whom will never be able to afford to fly in it.

6.2   INCOME AND CAR OWNERSHIP

Income can dictate who has access to good health, education, housing, leisure … and transport: whether one can afford a car, or the bus-fare, or is forced to walk. Figure 2.6 shows the close relationship between income (or, more accurately, GNP per capita) and car ownership across a range of cities in both developed and less developed countries. Extract 2.1 is from a report by the United Nations Centre for Human Settlement (UNCHS), and it illustrates that there are interesting variations to the pattern shown in Figure 2.6.

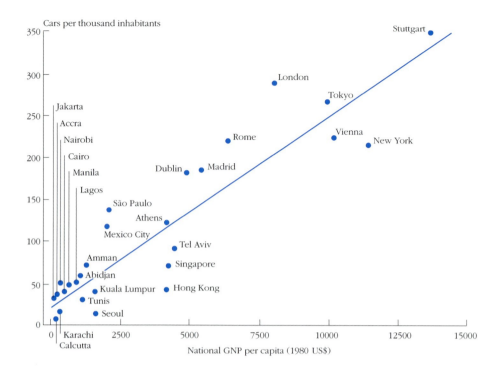

FIGURE 2.6 *Income and car ownership in various cities*

EXTRACT 2.1
UNCHS: 'Worldwide transport patterns'

The last ten to fifteen years [have] brought a continuing rise in the number of motorized road vehicles worldwide and continued growth in air traffic. For instance, road traffic in the European Union grew by 70 per cent between 1970 and 1985 and is expected to increase by a further 50 per cent between 1985 and 2000. The growth in automobile use and in air travel has been a major factor in growing levels of fossil fuel use and in greenhouse gas emissions. In the North, by the early 1990s, transport accounted for around 30 per cent of total energy consumption, with variation between countries. For instance, in the United States, it accounted for 37 per cent while in Japan it was 27 per cent and in the Netherlands 22 per cent. Virtually all cities have been transformed by motorized road vehicles …

The growth in the number of cars worldwide in recent decades has been far more rapid than the growth in the urban population. For instance, in 1950, there were around 53 million cars on the world's roads, three quarters of them in the United States; by 1990, there were more than 400 million and another 100 million trucks, buses and commercial vehicles. Around one third of these were in Europe, another third were in North America and the final third divided between the rest of the world.

Although in Africa, Asia and Latin America, the number of road vehicles per person remains far below the level in Europe and North America, certain countries in these regions have had the most rapid growth in the number of road vehicles – and a few of the wealthiest countries in these regions have levels of car ownership comparable to those of Europe. Number of passenger cars per 1000 inhabitants indicates the very large variations between some of the world's poorest countries with one or two passenger cars per 1000 inhabitants in 1985 and some of the wealthiest countries with 400 or more …

In many cities in the North, a significant proportion of the population now live in low-density suburbs within which no public transport can operate cost-effectively and in which at least two private cars per household are almost a necessity. Even in a small, compact country such as Denmark, each person (including children and the elderly) travels an average of 40 kilometres a day with this average projected to rise to 55 kilometres a day by the year 2010 …

A detailed study of 32 major cities in North America, Europe, Australia and Asia that looked at the extent of automobile dependence and the factors that helped explain this found that the 32 cities could be divided into five categories. Most US and Australian cities were within Categories 1 and 2 which have a high or very high automobile dependence and at most a minor role for public transport, walking and cycling. Most European cities fell into categories 3 and 4 which had moderate or low automobile dependence and an important role for public transport. However, Munich and Paris, both among the most prosperous cities in Europe, along with three of the most prosperous Asian cities (Tokyo, Singapore, Hong Kong) had a very low automobile dependence with public transport, walking and cycling more important than cars.

… in the mid 1980s, over two thirds of all motorized trips in Seoul, Bombay, Shanghai, Manila and Calcutta as well as Tokyo and Hong Kong were made by bus, rail or subway …

Cities in China such as Shanghai and Tianjin have among the world's highest rates of bicycles in use, although in both, these rates may be declining, with the rapid growth in the number of automobiles and with transport plans oriented towards increasing automobile use. Most cities in the South also have a high proportion of trips made by walking. But it is not only cities in relatively low-income countries that have a high proportion of trips made by walking or bicycling. For instance, in cities in West Germany in 1989, 27 per cent of trips were made by walking with 10 per cent by bicycle. In many cities in Denmark, France, Sweden, Germany and the Netherlands, a high proportion of all trips are made by bicycle; in Delft (the Netherlands) where special attention has been given to encouraging bicycle use, 43 per cent of trips are made by bicycle with 26 per cent made by walking.

Source: UNCHS, 1996, pp.23–5

Car ownership, as we have mentioned, is a great divider. In the West, people are increasingly regarded as deprived if they do not have access to a car; in other parts of the world, it is only the rich who have a car. Car ownership has been increasing throughout the western world, with a big leap in the 1970s, until now it is close to saturation point in many countries (see Figure 2.7). However, car ownership still tends to be lower in cities, usually in correlation with good public transport provision and high population density, as in London, New York, Singapore and Hong Kong (UNCHS, 1996, p.24).

FIGURE 2.7
World motor vehicle ownership, 1970–2010

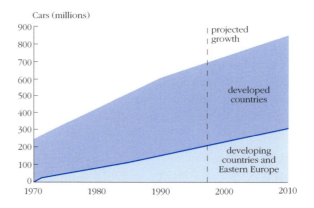

6.3   ATTITUDES TO TRANSPORT

The following quotation is about transport in São Paulo, which is one of the largest cities in Brazil:

> The bus system in São Paulo is of great importance in moving people around the city. Buses carry about 8 million passengers every day …
>
> There is a marked social division in the use of different modes. Even more so than in many other countries, to use the buses carries stigma. It is generally regarded as a sign of being among the poorer classes. It is said that a young person will put acquisition of a car – in whatever condition – before home, entertainment or other comforts …
>
> Although a number of bus priority routes have been implemented, introduction has become patchy and intermittent … Very low paid and insecure workers who fear losing their jobs if they are late for work may thus have to get up at 4 am to make a journey which may take up to three hours, and then face a similar journey at night.
>
> The metro system is a fragment of the total proposed – but is a showcase for the city … The system largely serves the better middle-class areas …
>
> The older railways of São Paulo contrast poorly with the Metro. Unreliability provoked a civil disturbance, including a fire, last year.
>
> (Maxon, 1998)

You will notice that São Paulo is frequently cited in this series as a useful example of a city in the developing world. For example, **Allen (1999)** describes how in the 1930s the French anthropologist Claude Lévi-Strauss had been thrilled by what he saw as the 'wildness' of the city. Today it contains some of the most startling juxtapositions of rich and poor in the world. The preference for car ownership in São Paulo is supported by Figure 2.6 above, where you will note that São Paulo is far above the average given its GNP per capita: car ownership is at a level more usually associated with income at double the rate. This depressing account of an old, poorly maintained and unreliable public transport system, which reinforces and is reinforced by hostile social attitudes, is perhaps especially stark in São Paulo, but it is by no means unique. General attitudes to public transport may not be as extreme in Europe, but individual expressions of dislike are not unknown.

ACTIVITY 2.5   You might like to consider how far attitudes to, and use of, a city's public transport are based on social assumptions rather than on actual experience of what it offers.

● How far are social attitudes shaped by what public transport is available?

● How do attitudes and experience contribute to inequality?

● How do assumptions about who uses what kind of transport, why and how, translate into policy? ◆

Extract 2.2 is taken from an article which, in its author's words, 'provides a comprehensive analysis of the situation and development of urban transport' in Beijing from 1949 to 1992. Like the great majority of writers on 'Third' World cities, Victor Sit sees transport as one of the major problems confronting such cities, but unlike many of his colleagues, he sees motorized transport as the answer, not the major part of the problem. The extract begins more than halfway through Sit's piece: his general position is that transport is about speed, and that congestion is amenable to the provision of more road space. In particular, the bicycle is presented as a cause of congestion, slowness and (he hints) social and economic backwardness.

EXTRACT 2.2
Victor Sit: 'Urban transport issues in a Third World and socialist setting'

The bicycle

The number of bicycles in Beijing has grown at an increasing rate … From 1949 to 1967 the annual increase of bicycles was about 50,000, but from 1967 to 1981, it was about 150,000. Since 1982, the annual addition exceeds half a million, showing a rising trend in spite of increasingly good economic performance and increased investment in public transport and a trend towards motorization. In 1990, a household on average owned more than two bicycles. Beijing is certainly the largest 'bicycle city' of the world, even surpassing Shanghai which has a larger population size, but slightly fewer bicycles (8.32 million).

In 1983 19.5% used public transport and 37% bicycles for commuting trips in Beijing. In 1988, the proportions had changed to 37% and 57.1%. Deteriorating public transport conditions have encouraged the use of the bicycle …

Data for 1983 in Beijing show that the average speed of public buses were 15 km/hr, and for the bicycle ride, 12 km/hr … the bicycle has become the predominant mode of transport, even if walking is also taken into account.

The predominance of the bicycle is related not only to substitution for poor public transport, but also to China's increasing capability in producing the bicycle leading to falling prices, as well as official encouragement of bicycle use. In the early 1980s, the average price of a bicycle of ¥400 was about two months' average income of a worker. In 1991, the average price of ¥300 was only about one month's income, a consequence of rapid expansion in production and improved income level … Journey-to-work by bicycle is encouraged through a monthly subsidy as it guarantees punctual attendance. The value of subsidy is such that the worker can buy a new bicycle every three years. As a bicycle can last for about ten years, the subsidy means not just 'free trips' but also a real monetary gain …

Increasing use of the bicycle contributes to congestion of the urban road network, particularly at road junctions. A 1986 survey that covers 178 main road junctions within the 650 sq. km of the planned urban areas illustrated the huge volume of bicycle flow during the morning peak hours of 7–8 a.m. Road junctions registering bicycle flows in excess of 20,000 during peak hour numbered 14 in 1986, a tremendous increase from 3 for 1978. Those with a flow of over 10,000 vehicles numbered 106, compared with 53 in 1978. The massive flow of bicycles has led to a declining average speed of motorized traffic and serious congestion problems. It is especially marked within the Old City, as only about 10% of the road network there provides separation for the motorized and non-motorized traffic. Even with separation, chaos at road junctions still prevails … Rough estimates made in 1987 show that problems of mixed traffic on roads and at road junctions had contributed 30–40% reduction in flow capacity of the road network. Bicycle parking has created yet another problem for the city districts. Within the Third Ring Road there were over 900 bicycle parking areas in 1988, with actual parking of 370,000 bicycles/hour and occupying 92,220 sq.m of land, 74.2% of which were on pedestrian paths and on roads … Thus, it has created not only problems of free movement on roads for vehicles, but also dangers for pedestrians.

Compared to buses and trolley buses, the bicycle shows a number of disadvantages. Its carrying capacity is low on a per seat basis. It occupies more road space, with slow speed and is unsuited for long distance trips. Bicycles occupy twice as much road space per passenger as a bus when stationary and ten times as much when in motion. Thus, the trend in Beijing since 1980 of moving increasingly towards the bicycle for commuting reflects an unhealthy situation, when considering the existing pattern of concentration of people and activities, commuting trips and the structure of the road network within the city and suburban districts.

Source: Sit, 1996, pp.265–6

6.4   ACCOMMODATING THE CAR

ACTIVITY 2.6   Transport has moved swiftly up the international political agenda in recent years: for example, it was high on the agenda of the Rio Earth Summit in 1992, the New York Earth Summit 2 in 1997 and the ministerial conference in Kyoto in the same year. Much if not most of the debate has focused on the perceived problem of the car, in the city as much as elsewhere. The debate is split between those who call for a much more restricted role for the car if environmental and social disaster is to be averted, and those who insist that other things will have to change in order to accommodate not just current levels of car use, but also the higher levels that are expected in the future. The predicted effects of current transport use are not contentious. It is agreed that something will have to change; the argument is about what that something is.

The question we want you to address for the remainder of this chapter is: how could the sustainable city cope with the predicted doubling of car use?

You should consider, among other things, the possibility that this debate is just another one of the panics that overwhelm the world from time to time – remember the worry about over-population in the 1970s? We are not commenting on the merits of the argument, but on the fact that issues rise and fall on the political agenda. Will the animus against the car go the same way? Could it be argued that science usually finds solutions, that life goes on? And how romantic is the transport environmentalists' case? Is it just hankering after some non-existent Golden Age?

The car is economically and socially important to life in the city: it provides employment; it generates huge government income in taxes; it can be a flexible and even convenient way of getting around; many people are proud of their cars – they are undoubtedly desirable objects. Cars can be sitting-rooms on wheels, providing privacy and even, should you wish, solitude.

Your case for keeping the car in the city will need to address not only the problems of air quality, noise and congestion, but also the amounts of space devoted to the car relative to other transport and non-transport uses; the needs of those (for example children) who will never have unfettered access to cars; and issues of equity. We look at all these matters in this chapter, although you may not agree with our analysis. Even if you think that car use should be severely limited, this exercise asks you to make the opposite case. ◆

What may appear to be an obsession with the car and its impact is by no means restricted to the developed countries: if anything, the alarm is greater in the less developed world, where indigenous concern with the profound changes ushered in by the car has been matched by the developed world's concern lest the developing world makes the same mistakes (or enjoys the same privileges – take your pick).

6.5   GENDER, RACE AND AGE

So who are the owners of cars in the city? The answer is that, wherever one looks, it is overwhelmingly men, especially middle-aged, relatively rich men. In cities in the developing world (with the odd exception like São Paulo) very few people own cars; even most middle-aged, relatively rich men are less likely to have cars, but they form a majority of those who do. Thus access to cars in a modern city is a complex matter. We have already discussed income, and we now turn to gender, race and age.

When you first encounter transport statistics, you will find that the most common unit of measurement is the household: traditionally, ownership and use of vehicles is described in terms of *households* having access to cars, not *people*. This means that, as car ownership becomes more prevalent in a city, planners are increasingly apt to assume that 'everyone' has access to a car, and to cease to make provision for those without – which will include many people in car-owning households. For instance, the man may take the household car away for the whole day just to journey to work and back; in a one-car household this can leave other family members (typically women and children) without access to the car, but not without important journeys to make.

Is the answer to this equity problem for everyone to have a car? We would suggest not: there is (usually) not room, for one thing, while environmental pollution, so closely allied to car use, is another concern (not to mention the prospect of young children driving). Plainly, the answer cannot be in terms of equal access to cars; it has to be about equal access, full stop. This requires a shift in perspective. One could begin by recognizing that to be car-less in the city is not necessarily to suffer disadvantage. In some cities it is perfectly possible to travel without a car at least as safely and at least as efficiently, and there are significant numbers of people who *choose* to live without a car. But for many people in cities, not to have a car is a deprivation, not least because the car is also a status symbol which creates and nourishes inequalities. Indeed, for some people it is a vital expression of their personality.

Another aspect of inequality created by increased car ownership is that it makes things more difficult for non-car users. The infrastructure put in place for the car very often disadvantages the rest of the city and its non-car users. The proliferation of cars has considerably altered the terms on which the use of the street is negotiated.

Income and the closely related trend of car ownership are not the only variables that produce inequality. Depending on where you look, movement in cities is male or female. On the whole it is men in the cars (and driving the taxis, buses and trains), and it is women on the buses and walking. Such inequality of access to the transport system, and thus to what it serves, militates against women's full participation in city life.

Men in regular nine-to-five employment are a minority of the population, but they have had a disproportionately large impact on the provision of transport in cities. It would appear that the bulk of their transport needs are satisfied by either the provision of public transport or of sufficient road space for their cars at certain times of the day, typically between seven and nine o'clock in the morning and five and seven o'clock in the evening; these are usually radial journeys, in and out of the centre from the suburbs. This has been the pattern in western cities for nearly half a century, and although there have been 'spreads' in peak times, and shifts away from city-centre employment, the pattern remains strong, especially in its influence on transport provision.

Most women, however, have more complicated regular transport needs. They are less likely to have a full-time paid job, so that their travel-to-work needs do not fit the 'peak' pattern, nor are they likely to be travelling long distances to work. The household-manager role is still overwhelmingly undertaken by women, and therefore it is women who are involved more in the socially necessary tasks of shopping, taking children to school, and visiting family and friends.

Race, unlike income and gender, is not a source of inequality in all cities: Dublin, Helsinki and Tokyo, for example, rarely have to grapple with racial problems, simply because the racial mix is not there. However, where racial tensions do exist, we can expect them to be played out in transport use, because transport happens in public space.

TABLE 2.2   *Modes of transport for metropolitan commuters (%) in South Africa*

	Bus	**Car**	**Taxi**	**Train**	**Walk**	**Other**
Black	21.5	11.2	41.9	14.9	8.7	0.8
White	3.4	88.3	0.2	4.0	3.4	0.7

Source: UNCHS, 1997, p.154

Consider Table 2.2, which shows transport use by black and white commuters in South Africa. The pattern of transport use shown in the table evolved in response to the restrictions of apartheid, the enforced relocation of black people, and 'security and racial considerations' (UNCHS, 1997, p.154). The high usage of taxis by blacks and the extremely low usage by whites indicates a quite different taxi industry in South Africa from that in, say, Europe.

Racial inequality thus manifests itself in transport use in several ways. In Europe and the USA, young black men are more likely to be stopped in cars by the police than young white men. In some city areas ethnic minorities will impose curfews on themselves, for fear of venturing onto the streets. When racial tension explodes, it is transport – cars, buses, taxis – that are among the first things to be destroyed. Often it is transport that provides the occasion for protest. For example, the civil rights protests in the southern states of the USA can be dated from the 1955 refusal of Rosa Parks to give up her seat on a bus in Montgomery, Alabama to a white person, as the law required her to do (see Extract 2.3).

EXTRACT 2.3
Charles Denby: 'Transport and political unrest'

A lot of tension was building up and nobody knew where or when it would break. And on December 5, 1955, their wasn't a soul who thought that when a working woman, a seamstress named Rosa Parks, refused to give up her seat to a white man in a bus in Montgomery, Alabama, that the break had come. Each concrete act took everyone by complete surprise, from the refusal by Mrs Parks to give up her seat to a white man, to the response to her arrest and court appearance, to the mass demonstrations led by the then unknown Rev. Martin Luther King Jr, to the black community running their own transportation system. It became Revolution, a word none of us ever used referring to an action defying the segregated conditions of life in the South. That mass action of revolt was the Montgomery Bus Boycott.

During the boycott, I talked with Rev. King, and he … said, 'You know – I can't tell you to save my life why Mrs Parks didn't move back when they told her to. She says she was tired. And I believe that; but I also know that she was active in the NAACP, never successfully. This time after they arrested her, all hell burst loose'.

He went on to say that there had been a few Black college youths on the bus from State Teacher's [sic] College and they found out that Mrs Parks was going to be tried on a certain day – I think it was Wednesday …

That Sunday, in practically every Black church in the city, the members were talking about the leaflet [mimeographed by the youths] calling for Blacks not to ride the bus on the day of Mrs Parks' trial, and to be at the court house to support her.

'And that', Rev. King told me, 'was the first I knew about it. All church members were asking their pastors what they should do, and practically every one of the pastors said they should stay off. I said the same thing to my members.'

He said about 80 per cent stayed off and walked that day … But after the Blacks boycotted the buses that Wednesday, and then went back to the bus stops on Thursday, something else happened. All the bus drivers – and they were all white then – would pull up to a stop and, where there were all Blacks standing there, went on by without picking up a single one of them.

The reaction of the Blacks was, 'What the hell! We walked yesterday … we can walk today.' And that, Rev. King said, was the beginning of the Montgomery Bus Boycott. And they kept on walking from that day on – for over nine months – until they won.

Source: Denby, 1978, pp.181–3

Not all different use of transport by race is due to discrimination by the majority. It can be part of the minority's cultural tradition. For example, within some minorities women either seldom go out, or only go out in private transport.

Age generally is another factor contributing to inequality in transport. In many cities children have very restricted access to transport: in Britain and the USA, for example, parents increasingly refuse to allow their children to travel to school or anywhere else on their own. Two surveys of children in England, the first in 1971 and a follow-up in 1990, showed dramatic falls in their independence. To take just one example: in 1971, three-quarters of all the children surveyed were allowed to cross a main road on their own; in 1990 only half were (Hillman, 1993, p.9). Open public space is increasingly denied to unaccompanied children under 12. This is therefore not so much a question of transport provision as of a cultural response to perceived danger, but it affects the transport system and the experience of it profoundly. On the one hand, there is the adult view, noticing the relative absence of children from the streets and the city's public transport; on the other, there is the child's view, knowing little of independent travel in their own area or of how to get out of the area.

At the other end of life, the problem is likely to be connected with decreasing personal agility, and attendant nervousness at coping with streets and systems which are not friendly to the halt and slow. Since older people are currently less likely to have a car, any failure of the public transport system to cater for them will greatly reduce their mobility. When older people do have cars (and in the West it is increasingly probable that they will own and drive cars, even in cities), this means that they are even less likely to try to use public transport, and so they contribute to inequalities in other ways.

7 *The rhetoric of freedom*

7.1   FREEDOM FOR WHOM?

Historically the city has been a place of freedom. It has attracted those formerly enslaved: it was where serfs went when they ceased to be serfs, where slaves went to escape their masters, where women went to evade censure. The anonymity of the city appeared to promise freedom, and often it delivered it. Transport in the city has more recently been an important element in all this. Not only did the train make the flight to the city so much easier, but once there, the buses and the metro contributed to the aura, and reality, of freedom. But nobody ever fled to the city in order to drive a car (the car is often the way to flee *from* the city); the freedom of the city has not been about free parking, nor have the keys to the city been car-keys. The car has rarely been an emblem of urban life, even in North American iconography. We can think only of ticker-tape parades in New York to welcome home heroes, and Grand Prix car-races in cities such as Monte Carlo and Melbourne.

Nevertheless, the rhetoric of freedom has been used very powerfully on behalf of the car, even in cities, even during the oil crises of the 1970s, and even while the cities of the world are being suffocated by cars and the public transport which is their lifeline is in almost universal decline. The tale of why this should be so is a sad one. First, though, let's consider what this rhetoric of 'freedom' consists of. The language is so familiar, so constant, that it can be hard to isolate it from less tendentious utterances.

The car represents freedom because:

- it is privacy, a place to be alone;
- you can pick and choose your companions;
- the car waits for you – waiting at a bus-stop is far less comfortable than sitting in a traffic-jam, however more environmentally kind the former may be.

At least as early as the 1950s supporters of private transport in the West have appropriated the language of freedom to justify claims for the car, and the same process of equating access to a car with personal liberation is now happening across the world. Part of the success of this argument has been due to the difficulty which supporters of non-car use have always had in using the same kind of language to stake their opposing claims. This failure is despite the fact that policy-makers have always known that the car reduced freedom for significant numbers of people (for example, for those without a car); the failure continues despite the fact that it is now very clear that the car reduces everyone's freedom, including that of the people in the car. Why should this success and this failure have happened?

Have you seen the film *Who Framed Roger Rabbit*? It was very successful on its release in 1988 (it took $150 million at the US box-office alone), largely because of its extraordinary mixture of cartoon characters (known as 'Toons' in the film) and real actors (most notably Bob Hoskins). The animation techniques were state-of-the-art, and the story moves along at a cracking pace. The point of mentioning it here is a little-noticed part of the plot. The film has a villain, Judge Doom, and we are offered two proofs of his villainy: first, that he hates Toons, and second, that he wants to destroy the much-loved and much-used Los Angeles streetcar system in order to build in its stead an automobile expressway. How many people who have seen the film know that Judge Doom's ambition is based on a real-life conspiracy – one which, unlike his, was successful, not only in Los Angeles but all over the United States?

Extract 2.4 below outlines why and how public transport systems across the USA were replaced by the private car. From the 1920s and starting in New York, General Motors targeted trolley-bus companies with a view to replacing them, first with buses and then with automobiles. They were eventually joined in this enterprise by Standard Oil of California, Mack Truck, Phillips Petroleum and Firestone (the tyre company). They created a front company, National City Lines, based on what had been a tiny bus company in Eveleth, Minnesota. By the time the US Department of Justice caught up with them all, after the Second World War, most of the damage had been done. Extract 2.4 is part of the transcript from a film made for US public television in 1997.

EXTRACT 2.4
Jim Klein and Martha Olson: 'Taken for a ride'

Narrator. When you're talking about public transportation in America, for the first part of this century, you're talking about streetcars. Trolleys ran on most major avenues every few minutes. Steel track and quiet electric motors made the ride smooth and clean and comfortable. The center of the road was reserved for streetcars, and the new automobiles had to move out of the way.

Bradford Snell, who has made a career researching the auto industry for 16 years: In 1922, only one American in ten owned an automobile. (Everyone else used rail.) … GM [General Motors] moved into Manhattan. They acquired interests in the New York railways and between 1926 and '36 they methodically destroyed the rails. When they finally motorized New York, General Motors issued ads throughout the country … [which] said, 'The motorization of 4th and Madison [Avenues] is the most important event in the history of community transportation'.

Narrator. In the mid-1930s, GM worked hard to create the impression of a nationwide trend away from rail. But there was no trend. Buses were a tough sell. They jolted. They smelled. They inched through traffic. City by city, it took the hidden hand of General Motors to replace streetcars with Yellow Coach buses.

In 1936, a company was founded that would grow to dominate American city transportation. National City Lines had no visible connection to General Motors. In fact, the director of operations came from a GM subsidiary, Yellow Coach, and members of the Board of Directors came from Greyhound, which was founded and controlled by General Motors. Over the next few years, Standard Oil of California, Mack Truck, Phillips Petroleum and Firestone Tire would join GM in backing this venture …

National City Lines grew quickly. By 1946 it controlled public-transit systems in over 83 cities. From Baltimore to St. Louis, Salt Lake City to LA [Los Angeles] …

Snell: The appearance was always that this was only a company that was owned by the Fitzgeralds. You know, these people had come from Minnesota with no money at all, and all of a sudden they were in control of this multi-million-dollar enterprise. But, in fact, the money was coming from the corporate sponsors …

Narrator: In 1946, the Justice Department began an antitrust investigation into National City Lines, General Motors and the other investors.

Justice Memo (voice over): Memorandum to the United States Attorney General: It appears that National City Lines and its manufacturing associates have entered into a plan to secure control over local transportation in important cities throughout the United States. If these companies are permitted to continue their program, they will soon have a stranglehold over the industry.

Snell: The key lawyers involved in the case told me there was not a scintilla of doubt that these defendants, General Motors and the others, had set out to destroy the streetcar system …

Narrator: The government's case was straight forward. National City Lines, General Motors and the other defendants were found guilty of conspiracy to monopolize the local transportation field.

Snell: These companies, that had probably eliminated systems that in order to reconstitute today would require maybe $300 billion, these companies were individually fined $5,000. And the individuals involved – like the treasurer of General Motors who had actively run Pacific City Lines, one of the subsidiaries, and was a major moving factor in all of National City Lines' operations – he was fined the magnanimous sum of $1 at the conclusion of the trial.

Source: Klein and Olson, 1997

7.2   INDIVIDUATION AND PRIVACY

Another example of a one-sided debate is that surrounding car-exhaust emissions. No one denies that cars pollute: it appears to be the one thing on which opponents and supporters of the car agree. However, there is a danger that eliminating car pollution will come to be seen, on all sides, as the removal of the problem of the car. Once there are 'clean' cars, so the argument goes, the car will be tamed. Partly in response to this reduction of the problem of the car to a concern over emissions and pollution (and partly because they have been largely responsible for influencing the argument in this way), car manufacturers such as Daimler-Benz are investing heavily in the development of a 'clean' engine that uses, for instance, fuel cells, compressed air or hydrogen.

- Thus the lack of opposition to the proliferation of the car can be traced in part to the activities of very powerful groupings in whose interest it is that car ownership is not curtailed.

- This has happened in different ways in different places, but there have been marked similarities in the broader political contexts that have nurtured these developments, chiefly in the promotion of a range of anti-collectivist attitudes and policies.

We want to connect some of these larger ideas with the resulting decline of public transport, but we are also aware that there are enormous intellectual and methodological problems not only in tracking ideas from group to group and person to person, but also in establishing how ideas motivate people's actions. Furthermore, we would not deny that the same ideas have been responsible for much that is positive in the world. The rhetoric of freedom, for example, was used to great effect in securing the release of Nelson Mandela, or in opening the Stasi's files to their victims in former East Germany. The mere juxtaposition of those events with the debate about cars highlights (for us at least) the inappropriateness of the use of that rhetoric in this case. That is not to say either that cars have not resulted in a significant measure of freedom for the individuals involved. For a great many families, life without a car is unthinkable, and would certainly be less rich in opportunity and enjoyment. This unthinkability is not because of a lack of imagination; it is because, for most people who now have a car, to give it up would involve profound changes in lifestyle. If people have chosen where they live, where they work and where their children go to school on the basis of having cars, then the short-term alternatives to having and using cars are truly hard to see.

The rhetoric of freedom has also been used negatively, to promote anti-collectivist attitudes and policies, including, and obviously going well beyond, transport issues. The apparent exigencies of the cold war predisposed many western governments to be suspicious of activities which brought large numbers of people together. While the destruction of public transport systems was already well advanced in the USA (see above), this anxiety was a timely opportunity for those who were interested in a similar reduction in, for example, British public transport. This is a very unfamiliar perspective on Britain's paranoia in the 1950s – a much diluted but still powerful parallel to the McCarthy years in the USA – but there is a great deal of uncollected evidence that official attitudes to the car and to public transport, as well as to television versus the cinema, or working-class gatherings (with the exception of sport), were deeply coloured by the fear of unrest and even revolution. We are not tracing this development with any precision here, but you may like to consider in this light Margaret Thatcher's well-documented dislike of mass provision of most kinds, including, famously, public transport.

A counterpart of this individuation of social need and provision has been the rise of the consumer society, dedicated to enshrining 'privacy' as a leading value in our lives. A transport-focused manifestation of these attitudes in the UK was the remark by Stephen Norris (the transport minister in John Major's Conservative government in the early 1990s) that it was unreasonable to expect people to sit next to the typical public transport passenger. Another aspect of this individuation has been a change in the negotiability of the street and of what one might call the privatization of space, which we discuss below.

The following quote is from *Travel Sickness*, a collection of essays about the need for a sustainable transport policy, and it sums up some of the contradictions inherent in the way the concept of freedom is applied to the transport policies of most governments around the world:

> If you give people freedom to move, more or less in whatever way they wish, and you also incorporate in that freedom a scantily punished ability to kill and maim other people, to consume earth resources in whatever way necessary to sustain this freedom, and to mutilate the environment with the roads and track and the effluence from the mechanized vehicles needed to move beyond a certain distance; if you have a small land area and a high population; if you do not intervene in the way land uses are put together or separated, and in the densities at which people live; if you have implicitly (often explicitly) different rules for the rich and the poor; if your system of government naturally follows a process of transport disintegration; if you sell off the public assets of your public transport system as an act of faith in a religion of market-worship … then you have a problem.

(Roberts, 1992, p.1)

8 Negotiability of the street and the privatization of space

In our discussion of issues of inequality and power, we mentioned that the car had altered the terms on which the street could be used. The encroachment of the car on city streets, by vastly increasing the demand for space, both for movement and, crucially, for parking, has affected its use for trading, for carnival, and for meeting people, as well as for other modes of transport like walking, buses and goods delivery. There are now roads in many cities on which pedestrians are not allowed: they may be called (urban) motorways, or freeways (freedom again!). This traffic not only pollutes the air and the streetscape, but it also imposes its own dehumanizing rhythm on all the movement and stillness of the city.

- The sound level of the horns and sirens used by emergency services is designed to alert the enclosed drivers of motorized road traffic rather than cyclists and pedestrians whose ears are not so protected.

- Bus-stops have been moved further and further away from the junctions where most passengers want to alight; this is in order to remove the obstacle that the stop presents to other (car) traffic.

- The sign-posting set up for car users rarely helps those on foot, even though it is typically erected on the pavement: it doubly alienates pedestrians from their own space, and may also be seen as evidence of the different local worlds which people inhabit.

We would argue that society's notion of what the street is for has been profoundly altered by car culture. One of the first effects of the car's takeover of a city street is to denude it of people. This creates a physical as well as a conceptual space on the streets, a space that is filled (or is perceived to be filled) by crime, insanity and other social deviation – and by more cars. Thus a downward spiral is created, a downward spiral of experience and expectation of what it is to be out on the street (except in a car). People come to resist venturing into public space. They seek bolts and entry-phones on the door, moats around the house, gates on the street. Of course, such aspirations are not new; the rich have long achieved them. What is new is the increasing numbers of people who both feel the need for them and are able to afford them. You may like to look again at Chapter 1's discussion of public spaces and private enclaves, which puts these issues in a wider context. We are not suggesting a simple causal model here, namely that cars lead in a straight line to gates at the end of your street. The connection between the desire for privacy and car culture is far more complex than that. However, we do want to stress that there *is* a connection that is also damaging to other freedoms of city life.

Despite this kind of analysis, the enemy in the city is generally agreed to be not the car, but other people. To the extent that the city has always attracted 'undesirable' people, the streets of the city have always felt unsafe to at least some citizens. Planners rush to assuage people's fears and build cities in the sky – cities where the street is only for driving on and parking under, and where drivers and passengers move from house to office, shop or theatre without once setting foot on the street. Even better, people can avoid the city altogether. They can live in the suburbs and turn their backs on the city, and they can shop in the mall on the perimeter and work in a landscaped greenfield site in the countryside. This flight from the city centre can be widely observed, and the ostensible causes are often described as racial tension, or poor educational facilities, or greener grass elsewhere. The new kinds of threat and depredation which the car has introduced have thus been blamed on these traditional shadowy scare-objects. Whilst these reasons might be valid, we want to stress the role that the car has played, both in causing the original discontent and in enabling the new lifestyle.

Back in the city, there has been an effort to mirror the car-induced impersonality of the street in other activities: the self-service petrol-station, banks' hole-in-the-wall cash-dispensers, and a whole range of services where the customer does not have to speak to anyone, or only does so at the very end of the transaction. This is not to say that the car has caused these developments: obviously businesses, especially in labour-intensive industries, have taken advantage of technological advances in order to cut costs. Now people accept such impersonality when they would not have done before. The car has fostered this absence of human interaction, and indeed has tended to glorify it, and so people tolerate and then actively seek it. 'Road rage' is perhaps just one of the consequences of this.

What is happening is the private appropriation of public space. Perhaps the best example of this is on-street parking. For reasons which we have never seen analysed, cars have been uniquely allowed to take over substantial proportions of our streets: often well over half the surface area of a city street is covered with dormant cars, whose only role at that moment (a moment which might well last 23 hours out of 24) is to be available to their user as easily and cheaply as possible. The justification for this usurpation is entirely focused on the individual's assertion of need; it takes no account of the community's need. Parking laws attempt to bring another perspective to bear, but the history everywhere of poor enforcement and the nigh-universal reviling of enforcement officers points to yet another connection between the freedom rhetoric we outlined above and anti-social behaviour.

This flight from the street might also be described as romantic. We would call it a flight from reality, except that it is a new reality which has been constructed, so that those wealthy enough (or sufficiently respectable-looking to the guards) are able to escape the now-unacceptable reality to which they have so signally contributed, and to which they continue to contribute each time they drive to the privatized parts of the city. Our final section concentrates on the horror and romance of travel.

9 The horror and romance of transport

Transport mirrors so much of the city – both the horror and the romance, as we said at the start of this chapter. Let's deal with the horror first.

Pollution is internationally recognized as an endemic problem of the modern city. We have already discussed the way in which the problem is presented, and therefore misunderstood, as being the *only* difficulty the city has with the car. Nevertheless we must acknowledge that pollution is a horror: in the UK three times as many people die from the effects of traffic pollution as are killed in road accidents. Children seem to be especially vulnerable to the deterioration of air quality, as the rising incidence of asthma worldwide testifies. Road traffic is a major producer of the chemicals that are causing global warming, which threatens all our lives and indeed the whole ecology of the planet. For example, in Mexico City on 17 March 1992, all schoolchildren under the age of 14 years were ordered to stay at home because air pollution had reached a record level of 398 micrograms of suspended particles per cubic metre. Forty per cent of the city's air pollution results from three million road vehicles which generate 5.2 million tons of contaminants (Gilbert, 1994).

In the UK in 1994 the Royal Commission on Environmental Pollution reported:

> The forecasts made by the [UK Department of Transport] in 1989 showed a doubling in the overall level of road traffic [in Great Britain] by 2025. Air traffic is forecast to increase still more rapidly. Even allowing for technical improvements in vehicle design, the consequence of growth on such a scale would be unacceptable in terms of emissions, noise, resource depletion, declining physical fitness and disruption of community life. In our view, the transport system must already be regarded as unsustainable … and will become progressively more so if recent trends continue. We believe this is an issue of such importance that it justifies placing significant constraints on the future evolution of the transport system … This will involve a gradual shift away from lifestyles which depend on high mobility and the use of cars.

(Royal Commission on Environmental Pollution, 1994, p.233)

Traffic congestion is pollution's twin and, like pollution, it is a city phenomenon. That it is crowded (at times) could almost be the definition of a city. Congestion – of people, buildings, entertainment, even traffic – can be the source of much of what makes cities good to live in, and of what makes them dangerous. At its worst, the traffic system breaks down entirely and there is gridlock – a series of major junctions become blocked and as a consequence nothing can move. What is done to relieve the congestion is perhaps the most important issue facing transport planners. They may well perceive the problem as too many vehicles in too little space. If the planners believe that the problem is too little space, their answer will be to create more, and the result will be even more vehicles, even more pollution, and the loss of whatever amenity was in the way of the extra road space – not to mention a patently unsustainable scenario. However, if what they perceive is too many cars, then the solutions will be very different: on the negative side there could be car-rationing, road-pricing and parking and speed restrictions; on the positive side there could be simply more space for pedestrians, buses, cycles, trains and trams – that is, an integrated approach to transport, relying on greater invention and innovation which moreover addresses sustainability.

The next horror is being a stranger in one's own city. People are familiar with the routes they know: they may not be the most direct routes, but they are theirs. There will be whole sections of their city which are unknown to them: they never go there, or perhaps all they know is a single road through it. If they turn left instead of right, they will be lost in the city they have lived in all their life. The city that the pedestrian knows is quite different from that of the metro user; the cyclist sees a different place from the car driver; while train passengers see a lot of back-gardens and derelict factories, places that they wouldn't recognize from the street. The person who commutes to the city and returns straight home every evening could work there all her life and not know anything beyond the streets between her office and the station. City-dwellers are notoriously apt not to know the sights to which tourists flock from all over the world, or they may know those 'sights' not as cultural beacons, but more as familiar landmarks.

Maps are there to help, but they may not tell what people want to know. Directional signs may crowd the pedestrian off the pavement, but the signs are for cars and lorries to follow. There are always lost people in cities: residents see them peering at maps on street corners, but the residents don't know the way either. They only know where they are going. Cities seem designed to confuse; people expect to be defeated by them. There is a common urban ritual in which arcane knowledge of short-cuts is shared.

People who use public transport (which is the majority in most cities) rarely have any choice about whom they share their travelling space with, and there are limits to the protection from other people's behaviour that can be afforded to passengers. While mugging and assault are obviously undesirable, they are also rare. But there is also the public display of what might be deemed private behaviour and which causes discomfort (amounting to a threat in some eyes) to travellers in the city: picking teeth, biting nails, couples kissing, playing music. There is also provocative behaviour: smoking in non-smoking areas, busking, changing clothes, drunkenness.

These are the least of some city-dwellers' problems: the city as a whole may offer great opportunities, but for those who cannot get out of their immediate area, parts of the city are inaccessible. Those who are poor and cannot afford a car, or the bus fare, are constrained however wonderful the roads and bus service may be. Those who live in an underresourced area of the city, with buses and trains only in the rush-hour, can find themselves marooned in broad

FIGURE 2.8
The horror of transport: the threatening ambience of the New York subway

daylight. Gender, race, age, degree of able-bodiedness and income, as we discussed above, all affect mobility.

Much of what we have fancifully called the 'horror' of transport can be traced to stress in the face of the unknown. First-time travellers know how stressful a new journey in a city can be, whether by car, foot or public transport. Unfamiliarity, especially in public spaces, brings with it anxiety, because unknown areas are assumed to be (and may in fact be) unsafe. Losing one's way, having to approach strangers for help, being late, missing one's stop, all involve lack of information, loss of control and loss of independence, and make for stressful experiences. Men may be especially prone to this source of stress: it appears that they are much less willing than women to ask for directions (Tannen, 1995). Even on familiar journeys there can be stress. For example: changing modes (from train to bus or metro, say) brings the possibility of missed connections; with driving there is often a competition for road space, and the damage to the cardio-vascular system which this continual fight-or-flight stance causes is well-documented; travelling after dark is worrying for many people, irrespective of mode; roadworks can disrupt one's journey; while new timetables can make it impossible to get to work on time. The two greatest fears behind transport-related stress are vehicle breakdown and personal-injury accidents.

Exasperation seems to be emerging as another corollary of city transport: in Dublin exasperated would-be passengers have hijacked buses; in London, train passengers have risked their lives by climbing out of faulty trains and walking along electrified lines to the station, despite urgent advice from the guard not to do so; and of course motorists' exasperation – expressed in physical assault on other drivers, or cyclists – is increasingly documented in newspapers around the world.

People can constrain *themselves*. They erect their own city walls and invent their own dragons: they never go over the river, over the tracks, into the next borough or *barrio* or *arrondissement*. They climb into their cars and travel through an area without seeing it; they take a metro under the city and can then completely ignore it. In wartime, people two streets away may well be too dangerous to travel among (as in Sarajevo, or Kigali in Rwanda), but in peacetime, too, people tend to demonize the 'other'. Two streets away in London, New York or Tokyo may be felt to be too dangerous to walk or even drive along.

However, to venture off one's patch can be an exploration – a *romance*. There is a long literary tradition, which presumably reflects readers' desires, in which travel is presented as a positive experience. An *Arabian Nights* flying carpet is not about the abolition of transport: it is about having the freedom to move – to

FIGURE 2.9
The romance of transport: Mott Street, New York, July 1988. Which would seem more exotic to a European visitor – the Chinese bank or the truck?

fly, indeed – without constraint of cost, class, gender, maps, technology or fear. Many stories are about a journey: from *The Old Curiosity Shop* to *Annie Hall*, transport features not only as a metaphor for the growth to maturity (and/or death), it also is there for real. *Tom Jones* is about our hero's gaining of knowledge and insight, but he really (in the fiction) does go to London, and really does have adventures on the way there. The fantasy of travelling from city to city on the *Orient Express* or on a Mediterranean cruise is more modern, and not fictional. Travel of the right sort – as close as one can get to the flying carpet, Orient Express kind – is nowadays an internationally recognized, status-bestowing reward for business success. The conference in Kyoto (unless one lives there) has far more allure than the one on the doorstep, irrespective of the subject matter: the travel probably compensates for it.

Is the trick for travellers to bring a similarly positive attitude to their own city's transport, to be a tourist in their own backyard – not just to visit exotic locations within the city (such as Chinatown in a western context), but to treat 'ordinary' locations as exotic? Often to stray just a few steps off one's normal route is to find oneself in what might as well be another city, so little is known of the surrounding streets, buildings, people, activities and rhythms. Even to take the same route through a city every day, as many city-dwellers do, is to make a different journey. While some things will remain the same from day to day, week to week, decade to decade, much alters by the minute: above all the people. They may well be engaged in the same activities each morning, but they will be performed in different ways and by different people each time.

Transport, if used in this way, makes the city an experience of variety and change, an opportunity to celebrate difference – travelling and arriving. The city can be defined by its concentration of desirable destinations, and it is transport that takes people to these destinations. But behaving like a perpetual tourist in one's own city is not the only way to experience it. People may prefer to read a book on the bus, or to sleep on the train, or to drive on automatic pilot listening to the radio. The enjoyment of the travel itself is a subversion of the ostensible purpose of a journey – to work, to school, to visit one's mother – and is rarely a factor in studies of travel behaviour.

It is not a new thing that we regard cities as places where bad things happen (see, for example, *Pride and Prejudice*, or Sodom and Gomorrah), nor that movement around the city is seen as risky. But there is also the example of Dickens, for whom the city was a source of inspiration, and movement through it a civilized pleasure.

Appendix

ACTIVITY 2.1: WHAT DOES TRANSPORT DO IN THE CITY?

Here is our list:

Transport …

- brings food and raw materials into the city
- takes the product from wholesaler to retailer to consumer
- takes waste out
- takes people to meetings and conferences
- allows people to have peripatetic jobs (sales reps, doctors, music teachers)
- facilitates social interaction by allowing people to get together
- delivers people to education, leisure, hospitals and culture
- dictates and responds to residential and shopping patterns
- provides employment
- builds its own political and administrative base
- contributes to and reflects a power hierarchy
- has military significance
- can isolate communities
- creates barriers to, and opportunities for, development
- provides temporary shelter
- is a hobby
- congests and pollutes
- inspires …

References

Allen, J. (1999) 'Worlds within cities', in Massey, D. *et al.* (eds).

Amis, M. (1998) 'Road rage and me', *The Guardian* (*The Week* section), 7 March, pp.1–2.

Birk, M.L., and Zegras, P.C. (1993) 'Moving toward integrated transport planning: energy, environment, and mobility in four Asian cities', at <http://solstice.crest.org/planning/curitiba/part4.html>, International Institute for Energy Conservation, Washington DC.

Blowers, A. and Pain, K. (1999) 'The unsustainable city?', in Pile, S., Brook, C. and Mooney, G. (eds) *Unruly Cities*, London, Routledge/The Open University (Book 3 in this series).

Denby, C. (1978) *Indignant Heart: A Black Worker's Journal,* Boston, South End Press.

Eliot, G. (1866/1972) *Felix Holt, the Radical,* Harmondsworth, Penguin.

Francis, J. (1851/1968) *A History of the English Railway: Its Social Relations and Revelations, 1820–1845*, vol.1, Newton Abbot, David & Charles.

Gilbert, A. (1994) *The Latin American City*, London, Latin American Bureau.

Graham, B. (1996) 'Flying on a British Airways Concorde airplane', at <http://www.kaiwan.com/~bruceg/bg_gboaa.html>.

Graham, S. (1997) 'Urban planning in the "information society"', *Town and Country Planning*, vol.66, no.11, pp.296–9.

Hamilton, K. and Potter, S. (1985) *Losing Track*, London, Routledge and Kegan Paul.

Hau, T.D. (1995) 'Transport for urban development in Hong Kong', in United Nations Centre for Human Settlement (UNCHS), *Economic Instruments and Regulatory Measures,* Nairobi, UNCHS.

Hillman, M. (ed.) (1993) *Children, Transport and the Quality of Life,* London, Policy Studies Institute.

Hood, C. (1993) *722 Miles. The Building of the Subways and How They Transformed New York,* New York, Simon and Schuster.

House of Commons (1997) *Minutes of Accounts Committee*, London, HMSO.

Klein, J. and Olson, M. (1997) 'Taken for a ride', film transcript at <http://www.self-propelled-city.com/advoc/bikezillaarticles.html#ride>, New Day Films, 314 Dayton St #207, Yellow Springs, OH 45387, USA.

Massey, D., Allen, J. and Pile, S. (eds) (1999) *City Worlds*, London, Routledge/The Open University (Book 1 in this series).

Maxon, R. (1998) *Transport in São Paulo*, unpublished report.

Mogridge, M.J.H. (1990) *Travel in Towns: Jam Yesterday, Jam Today and Jam Tomorrow,* Basingstoke, Macmillan.

Pile, S. (1999) 'What is a city?', in Massey, D. *et al.* (eds).

Roberts, J. (1992) 'The problem of British transport', in Roberts, J. *et al.* (eds) *Travel Sickness,* London, Lawrence and Wishart.

Royal Commission on Environmental Pollution (1994) *Eighteenth Report: Transport and the Environment,* London, HMSO.

Sit, V.S. (1996) 'Beijing: urban transport issues in a socialist Third World setting (1949–1992)', *Journal of Transport Geography,* vol.4, no.4, pp.253–73.

Tannen, D. (1995) *Talking from 9 to 5: How Women's and Men's Conversational Styles Affect Who Gets Heard, Who Gets Credit and What Gets Done at Work,* London, Virago.

UNCHS (United Nations Centre for Human Settlement) (1996) *An Urbanizing World: Global Report on Human Settlements 1996: Executive Summary,* Nairobi, UNCHS.

UNCHS (United Nations Centre for Human Settlement) (1997) *Transport and Communications for Urban Development: Report of the Habitat 2 Global Workshop,* Nairobi, UNCHS.

Wendell Cox Consultancy (1998) 'The public purpose: urban transport fact book: US urban public transport: 1990–1995 ridership by metropolitan area', at <http://www.publicpurpose.com/ut-met95.htm>.

READING 2A
Timothy D. Hau: 'Transport for urban development in Hong Kong'

Introduction

By the year 2000, rural-urban migration will result in half the world's population living in cities. Four-fifths of the world's megacities will be found in developing countries … the most effective and sustainable way of tackling [the resulting] urban transportation problem is with a combination of positive supply enhancements (in the form of road construction and public transport provision) and demand management measures. Further, the use of supply measures alone by transport planners would not be helpful in solving the congestion problem without the differential pricing of road use via peak/off-peak charges. It is hoped that Hong Kong's ability to hold its fort against the rising tide of traffic congestion that threatens many megalopolises may serve as a useful case study for other cities vigilant against being mired in gridlock.

Some cross country transport characteristics

… Although the city state only has three-quarters of Tokyo's population, Hong Kong emerges with the second highest population density in the world after Macao … As a leading Asian cub, Hong Kong (like Singapore) is ranked among the top 25 wealthiest countries and her citizens can increasingly afford the luxury of owning private automobiles … Hong Kong has only 4% individual car ownership compared with Japan's 28% (and Singapore's 10%) for the year 1992. With an average household size of 3.4, Hong Kong has a low household car ownership level of 14% … Hong Kong possesses one of the denser road networks per unit of land area around – about half that of Japan's (and a third that of Singapore's). Yet because of Hong Kong's diminutive size – on only a thousand square kilometres of land – the physical amount of road space is scarce …

GDP growth and auto ownership

Since motorization is correlated with a country's wealth, so the same can be expected to occur in Hong Kong, especially with such a low car ownership level at present. With the real GDP growth rate averaging about 7.5% … it is clear that in order to prevent hypercongestion from occurring the incessant demand for auto ownership cannot be allowed to grow unchecked without imposing some form of abatement …

The level of motorization in Hong Kong gathered momentum after the Second World War. With trade being Hong Kong's lifeline, export-induced economic prosperity resulted in the expansion of road transportation and an increase in motor vehicle acquisitions … [However,] three quarters of the road space [was] being used by only a quarter of the travelling population, namely motorists and taxi occupants …

The benefits and costs of ERP, 1983–85

The Hong Kong Electronic Road Pricing System (ERP) experiment in 1983–85 involved fitting a sample of 2500 vehicles with electronic number plates on the underside of a vehicle. This video-cassette sized transponder permits radiowave communication with the electronic loops embedded below the road surface. Road side microcomputers installed at selected charging points in turn relay the vehicle's identification code to a control center. Car owners (only) are then sent monthly billing statements (similar to telephone bills) listing the amount of actual road use subject to ERP … the ERP pilot experiment proved to be an overwhelming technical success at 99.7% reliability …

Lessons from the political failure of ERP

Yet despite the tremendous benefits to be obtained from [this] demand management measure, the proposal to implement a full-fledged ERP System based on the 1983–85 pilot scheme was rejected by the public. When confronted with the fact that eight-tenths of the population travel by public transport, a tenth by private cars and a tenth by taxis, it appears that Hong Kong possessed the ideal climate for the successful implementation of ERP. There are several reasons for ERP's failure in 1983–85, the main ones of which are summarized here. First, the enormous revenues to be had from ERP (with its eight to one revenue-cost ratio) aroused some to suspect Government's true motives and to regard ERP as the conception of a revenue-raising device … Second, the mounting of a transponder underneath a vehicle enables authorities to track citizens' movements, a most unwise decision in light of the signing in 1984 of the Sino-British Joint Declaration on the future of Hong Kong after July 1997. This fear of a 'big brother' government helped defeat ERP in 1985 when the Government sought public consultation for the implementation of a full system … Since then, technology has advanced to the stage where smart cards are used in electronic toll collection (as in Italy since 1989 on the autostrada). Smart cards are simply electronic purses similar to stored-value metro and phone cards. Invasion of privacy can no longer be used as an excuse to defeat road pricing schemes, as had been done in many places after Hong Kong's 1983–85 public relations disaster. Third, only private cars were charged, which created much ill-will on the part of motorists. After all, goods vehicles generate sizable congestion (as well as road damage) externalities …

Conclusions and recommendations

Without some form of peak/off-peak pricing, traffic will converge on preferred places and prime times until excessive delay results and inefficient trips undertaken. A rational transport policy for urban development must reflect the underlying principle of differential pricing; all other stand alone policies that do not reflect this principle are unsustainable over the long haul. Efficient prices will also signal the optimal level of road capacity to be invested. Thus efficiently pricing road use in the short run and optimally investing in road capacity over the long run should be the transport planner's dual objectives … The discussion here suggests that the congestion conundrum that Hong Kong faces is a generic urban problem.

Source: Hau, 1995, pp.267–88

READING 2B
Stephen Graham: 'Urban planning in the "information society"'

We are bombarded almost daily with rhetoric and hyperbole about the 'information society', in which capital, labour power, information and service and media products flow instantaneously around the planet, driving social and economic change.

Much of this excited commentary from the media, technology industries and futurists is based on the endless recycling of a series of received wisdoms.

First, our economy, society and culture are simultaneously becoming global and migrating into the 'on-line' realm of vast networks of interconnected computers and digital media appliances. Second, and as a result, anything can now be done anywhere and at any time, via wires or satellites strung across the globe. Third, and following on from this, the notion of the city is being undermined as workers start to decentralise to idyllic 'electronic cottages', from where they can, remotely, maintain intimate contact with work, family, friends and services.

Of course planners will already know instinctively that the information society rhetoric suggesting the 'end of geography' or the 'death of cities' is both profoundly misleading and dangerously simplistic.

The familiar life of towns and cities does not somehow die just because IT – information technology – capabilities are growing. A lot more holds towns and cities 'together' and sustains urban growth than just their ability to support concentrated, face-to-face communication.

In fact, nationally, the urban dominance of the UK economy, society and culture shows few signs of waning. More widely, with the emergence of 'mega-cities', global urban dominance is growing rapidly. All that is changing is that cities are becoming more and more bound up with the widespread shift to mediation by IT networks in all aspects of urban life.

All this means that the classic planning issues that stem from concentrated living in urban areas – physical development, congestion, pollution, environmental conflict, social and cultural divisions and inequities – will continue to fill planners' in-trays.

Rather than simply being replaced, transport demands at all scales are rising in parallel with exploding use of telecommunications. Both feed off each other in complex ways, and the shift is towards a highly mobile and communications-intensive society.

The links between interaction in urban places and

'electronic spaces' are therefore complex, subtle, and sometimes counter-intuitive. For example, the growing use of IT can actually make urban development more concentrated, as well as allowing dispersal from high-cost to low-cost locations …

In fact, in a highly volatile and competitive context, major 'global' cities are becoming key anchor points with the social 'milieux' and assets most suited to high-level corporate and decision-making functions and continuous innovation. Cities like London help global firms and institutions minimize risks, strategically control their interests over long distances, access many complementary services and infrastructures, and, above all, maintain highly skilled work forces whose on-going, face-to-face work can support continuous innovation, so keeping ahead of the competition …

IT networks and services are thus closely bound up in the manifold production of all types of new urban space and the restructuring of old urban space.

In fact the urban dominance of the Internet in the USA is actually *growing* rather than declining. The top 15 metropolitan core regions in the USA in Internet domains accounted for just 4.3% of national populations in 1996. But they contained 12.6% of the US total in April 1994 and by 1996 the figure had risen to almost 20% as the Internet became a massively diffused and corporately rich system.

As Moss and Townsend suggest, 'the highly disproportionate share of Internet growth in these cities demonstrates that Internet growth is not weakening the role of information-intensive cities. In fact, the activities of information-producing cities have been *driving the growth of the Internet* in the last three years' (emphasis added) …

But planners are not standing still in the face of these challenges. Many local agencies are now beginning to experiment with information infrastructures as planning tools to help shape new types of urban places, in what may be termed urban 'tele-planning' initiatives. A wide range of such strategies and plans are rapidly emerging, designed to propel places as sites for investment and creativity in the new media and communication landscape …

In California, new urban 'tele-village' or 'smart community' initiatives are rapidly emerging. These attempt to insert optic fibre grids and advanced information services into the fabric of the master-planned, liveable, and relatively high-density communities of the New Urbanism movement. In Los Angeles, urban tele-villages are being positioned near rapid transit stations within ambitious, integrated plans to manage transport and teleworking together in whole urban corridors, in a sophisticated effort to reduce automobile use …

On the economic front, most large, dynamic cities already have their campus-like 'technopoles', developed to house science parks, corporate research and development centres, and university business, science and engineering schools, usually in green, peripheral zones. Cities like Lille, Cologne and Sunderland have gone further and have developed high-profile 'tele-ports' which connect local industries directly to advanced services and satellite ground stations …

'Cultural industries' are being digitised too. In the centres of older industrial cities like Manchester, New York and Dublin, so-called 'information districts' are emerging, usually spontaneously. In these, small and micro digital media firms – thriving on the fusion of digital design, Internet services, advertising, music and art – concentrate into gentrifying inner urban neighbourhoods.

Such places are proving attractive through their ability to support the sort of ambient street life which attracts new entrepreneurs and supports intense, on-going, face-to-face contact. But they also have the high-capacity telecommunications grids to link globally. London's Soho, for example, has just been linked by a broadband network to Hollywood to support on-line film-editing and production …

Giant among the emerging generation of urban 'tele-planning' initiatives is the proposed Multimedia Supercorridor (MSC) in Malaysia. Here, at the heart of what, at least until the recent upheavals in stock and currency markets, were confidently regarded as the burgeoning 'miracle economies' of south east Asia, a whole national development strategy had effectively been condensed into a single, grandiose urban plan for a vast new urban corridor.

The aim of the MSC is – financial stability permitting – nothing less than to replace Malaysia's manufacturing-dominated economy with a booming constellation of service, IT, media and communications industries. This is to be achieved by turning a vast stretch of rain forest and rubber plantations into 'Asia's technology hub' by the year 2020 … Its scale and comprehensiveness are astonishing: 30 miles long and 10 miles wide, the corridor covers an area of 750km^2. In effect, it is a greenfield site as big as the whole of Singapore island. Massive new highway grids, rail networks, utilities and a very high-capacity optic fibre web provide the infrastructural matrix for the plan, at a cost of $2 billion to the State.

The key elements of the MSC plans include a network of 'smart schools', a new multi-media university, with surrounding research and development clusters, 'paperless' electronic government offices and campuses, on-line medicine and distance learning centres, advanced manufacturing centres, and tele-services and back-office zones for inward investors. Each is associated with its own plans for new urban districts, carefully integrating the IT spaces with a strategic planning framework providing associated housing (inevitably labelled 'cybervillages'), parks, leisure, transportation, and retailing uses …

It is increasingly clear that, almost unnoticed in the planning literature, a whole new set of geographical landscapes and planning practices are quickly emerging, driven by the economic, social, cultural and technological dynamics of urban development in this globalizing age. Urban 'tele-planning' initiatives demonstrate forcefully that new technology means neither some simple 'end' to the importance of space and place, nor the collapse of face-to-face contacts in towns and cities and the transport demands that derive from them. Rather, cities are *co-evolving* with new communications networks and applications, as they did with the telegraph, telephone, cinema and newsprint media before them.

In fact, the specific assets and characteristics of individual spaces and places are becoming *more important*. This is because powerful organisations, especially trans-national firms, are now able to scrutinise what each place offers much more intimately within the myriad of location options available, within a context of increasingly free flowing capital, finance, goods, technology, services and information.

As giant global networks linking places together become the norm, places will need to fight relentlessly to make sure, first, that their territories are represented on them (rather than being 'off line', excluded and marginalised), and, second, that they are represented in ways that bring the greatest possible developmental benefits and spin-offs, within complex international divisions of labour.

Reference

Moss, M. and Townsend, A. (1997) 'Manhattan leads the Net nation', at <http://www.nyu.edu/urban/ny_affairs/telecom.html>.

Source: Graham, 1997, pp.296–9

CHAPTER 3

City life and difference: negotiating diversity

by Linda McDowell

1 *Introduction*

This chapter has a dual focus. It examines the tension between *settlement* and *movement* that is a key feature of cities. We shall look both at the city as a fluid entity, at the flows of people to the city and the patterns of movement within it, and also at the ways in which people settle into different areas of cities. The chapter is closely linked to the previous one: we shall pursue the questions of movement that were addressed in Chapter 2, but from a different perspective. Our interest here is in people, in the urban inhabitants themselves, rather than in transport networks. But the two are, of course, inseparable, and some of the issues about fear, safety and pollution on city streets that were addressed in the previous chapter will be revisited in this one.

Two related sets of questions run through the chapter:

● The first set of questions is about urban migration or movement: about arrival in the city, about movement within its streets and between its residential areas. We shall look at the ways in which social differences between people, whether based on their income and wealth or on their class, gender or ethnicity, affect who may and who may not move within and between the public spaces of a city.

● In the second set of questions our attention shifts to the converse of movement: to stasis, settlement and immobility, to how people become fixed or trapped in space. We shall be addressing questions about who settles where in cities, about how people are sifted and sorted into different areas – made immobile if you like, living most of their everyday lives within the boundaries of a 'community' and in the private spaces of their homes.

While this distinction between movement and settlement is often mapped on to the spatial distinction or tension between public and private spaces, a key aim of this chapter is to rethink this distinction and, at the end of the chapter, to critically reconsider the term 'community'.

Throughout the chapter, it will help to keep in mind the paradoxical duality of cities: on the one hand cities are fluid collections of people who, through movement and migration, arrive in and move through city streets as strangers to each other; on the other hand cities are composed of a series of 'urban villages' – residential communities of settled neighbours – who live much of their daily lives in a relatively bounded locality, familiar with the networks of streets, shops and pubs in that area and with the people who inhabit and use them.

Elsewhere in the series **Allen (1999)** discusses the 'worlds within cities'; in this chapter we will be revisiting the themes that are outlined there. This chapter is also about proximity and difference, but instead of the main emphasis being on daily rhythms and patterns, we will begin to look at urban rhythms and patterns over a much longer time perspective. We shall be looking at events across the twentieth century, at the long-distance movements of money and people, through colonialism, trade and migration, that connect cities in different parts of the world into an interdependent network of unequal relations, as well as the movements and patterns of settlement that divide cities into distinctive areas or localities. The longer-term temporal perspective of this chapter is important. Cities traditionally are the places where strangers meet and also make their lives. Diverse strangers are brought into spatial proximity in urban centres, whether these people are drawn from the rural areas of the same nation state or from widely differing places as social and economic changes – from wars and famines to geographically uneven patterns of economic growth and decline – result in migration. Thus the great cities of the world challenge our notions of geographical distinctiveness. 'First' and 'Third' World populations live in the same cities, both in the 'West' and in the 'rest' of the world. Whether we take the example of London, New York, Cairo or Bombay, 'Third' World women undertake sweated labour in their homes, often in close proximity to the office areas where businessmen of different nationalities work for the same global corporations.

Finally, one further introductory comment about the public and private arenas of cities may be helpful. The distinction between, on the one hand, a crowd of strangers and, on the other, a familiar group of family, friends and neighbours is the basis of a key division in urban life, which roughly maps on to one of the most significant social and spatial divisions in cities – that between the public and the private. The ties that link urban strangers – the crowd or the 'public' – are the ties of the polity. The bonds between urban residents are not the close personal and intimate ties based on the private relationships between kith and kin and close personal friends. The urban crowd, the multitude of strangers, is instead united through the rights and obligations of common citizenship and through participation in the public life of a city. However, as we shall we, the public arena, be it the literal spaces of city streets and parks or the metaphorical public spaces of urban institutions and networks of power and influence, are not equally accessible to all urban residents. Those spaces which, without thinking, we call public spaces may in fact be no such thing.

Cities, then, are loci of power, difference and diversity. The social divisions among the urban masses, and the potential that city living offers for interactions between these diverse populations living in relative spatial proximity, mean that

● questions about space and privacy, about interactions and connections, about the potential for both social and spatial mobility on the one hand, or the prospect of achieving a place and a settled life in a new city on the other, are key urban issues.

While cities embody the exciting prospect of intense social relations with individuals and groups from a wide range of backgrounds with different attitudes, beliefs and customs, they are also arenas of potential conflict. An urban population has to deal with the issues that are raised by difference and diversity. Cities are also arenas of intolerance and persecution, in which strangers refuse to accept the validity of other ways of life. How to negotiate this diversity and to create social and political institutions which are based on the tolerant acceptance of difference rather than the refusal to accept the other's point of view is one of the key political issues facing cities at the turn of the century. How to negotiate an acceptance or tolerance of difference is the issue that we shall turn to in section 7.

2 *Movement, migration and urban diversity*

When was the last time you arrived in a strange city for the first time? Shut your eyes for a moment and remember what it was like as the multitude of new sounds and smells assailed your senses, and how, as you left the station or the airport, got off the boat or out of your car, a kaleidoscope of different scenes and places and crowds of strangers met your gaze. Even now, at the turn of the century, when cities and urban images are a common part of everyday life for most of us in the UK, there is a certain thrill or heightened expectation when we arrive at a London mainline station from the suburbs of a smaller town or a village: the hubbub, the noise of the crowds, the multiple sights and smells and the danger and excitement of the moving crowds of people banging and jostling past bring a smile to the lips and the expectation that something, anything, might happen to us in the city that day.

Arriving at a new city for a holiday or a short break from our daily routine is, however, a world away from the experience of a migrant who has travelled not from a town up the line somewhere but instead from another country, nation and society to start a new life in a strange land. In this case the kaleidoscopic scene may be threatening and impenetrable rather than pleasurable or challenging.

Urban growth and change, from the earliest cities of the ancient civilizations to the expanding cities of the Third World today, have been in large part the consequence of movement and migration. Indeed, it was not until the end of the nineteenth century in western Europe that there was significant urban growth resulting from the expansion of the internal population as death rates began to decline and the worst excesses of urban overcrowding, poverty and insanitary living conditions began to be ameliorated through public health interventions. But the great urban expansions that accompanied nineteenth-century industrial growth in western Europe and North America were achieved primarily through movement. In Britain, France and Germany, for example, the rural poor from the hinterlands of nascent cities and from colonial possessions moved in their hundreds of thousands to industrial cities in order to find jobs in the expanding factories, mines, shipyards and steelworks. Great social misfortunes such as famine or disease, large-scale economic changes such as agricultural restructuring, and smaller-scale personal tragedies drew people to cities. In Great Britain during the Irish famine, 300,000 Irish migrants arrived in Liverpool in a single year, 1847. While some settled there, others continued their journeys to other British cities or undertook the hazardous voyage across the Atlantic to settle in New York, Chicago and other US cities. The migration from Europe to the USA from the mid nineteenth century was considerable as American society began to take on its characteristic feature of diversity (see **Pile, 1999,** for a discussion of the impact on Chicago). In total an almost unbelievable figure of

80 million migrants entered the USA between 1840 and 1940. In more recent decades the migration flows into and between European countries and into the USA have been from a wider range of countries, as migrants from southern and eastern nations have moved north, or across the Pacific rather than the Atlantic, to the cities of the North.

At the same time, in the second half of the twentieth century, migration from the countryside has swollen the urban population of Third World cities. Whereas the giant cities of the early twentieth century were predominantly in the 'West', in the 1990s only Los Angeles was growing as fast as the cities of the South. **Massey (1999)** shows that the most populous cities of the world now include Mexico City and São Paulo, Shanghai and Beijing, Bombay and Seoul, although, as later chapters will show, political and economic power is increasingly concentrated in the global cities of the economically advanced nations. Technological innovation and economic restructuring have shifted the economic base of western cities from manufacturing to services, as the expansion of manufacturing employment in the cities of the Third World draws their populations into industrial employment. Millions of people across the globe have moved into the growing cities of the Third World or moved from the Third to the First World. Furthermore, in-migration from former colonies as empires are dismantled and from the poorest countries of the world has resulted in ever greater diversity in the cities of the advanced world. Migrants from Korea, China, Malaysia and Vietnam, from Colombia and Mexico, from the European periphery – Portugal, Greece and Turkey – and from the former socialist world have joined the descendants of those earlier migrants from, say, Ireland and eastern Europe who moved to the cites of the UK and the USA in earlier decades. For economic reasons or because of flood, famine, persecution or religious intolerance, these movements have increased the degree of variation among urban populations, especially in the inner areas of cities where, just as a century or so earlier, many migrants have made their initial homes in the city.

At the turn of the century the movements and migrations that have been so important in transforming cities have become even more complex as a new element of fluidity is added to their direction. As we saw in Chapter 2, the 'friction of distance' has been reduced as innovations in transport technology make it easier and faster for more and more people to cross space. Many of the border crossings that people made to reach a new urban home are perhaps less permanent as more people are able to move back and forth between their old and their new homelands, although many others are trapped in their new 'homes' either by poverty, fear of persecution or immigration controls. But the rise of the new technologies of information transfer that were considered in Chapter 1 have blurred the distinctions between 'here' and 'there'. Pilgrims, tourists, temporary workers and more or less permanent migrants are active in displacing older relations between culture and place. The cities of the West, for example, are remade and recast in the images of elsewhere, which themselves are affected by displacement and relocation, while technological change enables connections to be maintained between different cities. Iranian exiles in Los

Angeles, for example, can watch television programmes made in Baghdad. Most of the urban poor in the cities of the First and Third Worlds are not fax-users, e-mail receivers, jet passengers or satellite owners, yet the television set is almost ubiquitous, and so global images produced in the main in the West remake the spatial distinctions that once made cities in different parts of the world and the ways of life within them so distinctive.

ACTIVITY 3.1   Jot down a list of the ways in which contemporary patterns of urban migration, both temporary and permanent, might affect the structure of large cities. Then read the following short extract, which is taken from a chapter of a book by Farha Ghannam about Cairo in the 1980s, when President Sadat's liberalization policies opened the city to new forms of communication and to international migration to Cairo for both business and tourism. Compare your list to the effects of migration on Cairo noted by Ghannam:

> ... around 5,000 Egyptian families were moved during the period 1979 to 1981 from Central Cairo (Bulaq) ... This area, which over the years had housed thousands of Egyptian low-income families, is adjacent to the Ramses Hilton, next to the television station, around the corner from the World Trade Center, across the river from Zamalek (an upper-class neighbourhood), overlooks the Nile, and is very close to many of the facilities that are oriented to foreign tourists. The area then occupied by low-income families became very valuable ... [and] the old crowded houses were to be replaced by modern buildings, luxury houses, five-star hotels, offices, multi-storey parking lots, movie theatres, conference rooms and centers of culture. Officials thus emphasized the urgent need to remove the residents of this old quarter because many international companies were ready to initiate economic and tourist investment in the area.

> (Ghannam, 1997, p.122) ◆

In contemporary cities, then, the types of areas that once housed rural in-migrants are being rebuilt for the new migrations that are related to the current phase of global capitalism. New flows of international tourists and business people are restructuring old urban spaces such as Bulaq.

How did your list compare with the factors noted by Ghannam? This example is rather specific. I had in mind a city like London and included on my list not only urban redevelopment but new types of shops and restaurants, new art ventures, the availability of newspapers from all over the world, but also less positive features such as racist graffiti and housing problems.

Let's now turn back to a century or so ago, to pursue the implications of the tension between the movement/settlement–public/private dichotomies. In the next section we shall consider some of the main arguments about the impact of the movements to cities that were at the time predominantly permanent and consisted mainly of less affluent migrants.

3 *Strangers and the urban crowd*

The unprecedented scale of urban growth in the nineteenth century in western Europe and then in North America was reflected in, and related to, an enormous upheaval in social relations: in the ways men and women related to each other, in the jobs people did, in the freedom they had to escape the chafing bonds of the traditional social conventions that characterized smaller settlements. Movement to the city brought with it greater freedom and mobility, both in the sense of spatial freedoms to roam the newly built streets and highways and to enter public spaces like cafés and drinking houses, and of greater social freedoms and opportunities which were connected to the move to a new place. So, movement to the city reordered people's social and spatial lives.

In the growing cities of the nineteenth century, the public world became more significant than the private world of the home. More and more people, for example, made their living by selling their labour power – their strength or their mental skills – in the factories, shops and offices of these vast new cities. New divisions between the public arena and private spaces, between the workplace and the home, created new opportunities for public association for some of the multitudes of strangers brought together in cities. At work, but especially in leisure hours in pubs and clubs, theatres and libraries, in the new cinemas and public parks, many men and women were able to mingle freely with a far wider diversity of people than they would have met in earlier periods. Contacts between people were less rigid, less bounded by social convention. In the new public spaces people might meet by chance, rather than by design, and this new diversity had a significant impact on the imaginations of the new urban residents.

This development – of a strong public life, of public associations between strangers in cities – has had a profound impact on writing about the ideal city, in both a negative and positive way. Whereas certain theorists developed a pessimistic analysis of the meeting and mingling of strangers in cities, arguing about the impact of sensory overload, for example, or the loneliness that overcomes strangers in an urban crowd, a perhaps more interesting set of arguments are those from the urban optimists, among whom we might include the British writer and political critic Raymond Williams and the American urban sociologist Richard Sennett. Sennett's ideas about the uses of urban disorder are discussed elsewhere in this series (see **Allen, 1999**). Let's first take a brief look at the arguments of Raymond Williams.

Williams was an influential analyst of culture, and in an exploration of the rise of the modern movement he argued that the urban social diversity that characterized the rising metropolis was a key element in the development of new forms of artistic expression in western European and North American cities (Williams, 1989). In the period of rapid urbanization in the late nineteenth and early twentieth centuries, numerous experiments in written and visual expression – in art, music, photography, novels and poetry – grew out of the social mix between all sorts of people in the new public spaces of the city. Surrealism in art and literature, new forms of visual expression such as the cinema, experiments with the structure and language of novels: together these changes marked a new epoch in artistic expression so significant that it is labelled the 'modern movement' or 'modernism'. Thus the city was a site of great excitement and innovation, of the exploration of new freedoms and ways of living, and as such it was celebrated in the modern artistic movements.

Richard Sennett also connects the city, social diversity and cultural change. In his book *The Fall of Public Man* (1977/1986; the following page references are from the 1986 edition) he draws out the key significance of the development of the public sphere in the formation of a new capitalist, secular, urban culture. His work is important for our argument in this chapter as he suggests that 'The history of the words "public" and "private" is a key to understanding this basic shift in the terms of western culture' (p.16). The outlines of Sennett's argument are captured in the extract below. Before you begin to read it, try to define 'public' yourself, both in the sense of 'the public' and a 'public' space.

As I was thinking about this definition I was walking around Cambridge, where I live. I passed one of the key symbolic public sites of Cambridge University that is featured in all the publicity material. It is a square outside the Senate House where new graduates wait before receiving their degree certificate from the Vice-Chancellor. Firmly fixed on the railings is a large sign insisting 'No public access'. In this case the space which is a public area for students and other members of the university is not open to the wider public – the general population of the city. It seems, therefore, that the definition of public is spatially or socially specific. It depends on the nature of the space and who controls access to it. In your own examples of public spaces, did this context-dependent nature of public space become clear? How did you define 'the public'? Let's turn to Sennett:

> The first recorded uses of the word 'public' in English [in 1470] identify the 'public' with the common good in society … Some seventy years later, there was added a sense of 'public' as that which is manifest and open to general observation … By the end of the 17th Century, the opposition of 'public' and 'private' was shaded more like the way the terms are now used. 'Public' meant open to the scrutiny of anyone, whereas 'private' meant a sheltered region of life defined by one's family and friends.

> (Sennett, 1977/1986, p.16)

Sennett goes on to explain that as society began to be conceived in terms of a geography, people went out 'in public'. He continues his historical exploration of the term:

> The sense of who 'the public' were, and where one was when one was out 'in public', became enlarged in the early 18th Century … Bourgeois people became less concerned to cover up their social origins; there were many more of them; the cities they inhabited were becoming a world in which

widely diverse groups in society were coming into contact. By the time the word 'public' had taken on its modern meaning, therefore, it meant not only the region of social life located apart from the realm of family and close friends, but also that this public realm of acquaintances and strangers included a relatively wide diversity of people.

(Sennett, 1977/1986, p.17)

Pile (1999) discusses elsewhere in the series the emphasis on social heterogeneity in Wirth's definition of the city. Here Sennett is emphasizing the same feature of urban life: the contacts between diverse, complex social groups. In the paragraph below, Sennett explains the ways in which, and places where, strangers might meet in cities like Paris and London:

> As the cities grew, and developed networks of sociability … places where strangers might regularly meet grew up. This was the era [the late eighteenth and early nineteenth centuries] of the building of massive urban parks, of the first attempts at making streets fit for the special purpose of pedestrian strolling as a form of relaxation. It was the era in which coffeehouses, then cafés and coaching inns, became social centres in which the theater and opera became open to a wide public … urban amenities were diffused out from a small elite circle to a broader spectrum of society, so that even the laboring classes began to adopt some of the habits of sociability, like promenades in parks, which were formerly the exclusive province of the elite.

(Sennett, 1977/1986, p.17)

According to Sennett, this 'public and private geography' (p.19), despite prolonging itself into the nineteenth century, was changed from within, altering the balance between the public and private arenas. As cities grew and the social relations of the market became increasingly dominant, those who had the means began to shield themselves from the traumas of the hugely expanding cities. According to Sennett, 'gradually the will to control and shape the public order eroded, and people put more emphasis on protecting themselves from it. During the 19th Century the family came to appear less and less the center of a particular, non-public regime, more an idealized refuge, a world of its own with a higher moral value than the public realm' (p.19).

At the same time, in the rapidly expanding cities growing numbers of strangers were swelling the urban crowd, and they became increasingly indistinguishable one from another as the expansion of factory-produced goods and clothing blurred visible class differences, thus making an increasingly expanding range of opportunities available to the mass. At this time the meaning of the 'public' sphere began to change. If, as I suggested above, the private sphere became an arena of virtue, the public arena began to be associated with vice. Let's turn again to Sennett for an explanation:

> Out in the public was where moral violation occurred and was tolerated; in public one could break the laws of respectability. If the private was a refuge

from the terrors of society as a whole, a refuge created by idealizing the family, one could escape the burdens of this ideal by a special kind of experience, one passed among strangers or, more importantly, people determined to remain strangers to each other.

(Sennett, 1977/1986, p.23)

You might underline this last sentence, as it is an important part of Sennett's argument about the public arena in today's cities. Sennett believes that the determination to remain strangers is a continuing feature of public interactions in cities and that it is socially destructive. He yearns to return to the civil and civic values of the late eighteenth century.

Let's pause here and think for a moment about Sennett's reliance on the term 'strangers' and the 'public arena' in the extracts above. Do you think there are any problems in using these neutral terms? What would happen if we gave these anonymous strangers a range of social characteristics and then sent them out to explore the urban public domain? And what about the nature of the different public spaces?

ACTIVITY 3.2   Taking two different examples, describe the ways in which access to public spaces is affected by social attributes such as age, class, physical ability, and so on. Who is permitted in which spaces? Does the time of day make a difference, for example?

List the attributes for each space and compare them. Are the spaces gendered, perhaps, in the sense that men dominate one space or women the other? Perhaps an urban park is a good example, or a shopping mall. This latter example is a complicated one, as sometimes such malls are 'quasi-public'. A security firm may patrol the shopping centre at particular times, and 'move on' those who are merely hanging about and not buying. ◆

We are beginning to see that strangers are diverse. In your notes you may have differentiated between strangers on the basis of their age, perhaps, or their physical abilities, which mean that many elderly and infirm people find that a range of public spaces are inaccessible to them. Public spaces themselves have an identity too: particular public places are constructed with certain users in mind. They often have class or gender associations which act to make 'strangers' of the inappropriate sex or class feel 'out of place': men in a child health clinic perhaps, or middle-class women in the public bar of a working-class public house (and notice how inaccurate the double designation 'public' is in this case).

Let's unpack this idea of exclusive places in more detail. We need to work through the implications of this notion of differential access to different sorts of public places for non-anonymous strangers in order to assess how the interconnections between social and spatial identities restrict people's movement through urban spaces. We shall first take two empirical examples, one from the UK in recent times and one from pre-war New York, and then turn to a theoretical critique of the notion of 'the public arena'.

4 *Sexing the stranger*

4.1 WOMEN IN SPACE

Doubtless one of the differences that you began to think about in Activity 3.2 was the differential access to urban public spaces for men and women. It is hard to open a daily newspaper at present without some story of harassment or violence that a woman has been subjected to in an urban public space. Furthermore, in their everyday lives most women have to deal with a range of relatively trivial invasions of their privacy in the urban crowd, from personal comments and inappropriate gestures to more serious incidents of physical violence. Femininity therefore brings with it a range of restrictions in the use of public space. These restrictions may arise from actual experiences of urban violence or from women's fear of violence. Research has shown that many women, especially elderly women, are afraid to use certain spaces such as city streets and parks, particularly in the evening or at night because of their fear of attack, especially by gangs of youths (see Pain, 1991; Valentine, 1989). In fact, Home Office statistics of reported crimes show that it is young men who are actually most frequently the victims of random or unprovoked urban violence in public places.

Feminist critics have suggested that one of the reasons for women's relatively restricted access to a range of public spaces is because of the general assumption that women are in need of protection from the hurly-burly of the public arena. Women's construction as dependent on men, both economically and morally, or as lesser beings – as fragile or in need of protection – reduces their rights to freedom in public spaces (Pateman, 1988).

A clear illustration can be seen in the judgements made in cases of rape and harassment, when judges have sometimes argued that women should remain indoors for their own protection. At times when men who are thought to be dangerous are 'on the loose' or 'at large', there are often calls for curfews for women and girls. Feminist campaigns to 'reclaim the streets' or 'reclaim the night', together with counter-claims for curfews for men, challenge the assumed greater freedom of men to occupy open and public spaces. Interestingly, in cases of rape or murder, women are sometimes constructed as transgressors who, through their actions, have tempted men into wrongdoing. Thus the British judiciary seems to imply in such cases that a woman who has been out too late or in the wrong place deserves what happened; the element of fault, it is suggested, rests with the woman rather than the man who attacked her (Smith, 1989).

We see here an extremely interesting example of the way in which the spatial location of women is used to construct them as certain sorts of women. The widespread and dominant association of women with the private rather than the public spaces of cities is used to construct a binary distinction between 'decent' women and less decent or less deserving women. Thus for a woman in a public place, her spatial position is used to identify assumed social characteristics and especially sexual morality. Remember that a commonly used euphemism for prostitutes is 'streetwalkers'. So places themselves are gendered and sexed, and place and identity are co-constituted.

A good (or do I mean bad?) illustration of the spatial construction of sexual identity was provided in the search for the so-called 'Yorkshire Ripper' in Leeds in the 1970s. The police initially and incorrectly assumed that all the victims of Peter Sutcliffe, the murderer of ten women, were prostitutes, merely because they were out alone in public. This case was over twenty years ago, and social attitudes about gender relations have changed. For example, larger numbers of women are now in employment and so perhaps have greater financial independence. But to what extent have attitudes changed?

ACTIVITY 3.3

- Is the assumption that women should not go out alone, especially after dark, still a common one?

- Have there been recent cases in the town or city where you live where you have seen or heard police evidence or remarks by the judiciary that supports your answer?

- If you have a son and daughter, do you expect different behaviour from your daughter compared with her brother?

- Does your daughter have the same freedoms in public places as her brother? ◆

It is important not to forget that, in a more general sense, the movement to the city brought greater freedoms for many women. City living was and is important in the development of new forms of association between men and women, and especially in the history of women's struggles to achieve greater freedom. For example, Elizabeth Wilson (1992) argued in her book *The Sphinx and the City* that the anonymity and excitement of the late-nineteenth- and twentieth-century city was crucial in the rise of feminist politics, as urban living often brought freedom from patriarchal control, in both the literal sense of an escape from the father and in the more general sense. For the 'New Woman' of the twentieth century, challenging the Victorian sanctity of hearth and home, the city brought a welcome anonymity and new opportunities for public association with a wide variety of women from different backgrounds and the development of new types of sexuality and sexual associations. Feminist historians have documented in detail the lives and struggles of the early pioneers of women's education and employment, women involved in the suffrage movement, lesbians, and women who searched for new ways of independent living in cities.

4.2   GAY MEN IN THE CITY

The anonymous urban crowd, as well as being a place for women to escape the bounds of conventional familiar morality and patriarchal ties, was and remains a place of escape for many others seeking to transgress conventional boundaries. Cities offer opportunities to those whose lifestyle was once labelled deviant or perverse (Mort, 1995; Sennett, 1977/1986, 1991; Walkowitz, 1992). A growing literature, primarily by and about gay men but latterly including lesbians too, has documented the significance of urban bars and clubs and areas of gay residential gentrification in the establishment of alternative sexual identities during the twentieth century (Adler and Brenner, 1992; Bell and Valentine, 1995; Chauncey, 1995; Castells, 1983; Fitzgerald, 1986; Knopp, 1992; Lauria and Knopp, 1985).

In this work, too, social and spatial identities are mutually constituted, as, it is argued, gay men often do not feel themselves to be gay unless they are spatially visible. Certain areas of particular cities have become well-known as gay enclaves. Perhaps the best documented of these is the Castro area in San Francisco, but other areas in a range of cities are identifiable gay areas, including Bondi in Sydney and the gay village that centres on Canal Street in inner Manchester.

The significance of gay political movements in cities has often been dated as beginning with the Stonewall riots in New York in 1969, when the police forcibly closed down a gay bar. However, in his recent book, *Gay New York*, George Chauncey (1995) has argued that there was an important gay geography in New York in the half-century from 1890. He has uncovered in fascinating detail the making of a gay male world between 1890 and 1940. Using newspapers, oral histories, court records, books and letters, he reveals a surprisingly public world – for the contemporary reader at least – of gay life in rooming houses, the YMCA and cafeterias, at drag balls, in bath-houses and on the streets; he also traces the development of gay neighbourhood enclaves in Greenwich Village, Times Square and Harlem in the first part of the twentieth century. Each of these neighbourhoods had different class and ethnic characteristics and public reputations, and yet these worlds had been almost entirely overlooked by historians of the period. Before Chauncey's book, it had been widely argued that gay men lived a secret, internal world and that there was no visible 'queer geography' in cities. But Chauncey shows that before the Second World War there was indeed

> a highly visible, remarkably complex and continually changing gay male world in New York. That world included several gay neighbourhood enclaves, widely publicized dances and other social events, and a host of commercial establishments where gay men gathered, ranging from salons and speakeasies and bars to cheap cafeterias and elegant restaurants. The men who participated in that world forged a distinctive culture with its own language and customs, its own traditions and folk histories, its own heroes and heroines. They organised male beauty contests at Coney Island and drag balls in Harlem; they performed at gay clubs in the Village and at

tourist traps in Times Square. Gay writers and performers produced a flurry of gay literature and thereafter in the 1920s and 1930s gay impresarios organized cultural events that sustained and enhanced gay men's communal ties and group identity.

(Chauncey, 1995, p.1)

Chauncey argues that, in New York over this half-century, although the laws against homosexuality were draconian, they were enforced only irregularly. Gay men had to be careful, but 'like other marginalized peoples, they were able to construct spheres of relative cultural autonomy in the interstices of a city governed by hostile powers' (Chauncey, 1995, p.2). In these areas, networks of ties between people helped them to find jobs and apartments and deal with the practicalities of urban life, as well as find romance and friendship. These spatial associations were important in rejecting mainstream values of the time and in the forging of gay identity. The areas were so well-known that apparently many New Yorkers 'viewed the gay subculture's most dramatic manifestations as part of the spectacle that defined the distinctive character of their city. Tourists visited the Bowery, the Village and Harlem, in part to view gay men's haunts' (Chauncey, 1995, p.4). Whatever we may think about the voyeurism involved in these visits, it seems clear that spatial association was an important element in gay rights and politics of the time.

ACTIVITY 3.4   We have seen how spatial location in different urban public spaces is an important element in the differentiation of the urban crowd. Rather than being anonymous strangers, it is clear that who we are depends in some part on *where* we are. As we move through the city, we may take on the social attributes of the spaces that we enter. Can you think of instances where your own identity is changed by the spaces that you occupy? ◆

5 *Rethinking the public*

5.1   ACCESS TO PUBLIC SPACE

Let's recap our argument so far, before we go a step further and think about ways of defining and redefining 'the public'.

- Cities are areas of complex and cross-cutting movements, composed of migration to and between different urban spaces.
- Cities are places in which public spaces are significant.
- These public spaces are arenas for multiple encounters with proximate but unfamiliar strangers.
- Strangers, however, have social characteristics which have a differential impact on their access to public places.
- Location within particular spaces affects the social identity of the occupant; or, in other words, identity may be read off from the space being occupied.

Now, I wonder if you remember my earlier suggestion that there is an association between the public domain and the polity. In democratic societies it has long been argued that every individual is equal in the eyes of the law and the state. All adult urban residents in democratic societies are citizens. However, as we began to explore above, not everybody has *de facto* equal rights in public spaces.

Despite the gradual extension of the franchise to the working class and to women in most societies, a noticeable feature of urban practice and policy has been the separation, segregation and isolation of motley groups of 'others' – be they women, the working class, non-'natives' or the mob – who are less welcome in the civic spaces of the city. Thus the establishment of urban order and the participation of citizens in urban public institutions has tended to be restricted to 'decent' citizens – usually, though not exclusively, bourgeois men. As cities grew over the eighteenth and nineteenth centuries, this spatial division between the public and the private, between those with a legitimate right to the public arena and those who should be excluded, was enshrined in theory and in practice, in the state and in legislation, and in the definition of citizenship and civic values. Thus the urban civic arena – the public spaces and state institutions – was constructed as an exclusive rather than inclusive space, and the notion of citizenship itself took a spatially constituted form.

Women, children, young people and roaming crowds have long been defined as dangerous and in need of control. Anxieties about the free and public association of strangers remain strong in contemporary cities and are evident in the ways in which events such as political rallies and football matches and more aimless groups of young men on street corners are seen as frightening and are often subjected to surveillance and heavy police control. Indeed, in some US cities and, it seems, in a growing number of UK cities, young people are the subject of curfews during the late evening and overnight. Clearly, some citizens have greater freedom of public association than others. But even when people are not actually prevented from assembling in public places, conflicts may arise because of their different needs and demands in the use of these spaces. Let's take one final example to illustrate the conflicts that might arise when a diverse group of strangers attempt to use the same public space: in this case a public park.

ACTIVITY 3.5   Read the extract below, which is a description of a busy public park in Manhattan, New York City. The extract is from an article written in August 1989 by John Kifner in the *International Herald Tribune*:

> Tompkins Square is used by skateboarders, basketball players, mothers with small children, radicals looking like 1960s retreads, spiky-haired punk rockers in black, skinheads in heavy working boots looking to beat up the radicals and punks, dreadlocked Rastafarians, heavy metal bands, chess players, dog walkers – all occupy their spaces in the park, along with professionals carrying their dry-cleaned suits to the renovated 'gentrified' buildings that are changing the character of the neighbourhood.
>
> (quoted in Harvey, 1992, p.588)

As well as up to 300 homeless people who slept in the park during the night, it was also used by young women and children, Hell's Angels and bikers, the de-institutionalized, and people involved in substance abuse of various kinds. All these different groups found mutual and tolerant coexistence impossible and there was a great deal of tension, as well as frequent outbreaks of violence in the square.

- If you were the urban planner responsible for the park, how would you resolve these conflicts of use?
- What practical steps would you take to control or exclude certain groups?
- Which groups would they be?
- How would you ensure that the exclusions could be achieved? ◆

I am not sure that there are any easy answers. I wondered about the use of park wardens or even a security firm. The newspaper article above is quoted in a paper by David Harvey, an urban geographer, in which he illustrates the dilemmas that faced the city's planners in attempting to resolve the conflicts arising from these multiple users. He quotes also from an article in *The New York Times*:

> There are neighbourhood associations clamoring for the city to close the park and others just as insistent that it remain a refuge for the city's downtrodden. The local Assemblyman yesterday called for a curfew that would effectively evict more than a hundred homeless people camped out

in the park. Councilwoman Miriam Friedlander instead recommended that Social Services, like healthcare and drug treatment, be brought directly to the people living in the tent city. 'We do not find the park is being used appropriately', said deputy Mayor Barbara J Fife, 'but we do recognize that there are various interests'. There is, they go on to say, only one thing that is a consensus, first that there isn't a consensus over what should be done, except that any new plan is likely to provoke more disturbances, more violence.

(quoted in Harvey, 1992, p.590)

Unfortunately, Harvey's paper does not include details of what was done in the area, but clearly this example raises difficult questions about how to adjudicate between conflicting notions of appropriate access and use. In many of the small urban squares and parks in Manhattan, the homeless and the poor are now excluded by locked gates and/or security patrols (Zukin, 1995). Gated squares, accessible only to the urban middle class, have long been a familiar feature of the urban geography of London too.

5.2   CITIZENSHIP AND PUBLIC SPACE

It is now time to start to think harder about definitions of citizenship and human rights in cities. We also need to try and address the conflict that seems inherent in, on the one hand, the recognition that 'strangers' are in fact a socially diverse group of people with different abilities and needs, and, on the other hand, the strong belief that in a democratic society we should all have equal access to all the goods and resources that our society offers for public uses, including of course urban spaces and public places.

Let's try and spell out some of the questions that face us in thinking through this dilemma:

- Is it possible to hold on to a universal definition of human rights – that all individuals are equal – in cities marked by diversity and inequalities between individuals and between social groups?

- Has everybody an equal right (and opportunity) to go anywhere in a city? Should they have?

- If some people are to be excluded from urban public spaces, how should it be achieved?

These are important questions, as conflicts over the use of public spaces, as well as the right to live in particular areas in cities, often revolve around differential claims about the right of occupation. Liberal theorists argue that every individual as a member of the polity has an equal right to be in the public arena, but as feminists and others have pointed out, this is often denied in practice. We have already examined the assumptions about women's right to be out at night and about the rights of people of 'alternative' sexualities to dominate certain spaces.

In the operation of state-defined rules and in common practices there is also an assumption of moral worth, in which *de facto* as opposed to *de jure* rights of citizenship are defined as open to those who are deserving or who are capable of acting responsibly. In practice, therefore, as critics from both the left and the right have recognized (Sandal, 1996; Sklar, 1991; Young, 1990a and 1990b), citizenship is not inclusive but exclusive.

Notice here the parallels with the arguments that we have just addressed about public places. As we have already seen, a considerable number of urban residents are discriminated against in terms of their access to particular spaces. Young people, people of colour and 'counter-cultural' groups often find that they are harassed and moved along (McKay, 1995; Skelton and Valentine, 1997), while urban public spaces are becoming increasingly less accessible and privatized through, for example, the use of private security firms to patrol the spaces between corporate buildings. In addition, in a growing number of towns and cities in the UK, closed-circuit television (CCTV) has been installed in streets and squares, which means that our movements in these spaces are no longer anonymous. The crowd of urban strangers celebrated by Richard Sennett is no longer anonymous but increasingly identifiable to a growing number of organizations and institutions.

In my home town, CCTV cameras have been installed in all the main streets in the city centre, as well as in most department stores, the public library and the university library. So every day as I cycle to town, shop with my credit card or take out a library book, my movements are tracked both by the bank of cameras in the city's surveillance centre and by the electronic trails of purchases and borrowings recorded by the shops, banks and library. It seems as if the urban anonymity celebrated by some of the early urban theorists is increasingly elusive.

ACTIVITY 3.6

● How many of your regular daily movements are now electronically recorded?
● What is your view of the costs and benefits of this surveillance? ◆

For the growing number of people who have to pass a much greater proportion of their daily lives in public spaces than most of us – for example the growing numbers of homeless people who make their homes in public rather than in private interior spaces, living on the streets, in doorways, subways and other tunnels – this increasing surveillance of urban spaces causes great problems, as the opportunities to escape the vigilance of state institutions diminishes. (See **Massey, 1999,** for a discussion of the problems of the urban poor in Los Angeles.) It does seem hard to argue in the face of the information we have considered so far that an undifferentiated urban public, a polity of citizens with equal rights of access to and occupation of urban public goods and resources, including urban space, is a defensible concept. How, then, might we rethink our notion of public space?

Nancy Fraser (1990), a feminist political theorist, has argued that if we are to take these exclusions seriously, then we have to rethink our notion of public space, defining it as a set of multiple and differentiated public arenas to which some groups have access but from which others are excluded. Like the post-colonial theorists who argue that there are 'subaltern' subjects who are able to challenge the power and discourses of colonialism, Fraser suggests the notion of 'subaltern counter-publics', which are real and metaphorical public spaces where marginalized groups might articulate their needs. The term 'subaltern' has been used by a school of Indian scholars who use it to refer to colonial subjects. Originally used to refer to women's subordination in imperial India, it is now more widely used to encompass the excluded or powerless in general.

ACTIVITY 3.7   Some subaltern counter-publics perhaps exist already – that is, alternative or oppositional spaces where the powerless challenge conventional uses of space. Are there any examples where you live? ◆

The sorts of spaces I can think of include spaces of protest, such as 'new age' encampments or the tunnels and trees of road protesters.

The work of the geographer Don Mitchell (1995) on the history of the People's Park in Berkeley, which was claimed by counter-cultural groups in the late 1960s, shows how hard it is to preserve spaces like these in the face of the state and private property interests that seem increasingly reluctant to ensure open access to public spaces. But public spaces are important in that they are the spaces where the diversity that constitutes 'the public' is most apparent and where idealized notions of 'the public interest' are most evidently challenged.

However, let's leave these public and quasi-public spaces and turn in the final sections of the chapter to the private side of the public/private tension that distinguishes urban spaces. Does urban settlement rather than movement and the development of private and community spaces in cities raise similar sorts of questions about the inclusive and exclusive uses of space?

6 Settlement and a sense of place

6.1 THE CONSTRUCTION OF RESIDENTIAL COMMUNITY

As well as being a kaleidoscopic mixture of diverse strangers moving in more or less anonymous streams across urban space, cities are fixed entities, a collection of buildings that are relatively permanent (the average life-span for a house in UK cities today, for example, is about sixty years). As the first book in this series makes clear, cities reflect the social relations of an earlier era. For the diverse group of people – recent and less recent migrants and long-standing residents who make up a city's population – a small area of the city becomes their home, where they settle and construct a way of life that is reasonably permanent and which, in the main, takes place in a restricted spatial area. This is not to forget the uncertainties and impermanence of residence for many occupants of cities, especially the urban poor who make their homes in legal or illegal squatter settlements, shanty towns, *favelas* and *barrios* of western and non-western cities. As we saw earlier in the example from Cairo, governments often uproot the urban working classes, and less frequently the middle classes, for a range of urban development schemes. Nevertheless, as innumerable urban analysts have documented, urban dwellers develop a strong sense of place and attachment to their locality or community.

In his most recent work, Richard Sennett (1995) has suggested that the significance of these attachments has become increasingly marked in contemporary western cities. In the western cities to which migrants are moving at the end of the late twentieth century, there has been a great upheaval in the nature of the jobs that are available. The service economies that have replaced former manufacturing industries in these cities are marked by the expansion both of the best-paid and the worst-paid occupations. Consequently, many migrants, as well as the local working class, are forced to accept low-paid work in a range of 'servicing' occupations – in catering, cooking and cleaning, in the fast-food industry, as messengers or couriers, decorators and domestic workers. Although manufacturing jobs are still present in these cities, they are often in the informal or unorganized sector – in sweated labour in the clothing industry, for example (Sassen, 1991). There has also been an increase in the numbers of people who are only temporarily or casually employed and in those without work at all, increasing the extremes of income and wealth that have always been visible in cities.

In the passage below, Sennett vividly outlines the promises and problems of recent economic change for cities and their inhabitants, and, importantly for our focus in this section, he suggests why the changes seem to be leading to a deepening of people's attachment to place. This is how he begins:

Modern technology promises to banish routine work to the innards of new machines, and does so. It could therefore be argued, from a strictly technological point of view, that the division of labour is coming to an end, and with it one human evil of the old order. In reality, however, the new technology frequently 'deskills' workers, who now oversee, as the electronic janitors of robotic machines, complex tasks the workers once performed themselves. More brutally, the division of labour now separates those who get to work and those who don't; large numbers of people are set free of routine tasks only to find themselves useless, or at best under-used, especially in the context of global labour supply.

This spectre of uselessness [also] now shadows the lives of educated middle-class people, compounding the older experiential problem of routine among less-favoured workers: there are too many qualified engineers, programmers, systems analysts, not to speak of too many lawyers, MBAs, securities salesmen. The young suffer the pangs of uselessness in a particularly cruel way, since an ever-expanding educational system trains them ever more elaborately for jobs which do not exist … The masses, now comprising people in suits and ties as well as in overalls, appear peripheral to the elite productive core.

(Sennett, 1995, pp.12–14)

Sennett's gendered descriptions are probably accurate, as it is men in particular who have been most affected by the type of economic changes and urban restructuring that have taken place in large cities. It is, of course, also important to remember that these trends have had a differential impact, both over time and between cities in different countries.

One of the consequences of these changes, however, is that as work becomes less significant for many of us as the arena in which we construct our sense of self-identity and self-worth, then the city, in terms of the local area where we live and carry out our home life, may become correspondingly more significant. This is what Sennett suggests: 'as a result [of economic restructuring] people are seeking compensation from the places in which they live: personal standing locally if not on the job, a sense of cohesion and stability in the community which is absent in corporations which are continually repackaged and resold' (p.13). But Sennett warns us that this new sense of attachment to place may have paradoxical, even dangerous, consequences. He suggests that 'if neighbourhoods, cities or nations become defensive refuges against a hostile world, they may provide symbols of self-worth and belonging through practices of exclusion and intolerance' (p.13).

Here again, then, we meet the same issues of exclusion and tolerance that were so important in our earlier analysis of public space, and this is what we shall focus on for the rest of this chapter. However, we shall only be able to begin to address the wide range of issues about inclusion and exclusion, power and segregation that are involved in the distinction of urban residential communities and the construction of a sense of belonging in cities. These issues are discussed later in the series (see **Pile, Brook and Mooney, 1999**).

The most important point to note about residential segregation is that it is by definition an exclusionary act, whether the mechanisms are through prices (pricing the poor out of 'exclusive' areas), through state regulations that may restrict the occupation of publicly provided or state-subsidized housing to certain groups, or through more unpleasant restrictive practices based on prejudice and discrimination. While some urban residents are able to choose where they live in the city and, through individual or group actions, exclude others whom they consider undesirable, many other urban dwellers have to make do as best they can in the areas and accommodation to which they are allocated or that they can afford to rent or buy.

Let's briefly consider the ways in which the sifting and sorting mechanisms work in cities to produce communities of similar people. Such communities may be distinguished by their ethnicity, as in the many areas of cities that are named after their occupants (such as Chinatown, Little Tokyo and Koreatown in inner-city Los Angeles, which also has other spatially distinct areas occupied predominantly by people of Latin American origins, Vietnamese or African Americans). Rather than Los Angeles, however, let's return to Paris and the work of François Maspero and Anaïk Frantz, who are discussed elsewhere in the series (see **Allen, 1999**).

6.2   CROSSING PARIS

In the early 1990s, the French left-wing publisher François Maspero and his friend Anaïk Frantz, a photographer, decided to explore Paris by travelling across the region on the rapid express train, starting from Roissy airport. They got off the train at each station to explore the surroundings and the buildings, to meet the local people and to discover how and why they lived there, and what they felt about their locality. We shall look at three extracts from Maspero's (1994) record of their journey (*Roissy Express*), the first two from a stop in the outer suburbs, and the third from a stop further in towards the centre.

First, though, let's hear from Maspero on why they set out on their journey. Is it significant, do you think, that Maspero refers to himself and his companion in the third person?

> Being Parisians, they themselves had for years watched their bustling quarters being slowly transformed into museum-style shop windows; they had watched the departure of an entire class of craftsmen, workers and small shopkeepers – all the people who went to make up a Paris street. They themselves had hung on, but saw renovation force out the poorly off, old people, and young couples with their children who all disappeared as rents rose and flats were sold. Where did they all go? To the outskirts. To the suburbs. Paris had become a business hypermarket and a cultural Disneyland … [and in the suburbs] in a single kilometre, you move from one world to another.
>
> (Maspero, 1994, p.16)

In a few short sentences, Maspero sums up the connections that link our focus in this chapter: settlement and movement. The extracts also combine our dual focus, for Maspero and Frantz are middle-class and mobile, whereas the people they visit in the extracts below are settled, even trapped in their suburbs. Most of them are migrants to Paris drawn or forced here by the changing relations between metropolitan centres and peripheral cities constructed through imperialism or economic exploitation, but their current immobility is a consequence of economic and social changes, especially in the nature of work available in western cities that Sennett described in the quote in section 6.1 above. Paris, like other western cities, is an example of how migration – movement and mobility – is reformulating older geographical distinctions between what are sometimes referred to as 'First' World and 'Third' World cities. Growing numbers of Third World people now live right in the heart of the old imperial cities.

On their journey, Maspero and Frantz document the social diversity and difference between Parisian suburbs. As they are told by Akim, a 25 year-old French Algerian born in Aubervilliers, even neighbouring La Courneuve was a different world.

Before we go to Aubervilliers with François and Anaïk to visit Akim's father's flat, we shall start in the outer suburb of Aulnay-sous-Bois. Here our intrepid social commentators and urban explorers have got off the train at Villepinte, just two stops down the line from Roissy (see Figure 3.1).

FIGURE 3.1
The route of the suburban express train (RER) across Paris

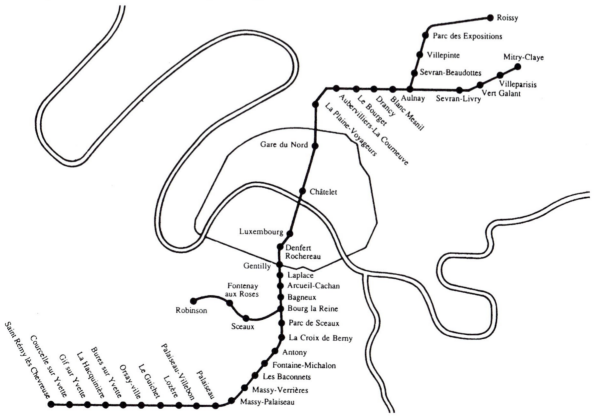

EXTRACT 3.1
François Maspero: 'Aulnay-sous-Bois'

They arrive on a big housing estate. Without realizing it, they have reached
the northern limits of Aulnay-sous-Bois. There are long lines of five-storey
'walls' of housing (a sixth floor would require lifts) but few high-rises.
They are tinted pink and pretty, seen from a distance. The grass is bare, the
trees are sickly-looking. Squares of soil and cement mark the site of dead
trees. Walls are dirty. They walk along the avenue between straight façades
of buildings. On one second-floor balcony, a man is watering a profusion
of plants and flowers in the quiet evening air. Anaïk wants to photograph
him. He refuses aggressively and, in reply to their compliments, says that
this year his plants aren't doing very well: it's the neighbours' fault. The
doors of the blocks of flats are cramped, as though making them narrow
costs less …

It is the time of day when children are out playing, their mothers come
home from shopping and the men tinker with their cars. The children
stand by the sign: photo-time. On the other side of the avenue, from
something half resembling an abandoned sports ground, half a public
garden, with a few concrete benches and beams, two girls call them over.
One of them has a child in a pushchair, the other a husky dog: her face is
very pale, and she is holding a slightly faded rose which, she says, comes
from her garden. She used to live just above the burned door. She's waiting
for her fiancé. She'd like to do what they're doing: have a walk round, take
photographs. Of people, and things. Be free. Earn a living like that. Do you
work for *Oxygène*? It's the local newspaper. They read it. This is a really
bad area. The Rose des Vents. The Aulnay 3000 estate: haven't you heard
of it? But it's famous. It's got a bad reputation. Rapes. Muggings. Drugs
especially. You should see the needles they find in the gutters. A boy died
from an overdose not long ago. It was in the papers. They've set up a
committee named after him: the Rodrigo Committee. But this is home.
Anaïk takes photos. Of them; the child; the dog. Then of them, the child
and the dog.

But this is home. At least there's fresh air. And space. You can breathe. Do
you come from Paris? Paris is suffocating. How do people live in Paris?
You're right to go and have a look round: there are some beautiful things
to see round here. Do you know the Vieux Pays? It's difficult to describe,
but it's nice. It's special, it's in the country. And Parc du Sausset? If you're
heading towards Garonor, go and see Parc Ballanger – there are geese and
goats. What would they do if they had the choice? One would like to live
in Châteauneuf-sur-Loire: she's been there for her holidays. The other
fancies the Charente. Down there you can trust people.

They go back to the foot of the 'walls'. Anaïk asks a lady if she may
photograph her, dressed in a long, glinting silk tunic, her head covered by
a scarf; she's holding a pigtailed little girl by the hand, waiting for her

husband to finish unloading the car. Madame Zineb has gaps between her teeth, the sort called *dents du bonheur*, 'good luck teeth', and a vertical blue line on her forehead. She has lived in Aulnay for twenty-one years. She was born in Tlemcen, Algeria. She has nine children. Two of her daughters are there, just home from work, and ask the travellers what they do. Do you work for *Oxygène?* Madame Zineb smiles: 'It warms my heart, you asking to take my photograph' …

October 1989 return to the 3000. Anaïk took Madame Zineb her photos, and they all had tea together. Madam Zineb talked about her husband, who used to be a builder and was unemployed. About her children, all born in France, some of whom lived at home: a son who had studied classical guitar and piano for seven years; a daughter who worked in a ministry in Paris and who would so like to have a job where she was free, like photography; another daughter, who was out of work but wanted to be a receptionist, a tourist guide or an interpreter. About the joy of her trip with her husband to Mecca. And about the hard times today, and the drugs taking hold of the young people. She'd like the government to clamp down more. Drugs are the new plague.

Source: Maspero, 1994, pp.31–5

FIGURES 3.2 and 3.3   *Aulnay, the 'Rose des Vents' (top) and a child living on the Aulnay 3000 estate*

119

ACTIVITY 3.8   Why do you think these outer suburban estates were built, and why have problems of poverty become so severe on the Aulnay estate in the 1990s? Think back to Sennett's description about the changing economic base of cities, and then read the next extract from Maspero's book. ◆

EXTRACT 3.2
François Maspero: 'The rise and fall of Aulnay'

Later they would find out more and realize that the 3000's first problem, its original curse, wasn't drugs, delinquency, intolerance, illiteracy or racism. The 3000's real problem was called Citroën. The estate was born in a state of euphoria in 1971: Citroën was opening a factory and needed workers – unskilled workers, the last of whom were mostly North Africans or Turks, who had been recruited specially to plug the local labour shortage and needed housing. The matter was entrusted to the public housing body and put in the hands of a town planner whose career notes say that he was interested only in aerial photos: viewed from a plane, problems of different ethnic groups living in overcrowded conditions, waterproofing, lack of soundproofing, and the quality of building materials obviously became hazy: only large masses matter. The planners thought big: 16,000 inhabitants were expected. The Communist town council, for its part, must certainly have dreamed of a radiantly happy proletariat living on a happy estate. One must also remember the background to the story: the shantytown period was barely over. There was a feeling of making a huge and praiseworthy effort to provide decent accommodation for all; not long before, the French bourgeoisie was still explaining that in any case, what was the point, the working classes still kept their coal in the bath – so you can imagine what the Turks are like. But it was time to think of applying widespread housing remedies that were less precarious, less shameful and ultimately more rational than shabby digs and shantytowns. These people were to get a home of their own. Bright apartments with everything that had been admired in previous decades in the name of 'modern comfort'. There was work nearby: 11,000 new jobs, 8500 of them at Citroën alone. Garonor was being built, and many other industrial zones were sprouting between the motorway and the airport. To separate the estate from the factory, they designed the Parc municipal de la Rose des Vents: it must all have looked pretty, seen from a plane. There was no need to provide transport to go job-hunting further afield. The birth of the 3000 on beet fields was particular in that, unlike other estates such as Sarcelles, no mixing of different social classes was intended. Apart from executives from neighbouring firms who had accommodation there, people stayed put. All the necessary social environment could be found on the estate: crèches, schools, community clinics, sports halls. A real little paradise.

Barely had the estate seen the light of day when crisis struck. As early as 1975, Citroën started laying people off. So did everyone else: Ideal

FIGURE 3.4   *What does the future hold for these young residents of the Aulnay estates?*

Standard's Aulnay plant closed down, making 2960 workers redundant in one go. In 1978–79 Citroën shed 1132 jobs, and more were to follow. How could people live on 'The Liner'? They had to get work elsewhere. A long, long way away, which required hours of travelling. Unemployment followed. Were they helped to pay the air fare home? They were now surplus to requirements. And as people were starting to talk a lot about tolerance thresholds, and housing departments and town councils everywhere were starting, more or less openly, to apply quota policies by refusing more than a certain percentage of foreigners or coloured people, an estate like the 3000, where the limit had been exceeded from the word go, served as a natural overflow for those refused elsewhere, whom socio-politico-administrative linguists, who never seem short of a verbal brainwave, describe as 'marginalized'. For many, the question was no longer how to leave the 3000 but how to stay. In Aulnay this year, 1989, 180 tenants have been evicted and 400 cases are under examination. It's not as if the authorities are stony-hearted: eviction is considered only when all makeshift remedies using social, humanitarian and charitable welfare have been exhausted.

Meanwhile, in an effort to defuse the situation, and also to keep different ethnic groups apart, the housing services blocked off 600 empty flats. Did they give up the profits for all that? No, because the operation was financed by the *FAS*, the welfare action fund. But from exactly which part of the *FAS's* budget was the money taken? From the part for immigrant housing. So who financed it? The immigrants themselves, would you believe …

What was to be done? In 1979, under Giscard's presidency, a big operation to renovate the estate was launched. It was barely eight years old ... The prefabs were defabricating and the prestressed concrete was destressing. They renovated. They repainted. It's crazy what gets painted in the Parisian suburbs. While they're at it, why not give everyone rose-tinted glasses? With the billions swallowed up, say the realists – are they optimists or pessimists? – there would have been enough to knock everything down and build *pavillons* for all ...

With the arrival of Mitterrand in 1981, efforts were stepped up. The 3000 was declared a 'sensitive zone' and a *ZEP*, an 'education priority zone': today there are ten school support groups for children of the estate. Never again will Citroën create unskilled jobs; nor will anyone else. But there are openings in the service sector, so if the children get good qualifications ... Meanwhile, at least an entire generation has lost its way: the youngsters who have been in limbo, and still are, who have never known what it is to have a real job. They are now twenty, twenty-five years old. They have missed the boat.

Source: Maspero, 1994, pp.37–9

The line between the outer suburbs and the inner ring of Parisian suburbs – known as *la petite couronne* – is still marked in places by the remains of a fortified wall which was constructed between the 1830s and 1870. Travelling in towards the centre of Paris, Maspero and Frantz got off the train at Aubervilliers (see Figure 3.1). Aubervilliers is a former industrial suburb where industrial work is now rare. Like Aulnay, it is also an area where many French Algerians live, in the main in high-rise estates built by the state. Read Extract 3.3 now, and compare it with the earlier extracts.

EXTRACT 3.3
François Maspero: 'Aubervilliers'

Lunch at Akim's father's flat. The travellers are happy to return briefly to a gentle family atmosphere. In truth it's not a big place: an *F2* in a high-rise in the 800. Two rooms linked by a kitchen, with a picture window running all the way along. Akim's father is fifty-nine, got divorced a few years back and has had three children with his new wife; and now he has just been pensioned off after nearly forty years at the factory. He arrived in France in the early 1950s. He knew Aubervilliers when it had fields and gardens, which he used to drive through on his moped to go to work at the Motobécane factory. It's a long time since he returned to Algeria – it's too dear, in any case. His life is here, though it's too much of a squeeze, unfortunately, with five in two rooms; but at the town hall they reply that young couples are given priority. The 800, says Akim, is a poor people's estate. Those who can go and live on more modern estates.

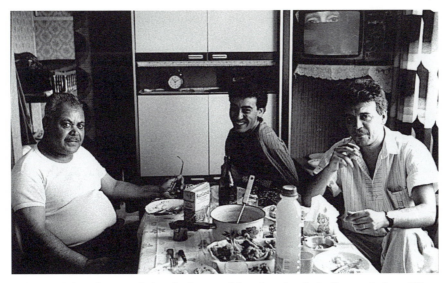

FIGURE 3.5   *Akim with his father and brother in their flat at Aubervilliers*

As though it were the most natural thing in the world, Akim's father has decided to welcome his son's unknown friends as if they were cousins arriving after a long journey: his wife has prepared a princely couscous. Round the table are his two eldest sons – Akim and his brother Sadi, a TV cameraman – and the three little ones. The eldest two talk about how worried they are about the young ones' education: if only they could avoid what they themselves have been through. The problem with the estates, says Akim, is that they don't let go of you easily: they're shut in on themselves, they offer a territory, a form of security. There are kids growing up on their Auber estate who have never really searched for other horizons. Even Paris exists only for the odd excursion. The gang ends up taking the place of a second family, of society, and everything takes second place to the role, status and prestige the youngsters can obtain within it: the most important thing is not to lose face in the eyes of the other members. It's tough living on the estate, but it's tough getting out too. Staying there and going back there is almost easier. So they just have to get by on 4000 francs a month, alternating between odd jobs and unemployment benefit. The gang, Akim goes on, isn't just linked to the economic crises. It comes from something even deeper. He thinks it goes in cycles: there are periods when you think gangs are going to disappear, that the new generation won't experience them, and then they surface again. 'I'll never come back to Aubervilliers,' says Sadi.

Round the table, a mild afternoon kind of feeling. Family photo-time. Akim's father wishes to explain that he doesn't understand how people can live in a country without adopting its customs, provided they're clean and civilized ones. He knows Tunisians and Moroccans who persist in observing the tradition of eating with their fingers, and he thinks that's stupid.

> François's lighter has gone out. The master of the house offers him a new one. 'Come back whenever you like,' he says when they say goodbye, at the foot of the tower block.
>
> Source: Maspero, 1994, pp.170–3

As these extracts illustrate, estates such as Aulnay 3000 and Auber 800 are not only traps for deprived people who are penned up there, marginalized from more affluent society, but also areas where they find security and feel comfortable in the social networks built up over time. Their residents are often powerless, excluded in all practical senses from the rights of urban citizenship, with little say in their future. Thus the ideal of citizenship, as we saw earlier, is again challenged by spatial divisions and mechanisms of exclusion that restrict the life chances of the least well-off.

In the final section of this chapter we shall turn to a consideration of how to rethink the definitions of community and citizenship in ways which recognize the difference that spatial location makes. The question that faces us is:

● How might greater democratic participation be extended to people who live in areas like the deprived Parisian suburbs visited by Maspero and Frantz?

7 *Diversity, social justice and community action*

To begin to answer the question posed at the end of section 6, we shall turn to the work of the political theorist Iris Marion Young, who is particularly interested in ideas about spatial justice and group participation in contemporary cities. Her arguments are discussed elsewhere in the series (see **Allen, 1999**). As Young's work is often quite abstract and, as she herself recognizes, utopian in its policy conclusions, it will help us to think through the implications of her argument if we counterpose it with an example of community action from present-day Birmingham. First, then, let's look at the Birmingham example and then we can use it to assess Young's claim about urban diversity.

7.1   BALSALL HEATH, BIRMINGHAM: COMMUNITY ACTION OR VIGILANTISM?

The example we are going to consider has many parallels with the case of Tompkins Square, which we considered in section 5.1. You will remember that there the main problem was one of conflicting uses of a public open space. Similar issues often arise in residential areas where some occupants may feel that their present or future neighbours are undesirable. They can also arise, of course, in the private use of the home, as families and households negotiate their own joint and individual use of the different rooms and spaces, often with conflicting needs or desires about what is appropriate behaviour at home.

ACTIVITY 3.9   Read the following paragraphs, which summarize several newspaper reports about conflict in Balsall Heath, an inner-city suburb of Birmingham. As you read them think back to your recommendations in the case of Tompkins Square. Are they relevant here? Or are different types of action more appropriate? ◆

Balsall Heath, in common with many similar areas in British cities, has a history of housing decline, class change and in-migration from different areas of Britain and also, since the Second World War but particularly from the 1970s, from other Commonwealth or former Commonwealth countries. Areas like Balsall Heath are often dubbed 'twilight areas' or areas of urban deprivation, but they are also home for many people who have a strong attachment to and pride in their area. Many have invested in owner occupation and in a range of improvements, and they resent not only the pejorative or negative labelling of their community but also, more problematically, the shifting population of less fortunate and less-rooted people who also find some sort of a home here. Such areas are thus cut through with social divisions – those of ethnicity, class, income, different types of lifestyle – and so with potential conflicts of interest between the residents.

In Balsall Heath another form of conflict became an increasing source of tension for the local residents in recent years. The area had become a place where sex was for sale: in some of the small Victorian houses where women sat under a light in the window, often in a state of semi-undress, and, more obviously, on the streets. Groups of women prostitutes, and some young men, would gather each evening, and during the day, on street corners in order to attract the attention of 'punters' who cruised the area in their cars. It was not only moral outrage, but also physical dangers to their children – from the passing cars, from the used condoms and syringes junked in the small front gardens – that stirred the local community into action. Headed by a group of mainly Asian men, a rota was established of men who would sit in full visibility on the street corners and patrol the local pavements each evening through into the night in an attempt to shame the 'punters' and destroy the trade for the 'working girls'.

ACTIVITY 3.10   What do you think about the action taken by these men in the local community? Is it a fine example of gender solidarity between local men and women uniting to protect their community, or is there another possible interpretation?

Melanie Phillips, a journalist, saw it in the former terms, declaring that the actions of these Muslim men had 'pointed social values in a more healthy and realistic direction' (Phillips, 1994, p.12).

● What is your reaction to these events?

● Should the views of the 'girls' be taken into account?

● Do you think all the local residents would support a group of predominantly Asian men in patrolling the area?

● Is there a more complicated story to tell here?

● Does Fraser's idea of counter-publics offer us any help? ◆

In an article in *The Guardian* published a few days after Melanie Phillips' comment, Maggie O'Kane (1994) chose to focus on the ways in which 'the local community' was divided along ethnic and gender lines. She interviewed several women residents who told her that they felt intimidated by the gangs of men patrolling the streets. Rather surprisingly, the attitudes of Asian women in the area, married or single, young or elderly, were not considered by either of the two reporters. However, as the work of the Southall Black Sisters, an Asian women's community action group in London, has shown, there are often tensions in Muslim families, not only between the different generations but also between young men and women, who have different attitudes about women's position in contemporary Britain. Some of the younger Asian women might be anxious to escape the patriarchal control of the older men in the area.

What is your conclusion about the action taken in Balsall Heath? Like the case of Tompkins Square, it raises difficult questions about who has rights to live and work in a neighbourhood. Let's try to spell out the advantages and disadvantages.

On the one hand, the community pickets were certainly successful in cleaning up Balsall Heath. Prostitutes, drug dealers, pimps and gangsters left the area as, according to Phillips, 'people power cleaned the streets'. But what is your view of this 'cleaning'? Do you perhaps agree with a local church minister, David Clark, who was quoted as saying: 'It's almost fascist; it's the sort of thing that was done to the Jews and the gypsies, when the police and local groups get together like this. It's treating people like objects, like the ethnic cleansing in Bosnia'. Do you have sympathy with this point of view?

There is no doubt that Phillips raises a difficult issue in her article. In a later article in which she returned to the issue, Phillips (1996, p.7) argued that: 'at the root of the controversy is a fundamental collision of values, between those who believe there are absolutes of right and wrong and those who believe that all values are relative and no-one should judge anyone else'.

ACTIVITY 3.11

- What do you think of Phillips' conclusion?
- Is she right that the key issue here is a collision of values?
- Should pimps and drug dealers have no rights?
- Can and should we distinguish between 'innocent' women and prostitutes?
- Are there absolutes of right and wrong in making decisions about urban issues?
- Who should decide who has access to which pieces of urban space?

We seem to be faced by innumerable questions here, to which there are few easy answers. This case study makes clear, however, that a serious look at spatial issues means that

- what seem like simple questions about equality and citizenship collapse into complex distributional questions. When we try to address the issues of whether a spatial social justice is possible, we quickly realize, as we have seen here, that any notion of a singular 'public interest' cannot be sustained.

So what is to be done? Let's now turn to the work of Iris Marion Young, who has suggested that if we accept diversity between people, then we must reject the liberal view that all individuals are equal and construct a new, group-based notion of social justice. She also suggests, however, that we must reject the notion of community. Here at last, then, we are perhaps finally approaching a way of dissolving the tension between difference and community with which we began this chapter.

7.2   A POLITICS OF DIFFERENCE

Let's start with Iris Young's succinct summary of the problems with the ideal of community. Read the following extract, keeping in mind the example of Balsall Heath as you do so:

> The ideal of community privileges unity over difference, immediacy over mediation, sympathy over recognition of the limits of one's understanding of others from their point of view. Community is an understandable dream, expressing a desire for selves that are transparent to one another, relationships of mutual identification, social closeness and comfort. The dream is understandable, but politically problematic, I argue, because those motivated by it will tend to suppress their differences among themselves or implicitly to exclude from their political group persons with whom they do not identify. The vision of small, face-to-face, decentralized units this ideal promotes, moreover, is an unrealistic vision for transformative politics in mass urban society.
>
> (Young, 1990a, p.300)

As you can see, in this passage Young identifies some of the problems that we have already identified when assessing local action in Balsall Heath. Taking action on the basis of an assumed coincidence of interest among all the inhabitants of a small area in a complex city like Birmingham means that alternative points of view are necessarily ignored. As Young (1990a, p.302) points out, 'community' action is problematic because it is often based on a 'desire for social wholeness and identification that underlies racism and ethnic chauvinism on the one hand and political sectarianism on the other'.

But is there an alternative? Instead of a city broken into separate spaces (communities?) – areas divided by race and ethnicity or social class where people feel comfortable in their face-to-face interactions with people like themselves – Young yearns for a diverse city that is not organized to suppress difference and oppress 'others' (that is, whoever differs from the dominant norm) but is open to 'unassimilated otherness' (p.301).

Let's explore what Young might mean by this concept. In the extract below, Young first expands on her criticisms of an idealized notion of community. This is what she argues:

> Insofar as the ideal of community entails promoting a model of face-to-face relations as best, it devalues and denies difference in the form of temporal and spatial distancing. The ideal of a society consisting of decentralized face-to-face communities is undesirably utopian in several ways. It fails to see that alienation and violence are not only a function of mediation of social relations but can and do exist in face-to-face relations. It implausibly proposes a society without the city … [It sets up] an opposition between authentic and inauthentic social relations … [it posits] the desired society as the complete negation of existing society [and] it provides no understanding

of the move from here to there that would be rooted in an understanding of the contradictions and possibilities of existing society.

(Young, 1990a, p.302)

Make sure you understand this passage, as it is a complicated argument. Remember that Sennett argued that an opposition between the family and the crowd became important in the nineteenth century when it was assumed that the highest virtues or moral values were to be found in the private sphere of the home and the household. This private/public opposition is paralleled here in the claim that face-to-face communities are somehow more 'authentic': a claim, you will note, that Iris Young is attempting to displace. She believes that in the complex, diverse societies of contemporary cities, something else is required. In her own words, that 'something else' should be 'social relations without domination in which persons live together in relations of mediation among strangers with whom they are not in community' (p.303).

This is a dense sentence and needs careful thought, especially as we are so used to thinking about the concept of community in a positive sense: in association, for example, with ideas of mutual aid and co-operation – the local self-help that Phillips valued so highly in Balsall Heath. Indeed, community is often counterposed to individualism – the emphasis on a solid, self-sufficient unity, each person for him- or herself – and accorded higher value by many interested in greater social equality in cities. But, as Young notes,

> identification as a member of a community often occurs as an oppositional differentiation from other groups, who are feared or at best devalued. Persons identify only with some other persons, feel in community only with those, and fear the difference others confront them with because they identify with a different culture, history, and point of view on the world.

(Young, 1990a, p.311)

Thus the basis of community, Young suggests, 'stems from the desire to understand others as they understand themselves and from the desire to be understood as I understand myself' (p.311). In cities, in the division of urban space into different residential 'communities', this identification occurs within homogeneous groups, based on the exclusion of a multiple group of 'others' on the basis of their difference from, and lesser value than, those with whom identification or community is felt.

Here too we can see the parallels between some of Sennett's claims and Young's arguments. Elsewhere (see **Allen, 1999**) Sennett has celebrated the multiple connections and complex loyalties in some of the ethnically diverse communities of Chicago in the early twentieth century. It might help you to think through Young's argument and untangle her rather complex prose if you test her claims against examples of what you consider to be a community or community action – for instance a locality in a town or city, or perhaps an interest group or an organization – as well as the case study of Balsall Heath.

I chose as my example the ways in which the residents of an 'exclusive' area of high-status housing have excluded ethnic minorities from their estate. Perhaps rather surprisingly, given that she is interested in the city, Young herself provides us with a non-urban example to clarify her arguments. She suggests that in the early days of the Women's Movement in the 1960s, 'strong pressure within women's groups for members to share the same understanding of the world … often led to homogeneity – primarily straight, or primarily lesbian, or primarily white, or primarily academic' (Young, 1990a, p.312). The anger felt by many working-class or black women at their exclusion from the Women's Movement led to bitter fractional disagreements in the 1970s. In more recent years a more tolerant perspective has developed that celebrates the multiple ways of 'being a woman'. This recognition of diversity seems to be leading to a movement based not on its members' understanding one another as they understand themselves, but rather on an acceptance of the differences and distance between them without leading to exclusion.

Is this a possible or merely a utopian aim in cities, when the distance between people is often physical and reflected in actual or metaphorical spatial boundaries, such as the walls around gated communities? Young does not deny that there are advantages in spatial proximity and the creation of communities of socially homogeneous groups: 'In a racist, sexist, homophobic society that has despised and devalued certain groups, it is necessary and desirable for members of these groups to adhere with one another and celebrate a common culture, heritage and experience' (Young, 1990a, p.312).

As we have seen, urban communities like Balsall Heath are a response to exclusion and deprivation. But as we have also seen, the desire for unity often leads to intolerance of differences within the group – the problems of young Asian women in this case. Another example, perhaps, is the personal difficulties sometimes experienced by gays and lesbians in some militantly 'out' groups.

Young suggests that we abandon the privileging of face-to-face communities as an ideal, although she does not deny the value of mutual friendship and co-operation in relatively localized city spaces. However, in the complex urban societies of the present day, a richer and more diverse set of social relations and connections is possible, as the media that link people across time and space expand. Think of how many people you know reasonably well in your own town or city. How many of these links and friendships are based on co-residence in the same locality?

Young suggests that we must accept the character of cities as they are today – complex centres of goods, resources and culture, containing large numbers of strangers with particular characteristics, backgrounds, needs and abilities, who do not, and cannot, 'understand one another in a subjective and immediate sense, relating across time and distance' (Young, 1990a, p.317). You will remember from the Tompkins Square and the Balsall Heath examples that cities also comprise strangers who do not understand or tolerate each other even when they are in proximate relations. What is needed, therefore, is

● to think about what a politics of difference or diversity might entail, a politics to ensure that cities continue to offer anonymity and the prospect of independence and tolerance for people deemed deviant.

The prospect of greater freedom and the chance to transgress conventional norms has long been an important aspect of city life. Think back to our earlier debates about gender and sexuality, for example. The city is an arena for tolerance, as well as intolerance, providing opportunities for people who are different from one another to accept and understand the points of view and ways of life of the numerous strangers and 'others' who live in the urban spaces.

But is the ideal of a politics of difference as utopian in its own way as the ideal of community? This is a difficult question to answer, but surely we all agree that tolerance is preferable to reactive and fearful exclusion, and that celebration of the diverse cultures of cities is preferable to suspicion and rejection of anything we find strange or uncomfortable? Indeed, it is possible that we might all change for the better, influenced by increasing urban diversity. 'British' urban culture, for example, is certainly more diverse and interesting at the end of the century than it was in, say, the 1950s: think of the influence of music from the southern USA, Africa and Bangladesh, for example, or of the cosmopolitan clothes and style, or the new languages and literatures developing in many metropolises. The sociologist Stuart Hall (1990) has suggested that one of the consequences of the meeting and mixing of diverse populations in cities may be what he terms 'translation', as our sense of ourselves and group identities in cities change in positive ways. This idea seems more optimistic to me than Young's call for the celebration of 'unassimilated otherness'. Translation implies mutual respect and influence, but recognizes and celebrates – using Young's terminology – the exciting impact that urban movement and migration have on the population.

Young herself recognizes that her call for tolerance and the acceptance of the multiple differences – in values, attitudes and ways of life – that exist among the diverse populations of cities may be an idealistic one, but she is prepared to go further and suggest practical ways in which a politics of difference might be implemented in cities as they are at present. Let's end this section with a look at her suggestions and the questions they raise.

Young suggests that the way to ensure tolerance in cities is to aim for as complete as possible democratic representation of the different groups who live in the same city (and remember the reasons why she prefers the term 'group' to 'community'. The most difficult question that immediately arises is: should *all* groups be represented? Indeed, Young (1990a, p.320) herself asks: 'what defines a group that deserves recognition and celebration?'. Consider, for example, extreme right-wing groups which preach racial intolerance, or white supremacy or terrorist groups which are prepared to sacrifice lives to achieve their aims, both of which groups are commonplace in many cities in different parts of the world at present. If we accept that such groups are united by their intolerance of others, then surely it is not ethical to argue that they should have the same democratic rights as other groups?

131

Young has a partial answer to this ethical problem of how to judge between groups. She suggests that we might start by rejecting racism, sexism, class prejudice and homophobia – all of which are the bases of visible social divisions in cities. Such a rejection is a necessary basis for the development of greater tolerance of attitudes and ways of life that may differ from our own but which do not discriminate against, or do harm to, others, although how to ensure anti-sexist or anti-racist behaviour is another vexed issue.

If we turn now to the decision about who should participate and which issues are permissible in urban decision-making fora, Young suggests that this should be decided by the adherence to three principles. Individual and group actions and their consequences must be judged:

● not to do harm to others;

● not to inhibit the ability of individuals to develop and exercise their capacities within the limits of mutual respect and co-operation; and

● not to determine the conditions under which other agents are compelled to act (Young, 1990b, p.251).

Young goes on to suggest that representative bodies, elected at the smallest spatial scale, are a necessary element for the participation of all the diverse and multiple groups living in cities. However, a strategic city-wide elected body is also an essential element of a politics of diversity, since judgements between the differential claims have also to be made at this spatial scale, as well as the decisions about which groups are included and which are to be excluded when the local fora are established.

Finally, then, we have seen in this section that Young is a pro-urban philosopher. She sees cities as:

● potentially liberating arenas within which variety, tolerance and diversity might be recognized and even celebrated, and in which the fluidity and mobility of the population and the option to participate in a wide range of activities in accessible and open public spaces facilitate the recognition and acceptance of different ways of living.

However, we have also seen in the examples we have considered here that cities are:

● locations riven with, even constituted by, a series of tensions. Although there is some tolerant acceptance of multiple ways of being, intolerance is also rife. In addition to open and accessible spaces and places through which the diverse urban population might move, there are also spaces of exclusion in cities, and populations which are trapped and discriminated against.

We have noted, through the examples, that individuals are marked by social characteristics including class, ethnicity, gender, age and sexuality, and that these are themselves partly spatially constituted. Women and men are at 'home' in different spaces, dominating some and discriminated against in others. We have seen, for example, that the tension between public and private spaces is a particularly important factor in the socio-spatial constitution of gender identities. But all groups are differentially placed in terms of their access to, and power to control, different spaces, whether they are open or public spaces or the less public spaces of localities and residential communities where the option to exclude others is important.

All urban spaces reflect and affect social divisions and relations of power, and are constituted by a series of tensions, especially the tension between settlement and movement and between, on the one hand, the desire to exclude threatening or challenging 'others' and, on the other hand, the opportunity to celebrate the diverse experiences, peoples, values and ways of life that constitute the contemporary city. While Young celebrates the variety, heterogeneity, publicity and eroticism of the city – the four features that she regards as defining the specificity of cities – for many urban residents in contemporary cities fear of crime, noise, danger, dirt and pollution lead them into protective strategies and reactive behaviours and a desire to exclude those others who may be perceived as different from, or as a direct threat to, the areas in which they live. Both reactions are part of urban existence and represent the dual focus that this chapter has addressed.

In the next chapter we turn to an aspect of urban life that is often neglected: the place of nature in cities. Young's view of urban life reflects, I think, a particular version of the city – the advanced industrial city in western societies at the turn of the millennium. These are cities in which we too easily assume that nature is either irrelevant or non-existent, but, as Chapter 4 will demonstrate, this tension between the 'artificial' city and 'natural' nature also needs unsettling.

References

Adler, S. and Brenner, J. (1992) 'Gender and space: lesbians and gay men in the city', *International Journal of Urban and Regional Research,* vol.16, no.1, pp.24–34.

Allen, J. (1999) 'Worlds within cities', in Massey, D. *et al.* (eds).

Bell, D. and Valentine, G. (eds) (1995) *Mapping Desire*, London, Routledge.

Castells, M. (1983) *The City and the Grassroots*, Oxford, Blackwell.

Chauncey, G. (1995) *Gay New York: The Making of the Gay Male World, 1890–1940*, London, HarperCollins.

Fitzgerald, F. (1986) *Cities on a Hill*, New York, Pantheon.

Fraser, N. (1990) 'Rethinking the public sphere: a contribution to the critique of actually existing democracy', *Social Text*, vol.25–6, pp.56–80.

Ghannam, F. (1997) 'Re-imagining the global: relocation and local identities in Cairo', in Oncu, A. and Weyland, P. (eds) *Space, Culture and Power: New Identities in Globalizing Cities*, London, Zed Books.

Hall, S. (1990) 'Cultural identity and diaspora', in Rutherford, J. (ed.) *Identity: Community, Culture, Difference*, London, Lawrence and Wishart.

Harvey, D. (1992) 'Social justice, postmodernism and the city', *International Journal of Urban and Regional Research*, vol.16, no.4, pp.588–601.

Knopp, L. (1992) 'Sexuality and the spatial dynamics of capitalism', *Environment and Planning D: Society and Space*, vol.10, no.6, pp.651–70.

Lauria, M. and Knopp, L. (1985) 'Towards an analysis of the role of gay communities in the urban renaissance', *Urban Geography,* vol.5, no.3, pp.152–69.

McKay, G. (1995) *Senseless Acts of Beauty*, London, Verso.

Massey, D. (1999) 'Cities in the world', in Massey, D. *et al.* (eds).

Massey, D., Allen, J. and Pile, S. (eds) (1999) *City Worlds*, London, Routledge/The Open University (Book 1 in this series).

Maspero, F. (1994) *Roissy Express: A Journey Through the Paris Suburbs*, London, Verso.

Mitchell, D. (1995) 'The end of public space? People's Park, definitions of the public and democracy', *Annals of the Association of American Geographers*, vol.85, pp.109–33.

Mort, F. (1995) 'Archaeologies of city life: commercial culture, masculinity, and spatial relations in 1980s London', *Environment and Planning A: Society and Space*, vol.13, no.5, pp.573–90.

O'Kane, M. (1994) 'Cruising, abusing or on the game', *The Guardian*, 23 July, p.25.

Pain, R. (1991) 'Space, sexual violence and social control', *Progress in Human Geography*, vol.15, no.4, pp.415–31.

Pateman, C. (1988) *The Sexual Contract,* Cambridge, Polity.

Phillips, M. (1994) 'When a community stands up for itself', *The Observer,* 17 July, p.25.

Phillips, M. (1996) 'Bye hooker, bye crook', *The Observer,* 4 February, p.7.

Pile, S. (1999) 'What is a city?', in Massey, D. *et al.* (eds).

Pile, S., Brook, C. and Mooney, G. (eds) (1999) *Unruly Cities*, London, Routledge/The Open University (Book 3 in this series).

Sandal, M. (1996) *Democracy's Discontent: America in Search of a Public Philosophy,* Cambridge MA, Belknap Press.

Sassen, S. (1991) *The Global City,* Princeton NJ, Princeton University Press.

Sennett, R. (1977/1986) *The Fall of Public Man*, first published 1977, New York, Knopf Inc.; reprinted 1986, London, Faber and Faber.

Sennett, R. (1991) *Flesh and Stone*, London, Faber and Faber.

Sennett, R. (1995) 'Something in the city', *Times Literary Supplement,* 22 September, pp.13–15.

Skelton, T. and Valentine, G. (eds) (1997) *Cool Places: Geographies of Youth Culture*, London, Routledge.

Sklar, J. (1991) *American Citizenship: The Search for Inclusion*, Cambridge MA, Harvard University Press.

Smith, J. (1989) *Misogynies,* London, Faber.

Valentine, G. (1989) 'The geography of women's fear', *Area*, vol.21, no.4, pp.385–90.

Walkowitz, J. (1992) *City of Dreadful Delight*, London, Virago.

Williams, R. (1989) 'Metropolitan perceptions and the emergence of modernism', in *The Politics of Modernism,* London, Verso.

Wilson, E. (1992) *The Sphinx and the City,* London, Virago.

Young, I.M. (1990a) 'The ideal of community and the politics of difference', in Nicholson, L. (ed.) *Feminism/Postmodernism*, London, Routledge.

Young, I.M. (1990b) *Justice and the Politics of Difference*, Princeton NJ, Princeton University Press.

Zukin, S. (1995) *The Cultures of Cities*, Oxford, Blackwell.

CHAPTER 4
Cities and natures: intimate strangers

by Steve Hinchliffe

1 *Introduction*

Cities are shaped, at least in part, by movements and settlements. As we have seen in the previous chapters, these movements and settlements are formed by, and help to produce, cities of difference and diversity. Furthermore, all the chapters in this book have so far agreed that movements and settlements help to make many cities exciting and frustrating places. This chapter plots a similar course, except that here we will try to think of cities as not just socially and culturally diverse places – a definition of cities based largely on the people that pass within, through and between them. We will be considering other movements and settlements that make up cities: we will encounter some of the animals, plants, materials and energies that pass between, flow through and linger within cities. My aim will be to suggest that cities are often characterized by environmental as well as social diversity. Just as we saw in Chapter 3 that cities of social difference have often produced possibilities for new political movements, so the heterogeneous collections of species and materials that I speak of here have the potential to *refigure* what we mean by cities and by urban-environmental politics. Far from arguing that cities are necessarily anti-nature, cities contain the possibility for some exciting changes to human relations with nature.

Before we begin the chapter proper, it is worth thinking about how we tend to relate cities and nature. Cities are sometimes imagined as places where nature stops – stops in the sense that, if you imagine travelling overland into a city, you might suppose you are leaving behind the open and/or green spaces of the countryside. You might think of the city you are entering as more cramped and closed in, a place where nature is largely absent. A related thought is that cities disrupt natural rhythms. We sometimes hear how cities are thought to have turned people (and other species) into different kinds of being. We often hear how cities can alter the weather, affecting temperatures and moisture regimes. When songbirds sing throughout the night in city centres, even the course of day and night, and the rhythms of the seasons, seem to have been altered.

Indeed, we might ask: what is natural about cities? Given our common-sense understandings of the terms 'cities' and 'nature', the immediate response of many of us to this question would be 'not much'. We might even say that the terms nature and city shouldn't be put together like this. They belong to different categories altogether. After all, people often think of cities as human, social and cultural achievements. It was Yi-Fu Tuan, the cultural geographer, who suggested that cities might be ranked according to how far they had departed from farm life, agricultural rhythms and cycles (see Tuan, 1978, and **Pile, 1999**). Elsewhere, Tuan (1984) maintains that if the natural is included in a city, then this is either thought of as an inconvenience which should be overcome (think of flooding rivers), or it is largely a human-made token (like a garden, a park or a pet dog). In this sense, cities are often understood to be anything but natural. In section 2 we look in more detail at this apparent divide between cities and nature.

One of the aims of this book is to challenge our common-sense ideas about cities. With this unsettling aim in mind, we will look for ways in which the division that places cities and nature in separate realms is itself constituted through a diverse range of movements. In bringing connections, disconnections, flows, networks and other movements to the fore, we might be able to challenge the idea that nature stops at the city boundaries. In sections 3 and 4, therefore, the notion that cities are important sites for non-human animals and plants will be introduced. The aim will be to suggest the extent to which humans share their urban spaces with a vast number of invited and uninvited guests. The point will be to reverse the idea set up in section 2 that cities and nature are worlds apart.

We will need to go further than this, though. Saying that cities *contain* nature is important, but it is not saying as much as we would perhaps like. In particular, we need to engage with the potential benefits of understanding cities as other than spatially bounded or discrete entities. We need to remember the flows between cities, and the other movements across boundaries. We also need to apply a similar analysis to nature, and try to imagine not only the itinerant species that share or inhabit certain urban spaces, but also the other *city-nature formations* that move within, between and through urban lives. I will use this not altogether obvious term 'city-nature formations' to express the varieties of spaces, movements and changes that occur and are produced in and through the cities that are considered in this chapter. In doing so, I want to communicate the idea that city-nature formations are not simply an addition of the urban and the natural. Rather, I will suggest that the urban and the natural are inseparable. In this sense, another answer to the question 'what is natural about cities?' will be put forward. This time, having blurred the boundaries between what we formerly meant by nature and cities, the response will be that there is nothing unnatural about cities.

Another reason for going beyond a simple statement that cities contain nature is by far the most important. By suggesting that cities and natures are somehow and sometimes more indistinct than we might have imagined, we open up interesting and vital questions about various distributions and arrangements of city-nature formations. These distributions will have consequences, so one of the aims of this chapter is not simply to confound conventional modes of thinking and ordering cities for the sake of academic satisfaction. Rather, and in addition, the chapter will be concerned to draw out some of the practical implications of rethinking city-nature questions. These issues will be picked up in the final section of the chapter.

Here are the issues and themes that you should bear in mind as you work through the chapter:

- Boundaries between the urban and the rural, and between cities and nature, are difficult to draw. The concepts of cities and nature are not as settled as we might at first think.

- Nature moves within and through cities, so much so that it becomes possible to talk of city-nature formations.

- The arrangements and distributions that constitute city-nature formations can be recognized as a basis for thinking about urban environmental politics.

2 The separation of the urban and the natural

In many cities, and for many people, the rat is often a symbol of disease, of unease. In the past, in many towns in Europe, the rat-catcher's job was to run the rodent out of town. Failure to remove the rat from cities provided conditions in which plague and pestilence could thrive. Today, the image of rats waiting to surface from the sewers is a reminder of the derivation of the word rodent – from the Latin *rodere*, to gnaw: rats are often depicted 'as a sinister force, gnawing at the foundations of human control' (McWhirter, 1996, p.109). To this day rats are often portrayed as figures to fear in cities; they undermine the senses and practices of human order. They are useful for our purposes in this chapter in that they hint at something about the ways in which cities are often understood. As the quote above suggests, rats gnaw at the foundations of city life. By that, I take it that the author isn't simply referring to the materials that underlie much of the urban fabric – the wooden piles, concrete trenches, or the more flimsy lean-to arrangements of the permanent-temporary buildings of shanty towns. I also take it that the visual presence of rats undermines our understandings of what it means to be urban. Rats emerging from a sewer, or even through a bathroom toilet, unsettle the fiction that cities are simply human affairs.

THE RAT CATCHER.

FIGURE 4.1
*The rat-catcher:
a Victorian engraving*

This myth, that cities are purely human and cultural achievements, is often based in what we can call a 'foundational story'. These are the stories that are told of how a city came to be founded, and what sorts of achievements, people and practices the city is built upon. They are stories of progressive settlement. As I will suggest, these stories often include a reference to human mastery over nature, including, of course, running the rats and other bearers of disease out of town (or at least out of sight).

2.1   TELLING FOUNDATIONAL STORIES

You may already be familiar with the idea of a foundational story. As we shall see, these stories exist in different forms, but they tend to be told as fairly straightforward histories. That is, they privilege a version of time that is linear, and follow a fairly predictable, developmental pathway through history (**Massey, 1999a and 1999b** discusses some reservations about linear stories.) They often privilege lots of other things as well. For example, we often hear of the founding fathers (not mothers) of cities. Of course, these stories often hide as much as they reveal about cities. Nevertheless, they are commonly told. You will most probably find examples of foundational stories displayed or held in the libraries or museums of the cities that you visit or in which you live.

The question I want to ask of urban foundational stories is: what do they tend to say about the place of nature? The foundational stories that we often tell of cities are ones where the origin is something akin to 'nature', and the destiny or outcome is something far removed from that primordial state. So nature is progressively expelled from urban space as the city grows and develops through time. (You might like to refer here to Tuan's ideas on the distancing of cities from nature – see **Pile, 1999**). The 'original' landscape and ecology is displaced by the urban landscape. Simplifying somewhat, it might be possible to outline two versions of this story, each of which is partially determined by the status given to origins.

The first type of urban foundational story is the *civilization story*. Here, primordial nature is something to be feared and to distance oneself from. Modern, civilized cities, so the story goes, are settled places where the natural is held off, outside the city's boundaries and beyond day-to-day lives. Holding nature at bay is a spatial achievement, and is often portrayed as something to congratulate ourselves about. It is often short-hand for the mark of being 'civilized'.

The second type of urban foundational story is the *degradation/pollution* story, where cities are seen to despoil the landscapes and ecological relations that once existed on that site. The 'origin' in this story is not so much feared; rather it is revered. Again, 'true' nature exists outside the city, but in this case there is guilt rather than pride. In addition, there is a twist to this second version of the foundational story. Nature may yet visit revenge on the city, and the progeny of human disregard for nature, in the form of hazards and disasters, is always waiting to re-invade the purified spaces of city life, although it will not do so in an even-handed fashion.

ACTIVITY 4.1   Now read Extract 4.1, 'Mexico City trashed', which is written by Joel Simon, a journalist. When you have read his account of Mexico City's history and environmental problems, answer the following questions:

● Compare the ways in which the author describes the Valley of Mexico prior to and after the Spanish colonization. What kind of foundational story is the author telling?

● In the article, the author describes some of the acute environmental and health-related problems that are experienced by many of the people presently living in Mexico City. It is fairly clear who and what the author blames for these contemporary problems. Can you think of any other causes that are missing from the author's account? ◆

EXTRACT 4.1
Joel Simon: 'Mexico City trashed'

The first question most visitors ask when they arrive in Mexico City is: 'Can I drink the water?' The second is often: 'Can I breathe the air?' The answer, in both cases, is probably no. Like no other place in the world, Mexico City is condemned to live in its own waste.

The roots of Mexico City's environmental crisis lie in its unique geography. The city sits in a valley at 7,400 feet, surrounded by snow-covered mountains as high as 18,000 feet. Two million years ago, a volcanic eruption sealed off the valley's natural drainage in the south. Rainwater backed up, filling a series of five interconnected lakes, and creating a completely self-contained ecosystem.

In 1325, after wandering in the northern desert, the Aztecs arrived in the Valley of Mexico. They thought they had found paradise. Everything they needed for survival could be produced in a single valley. Pine forests rose on the slopes of the rain-drenched volcanoes, while cactus and maquey … grew in the more arid north. Mountain creeks were diverted to irrigate the cornfields. The lakes themselves were not only a source of food, but in a country without horses or any other pack animals, they provided the basis for a transportation revolution. By the time of the Spanish Conquest, an estimated 200,000 canoes plied the waters, transporting food and other cargo long distances with astounding ease.

The Aztecs' reverence for their watery environment was so great that they built Tenochtitlan, their capital city, on an island in the middle of one of the lakes … When the Spanish conquerors first set eyes on the Aztec capital in 1521, they were awed by its splendor. They compared it to Venice, but in truth, it was larger and grander than any city in Europe at the time. Although the Spaniards admired the beauty of the Aztec city, they were land-based people who preferred to travel on the horses they had brought from Spain, rather than in Indian canoes. Therefore, they filled in the canals which had crisscrossed the city, and dismantled the dikes and causeways which had protected it from floods.

The introduction of horses and iron axes also allowed the Spaniards to cut down huge tracts of forest. They used the lumber to build palaces and churches in the new city. The advent of grazing animals, unknown in the new world, unleashed an unprecedented wave of erosion. Also, winds blew the unprotected topsoil off the mountains. By the beginning of the 16th century, Mexico City had its first smog emergency as an enormous cloud of swirling dust called a *tolvanera* descended from the mountains and covered the city. To this day, dust blown from the deforested hillsides remains a major source of air pollution.

The destruction of the forest also caused another unexpected problem. Rainwater, which used to gurgle down the mountain in small creeks, began to run off the exposed soil in a muddy torrent. As sediment washed into the lakes, the water level began to rise and Mexico City found itself battling more and more serious floods. The worst flood lasted from 1629 to 1634. The entire city was underwater and, as in Aztec times, the Spaniards were reduced to moving through the streets in canoes.

Rather than building dams to better control the flooding, the Spaniards declared war on the water. They decreed that the lakes, which still supported a large Indian population, would have to be destroyed. Thus began one of the most monumental projects of environmental engineering in human history – the draining of the Valley of Mexico. In 1608, the first phase of the project was completed. Sixty thousand Indians, under the supervision of Spanish authorities, completed a tunnel through the northern mountains, providing an outlet for the waters which had been trapped in the valley for 2 million years …

If the lakes had not been drained, Mexico City would not have reached such mammoth proportions. It was on the dried-up lake beds where rural refugees, driven from the countryside, built urban slums. As the city grew from a relative backwater to an urban monstrosity, the ecological deterioration of the valley accelerated until it reached the point where the very survival of one of the world's largest cities hangs in the balance …

Air pollution has made Mexico City infamous, but it is only one of the environmental woes afflicting the city. Decisions made five centuries ago by the Spanish conquerors still reverberate today. Mexico City has completely destroyed all the natural cycles which once governed life in one of the world's most unique ecosystems. For as long as the city lasts, it will be paying the cost.

Source: Simon, 1996, pp.172–5

This piece, like other foundational stories, should not be read uncritically. In reading the article and thinking about the questions in the activity, you may well have started to become uneasy with this account. When I thought about it, I managed to identify and develop four problematic issues with this form of narrative (there may well be others). The following sub-section offers some of my thoughts on the questions in Activity 4.1 and integrates them into a more general critique. When you have read the analysis in section 2.2 , refer back to your answers to the activity.

2.2   FOUR ISSUES CONCERNING URBAN FOUNDATIONAL STORIES

2.2.1   Idealizing the past

Simon's historical account of Mexico City in Extract 4.1 is a close approximation to a degradation/pollution urban narrative. Like many of these stories, it holds more than a passing resemblance to a creationist myth. The innocent, harmonious paradise of the Aztecs is compromised and rendered chaotic by the colonizing people of Europe. Through choice or enslavement, the people living in the Valley of Mexico were relatively quickly evicted from their Eden. Through greed, arrogance and temptation, the paradise became a 'monstrosity'. The morality tale is fairly explicit: it is a foreboding story to tell to would-be urban environmental engineers. Actions today reverberate into the future, and if we are to make good the fall, we had better learn to live with nature (in its 'original' paradisiacal form) or face the consequences. It's a 'fire and brimstone' story, straight out of the Euro-American providential tradition. When faced with stories like this, we should always be wary of the work they are doing, the traditions they draw on, and the possibility that there might be other ways of telling the story.

For example, consider the account of the meetings of the Spanish and the Aztecs in **Massey (1999a)**. There, it is the coming together of worlds that makes the city. To be sure, the meetings between the two worlds changed after 1521, and the terms on which the meetings took place were undoubtedly unequal. In Massey's account, however, we get the sense that the sixteenth-century Valley of Mexico marked a time-space when two culture-natures met up. Importantly, both the Spanish and the Aztecs were imperial cultures, and both had designs on the landscape. Without wanting to gloss over what must have been a cruel and violently unequal exchange, the Spanish designs very soon got the upper hand. The point for our purposes here is that we can spot elements of this story in Joel Simon's account, but the author doesn't address it in quite these terms. Indeed, it seems that Simon is suggesting at some points that the sixteenth-century meeting was a coming together of European culture and Aztec nature. The latter is made to sound ideal and viewed in a wistful, almost romantic fashion, whilst the former is symptomatic of the fall from grace of human culture.

In short, the point is that we need to be aware of who is telling the foundational story and where it is coming from. Viewing historical city-nature formations as somehow more natural than those of the present might not only overlook important details (including the scale of the Aztec city-nature formations); it could also obscure opportunities for more just formations in the future. This would certainly be the case if we were left with the impression that cities are no longer part and parcel of what we call nature. This brings us to the second reservation.

2.2.2 Undercutting the future

In suggesting that nature has been annihilated ('destroyed') in Mexico City, and stating that contemporary environmental woes were caused by decisions taken five centuries ago, the author does not provide readers with much hope. Indeed, we might say that the author is suggesting that the inhabitants' fate was sealed centuries ago. Mexico City ('for as long as it lasts') and its inhabitants (for as long as they last) will, it seems, be paying the cost of environmental destruction – there is no escape. Herein lies a problem: there is no room for city-nature politics where there is no hope.

Of course, the socio-environmental injustices in this and other cities cannot be 'storied away', but neither do these kinds of foundational story, and the social causes they imply, offer much in the way of imaginary resources for rethinking and reactivating urban environmental *movements*. I have emphasized the word 'movement' because Simon's story of Mexico City's settlement is, perhaps, too static. To borrow the terminology of Paul Gilroy (1993), this story speaks too much of roots, and not enough is said about routes (including interconnections, pathways, flows and possible 'exits').

A question that you might think about for the remainder of this chapter is: how can we tell more hopeful stories, which at the same time avoid a tendency to make urban environmental problems disappear from our accounts? Another way of putting this is to restate one of the aims of the chapter. We are trying to reactivate city-nature formations in order to represent and live in cities in ways which do not preclude possibilities for positive changes. The next two sub-sections take this point further by problematizing the spatial assumptions that adhere to urban foundational story-telling.

2.2.3 Anti-urban sentiments

In the idealization of origins and the condemnation of destinies, there is a powerful anti-urban sentiment that underlies many of the pollution/degradation urban foundational narratives. Cities, at least on the scale to be found in the twentieth century, are often depicted as the antithesis of environmentally sustainable futures, green living and the survival of 'nature'. This anti-urbanism has a long history, but contemporary environmentalisms have tended to revoice these sentiments. As the US cultural critic Andrew Ross (1994, p.101) points out, North American anti-urbanism 'perpetuates the monstrous image of the city as a greedy parasite, an overpopulated and dangerously polluted concrete jungle which has long transcended its appropriate biological limits'. We will challenge this anti-urbanism in various ways throughout this chapter, but for the moment it is worth noting the problems with such a view of cities.

First, anti-urbanism is often based on a mythical origin, and is told through a foundational story. In this way, contemporary cities are compared unfavourably to an imagined golden age or paradise. Herein lies a rural nostalgia which, as Ross (1994, p.101) points out, 'barely submerges its racism'. For Ross, North American anti-urbanism, and the 'white flight' from cities to green and rural suburbs in the post-war period, was part of a search for a white, middle-class American dream. The mistrust of the city, its racially marked inhabitants, its politics and its landscape, was therefore part of a foundational urban narrative and the selection of convenient histories.

Second, anti-urbanism is politically ineffective. You may now be familiar with the idea that cities are, in many ways, far from declining. They are home to more and more of the world's human population. **Massey (1999a)** has rightly drawn attention to the statistic that around half of the world's population lives in mega-cities. To write cities off as having no future seems, at the very least, impractical. If one adds to this the possibility that cities provide the most probable sites for a political mobilization around issues to do with environmental justice and nature politics, then it clearly seems counter-productive to perpetuate an anti-urban project.

More generally, anti-urbanism makes a rather simplistic distinction between the urban and its other, the rural. The following sub-section takes issue with this spatial purification.

2.2.4   Oversimplifying urban space

In spatial terms, the foundational story of Mexico City tells a very simple, two-dimensional map-story. In this map, nature is either present or absent. Anything in between this presence-absence, on-off type of mapping is given short shrift. Prior to 1521, nature is present, streams gurgle and, as a 'backwater' (despite Tenochtitlan being larger than most other European cities at the time), people live in a landscape resembling the garden of Eden, with nature. (This romantic paradise is, of course, wildly overstated.) After the Spanish colonization, nature is removed by European-style agricultural, forestry and engineering techniques. So, in the modern Mexico City, nature is truly absent – 'completely destroyed'. In this map of contemporary Mexico City, nature exists outside of the city – in the non-urbanized, rural landscape – and in the city's past. This is a simple containerization of natures and cities. In other words, natures and cities are separated by clear boundaries: they, or at least the urban category, form bounded spaces, and in those bounded spaces we have a certain uniformity. That is, the boundary between the country and the city marks a separation between, on one side, a purely urban landscape and, on the other, a purely rural landscape. Conceptually, at least, the spaces are considered to be purified. In the view from above, the boundaries and territories are clearly demarcated. Figure 4.2 illustrates this spatial view of the city and nature.

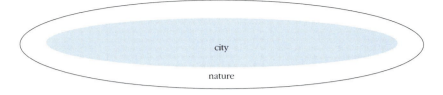

FIGURE 4.2   *The containerization of the urban and the natural*

This mapping of urban and non-urban spaces is problematic. It is divisive in that it tells us that urban lives and objects are unnatural, whilst their rural corollaries are natural. It is also symptomatic of a morality tale, suggesting that cities are somehow 'un-environmental'. It fixes spaces, ignores variation and difference, and in so doing misses some important movements within and between 'urban' and 'non-urban' spaces.

In the following sections we will look into ways in which such foundational city stories can be retold, and how, through this retelling, we can open up opportunities for remaking cities. We will draw on different spatial types to do this. Some, like the routes we have mentioned, have already been introduced, but we will be seeking to work with as many spatial types as possible in order to put city-natures on the move and to avoid the static spaces of nature and city that we have encountered so far in this chapter.

Here are some of the points that you should take from the foregoing discussion:

● Foundational stories work with a particular, and rather simplified, spatial map of cities and natures. In short, cities displace nature.

● The anti-urbanism that is a feature of some foundational stories is problematic. It tends to focus environmental political attention and action on the non-urban.

3 *Nature in the city*

In the previous section we ended by criticizing the notion that cities exist through the exclusion of natures. There are some fairly straightforward ways in which this containerization might be challenged. If you think about a city most familiar to you, it may well be possible to list the ways in which all kinds of 'natures' can be said to be present. I have already mentioned city rats. The rat-catcher's job is seemingly never ending, and if any animal challenges the contained spaces that were described towards the end of the last section, it is the rat. I now want to look at some of the other non-humans that live together with people in many of the world's cities.

3.1 PLANTS IN THE CITY

ACTIVITY 4.2 Look at Figure 4.3, which compares the populations of a number of European cities and villages (all located within 48–55°N and 4–23°E), with the number of plant species recorded within the settlement (taken from Pysek, 1993, p.92).

● What is the generalized relationship between city size and number of plant species?

● How would you account for this relationship? ◆

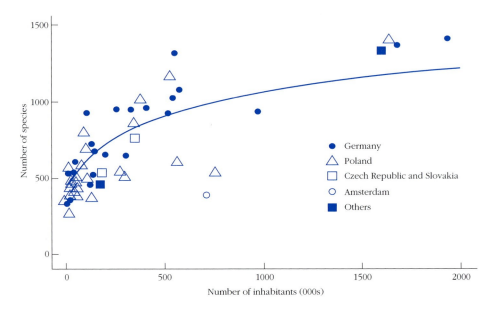

FIGURE 4.3 *The relation of floristic and vegetation diversity to city size*

Pysek's data seem to suggest that the larger the city, the higher the number of plant species. There are exceptions, notably in Amsterdam (the outlier shown as an empty circle), which seems to have fewer species for its size than we might expect (Amsterdam has very few parks or open spaces). Despite this, and given the fact that there may well be other ways of constructing these data, the relationship does suggest that, on this criterion, cities are far from being unnatural. Indeed, in the European cities that are used here, the surrounding countryside may be characterized by much more severe human control than is the case for the urban landscape. Further, the wide variety of cultural routes (networks, meetings, introductions and fostering of relations) that go to make up cities, along with diversities of materials and topographies, tend to result in greater varieties of urban plant life. Whatever can be said about city-natures, they are unlikely to be classed as monocultures.

By drawing attention to these city-nature formations, and in suggesting that countryside-nature formations cannot be differentiated from their urban counterparts on the basis of variety of species, we enter into some interesting debates. The vast movements of plants within and into cities is one way of suggesting that cities are part of, and not distinct from, ecological relations. We should perhaps be wary of planning, development and other environmental policies that ignore these city-nature formations. The latter are undoubtedly constituted through movements which integrate rural and urban landscapes. To protect a green belt at the expense of city allotments, or to develop land on the outskirts of a town in order to cash in on a rural-nostalgic dream, may miss the value of city-natures and their interrelations. We will pursue this theme as we turn to look at urban animal life.

3.2   ANIMALS IN THE CITY

In UK cities it has become common in recent years to see or hear of rats, urban foxes, hedgehogs, and other species which carry the 'nature' label with them, criss-crossing central urban spaces. While the species that are worth remarking upon might change, this movement and cutting of contained space is by no means a recent phenomenon.

There are a number of interesting issues that start to emerge from these movements and presences. As a way of raising and discussing some of these issues, and moving on to their significance, we will engage with two very different authors who have written about city-nature formations. The first is W.H. Hudson, who was writing in the late nineteenth century and was far more interested in nature than he was in cities. The second is Elizabeth Wilson, who, writing at the end of the twentieth century, is probably more interested in cities than nature. Nevertheless, both are, at least in part, interested in the coming together of city-natures.

ACTIVITY 4.3   Read Extract 4.2, which is by the famous naturalist, ornithologist and traveller, W.H. Hudson. The extract comprises two passages from Hudson's book *Birds in London*, which he wrote at the end of the nineteenth century. As you read the text, look for the elements in Hudson's writing that are suggestive of an urban foundational story. Second, and in some ways conversely, try to draw out the ways in which the foundational story and the spatial simplifications of such a narrative are overturned. ◆

EXTRACT 4.2
W.H. Hudson: 'Birds in London'

The subject [of London bird life] was often in my mind during the summer months of 1896 and 1897, which, for my sins, I was compelled to spend in town. During this wasted and dreary period, when I was often in the parks and open spaces in all parts of London, I was impressed more than I had been before with the changes constantly going on in the character of the bird population of the metropolis. These changes are not rapid enough to show a marked difference in a space of two or three years; but when we take a period of fifteen or twenty years, they strike us as really very great. They are the result of the gradual decrease in numbers and final dying out of many of the old-established species, chiefly singing birds, and, at the same time, the appearance of other species previously unknown in London, and their increase and diffusion …

Of the species which have established colonies in London during recent years, the wood-pigeon, or ringdove, is the most important, being the largest in size and the most numerous; and it is also remarkable on account of its beauty, melody, and tameness. Indeed, the presence of this bird and its abundance is a compensation for some of our losses suffered in recent years. It has for many of us, albeit in a less degree than the carrion crow, somewhat of glamour, producing in such a place as Kensington Gardens an illusion of wild nature; and watching it suddenly spring aloft, with loud flap of wings, to soar circling on high and descend in a graceful curve to its tree again, and listening to the beautiful sound of its human-like plaint, which may be heard not only in summer but on any mild day in winter, one is apt to lose sight of the increasingly artificial aspect of things; to forget the havoc that has been wrought, until the surviving trees – the decayed giants about whose roots the cruel, hungry, glittering axe ever flits and plays like a hawk-moth in the summer twilight – no longer seem conscious of their doom.

Twenty years ago the wood-pigeon was almost unknown in London, the very few birds that existed being confined to woods on the borders of the metropolis and to some of the old private parks … Tree-felling caused these birds to abandon the parks sometime during the seventies. But from

1883, when a single pair nested in Buckingham Palace Gardens, wood-pigeons have increased and spread from year to year until the present time, when there is not any park with large old trees, or with trees of a moderate size, where these birds are not annual breeders. As the park trees no longer afford them sufficient accommodation they have gone to other smaller areas, and to many squares and gardens, private and public. Thus, in Soho Square no fewer than six pairs had nests last summer. It was very pleasant, a friend told me, to look out of his window on an April morning and see two milk-white eggs, bright as gems in the sunlight, lying in the frail nest in a plane tree not many yards away …

Even in the heart of the smoky, roaring City they build their nests and rear their young on any large tree. To other spaces, where there are no suitable trees, they are daily visitors; and lately I have been amused to see them come in small flocks to the coal deposits of the Great Western Railway at Westbourne Park. What attraction this busy black place, vexed with rumbling, puffing, and shrieking noises, can have for them I cannot guess. These doves, when disturbed, invariably fly to a terrace of houses close by and perch on the chimney-pots, a newly acquired habit. In Leicester Square I have seen as many as a dozen to twenty birds at a time, leisurely moving about on the asphalted walks in search of crumbs of bread. It is not unusual to see one bird perched in a pretty attitude on the head of Shakespeare's statue in the middle of the square, the most commanding position. I never admired that marble until I saw it thus occupied by the pretty dove-coloured quest, with white collar, iridescent neck, and orange bill.

Source: Hudson, 1898/1969, pp.5–6 and 89–92

When Hudson says that he was compelled to stay in town during the summers of 1896 and 1897, he immediately conveys an anti-urban sentiment. He sees the city, and his time in it, as 'wasted and dreary'. Later in the extract he refers to the city as having an artificial aspect. He talks of 'the havoc that has been wrought' as the city has grown up. In these moments, Hudson, as I read him, is expressing an anti-urbanism that is closely tied to an underlying urban foundational story which highlights decay and pollution. In a sense, though, Hudson's 'confinement' allowed him to see beyond the 'artificial' settlement of London. Walking in the city in the presence of nature helps Hudson to daydream. He sees the city anew. The dreary statue of Shakespeare, which elsewhere he describes as mere 'furniture', has come alive in the presence of a wood-pigeon. Squares, which were formerly bounded spaces, become spaces of movement and of life. The delight in reporting the observation of a bird's eggs in the centre of the city is something of an expression of hope for Hudson. The return of the wood-pigeon to the landscape of London has marked a possibly brief, but important, interlude in the expansion of the city at the expense of nature.

Despite this, there is a strong feeling in Hudson's writing of 'making the best of a bad job'. The foundational story of nature displacement emerges clearly from his text. The incursions of a selection of bird species, often aided by some 'enlightened' urban planning in the form of parks and open spaces, only vaguely offset the sense of loss in Hudson's writing. While every effort should be made, in Hudson's opinion, to encourage species to move through urban space, it will only ever be a pale imitation of 'true' nature. Never, in Hudson's opinion, can the parks of London approach the 'richness and harmony seen in … [a] perfect transcript of wild nature' (Hudson, 1898/1969, p.329).

We have already seen how this idealization of origins can be overstated and might ultimately be politically constraining. This is not to say that we should be too romantic about the presence of a breeding pair of wood-pigeons in the centre of the city. Clearly, it is a difficult task to construct politically useful city-nature stories. In this sense, Hudson's writing can perhaps best be considered as a useful start in trying to open up city-nature formations, and to avoid thinking of cities and natures as worlds apart.

ACTIVITY 4.4   It might be useful to express Hudson's view of the spatial relationship between cities and nature in another simple diagram. Bearing in mind that Hudson saw nature as both inside and outside the city, how would you redraw Figure 4.2? When doing this, try to incorporate 'natural' spaces, like parks, into the city. ◆

Figure 4.4 is my attempt to represent Hudson's view of nature in the city. The city is represented by the grey tinted area, while the blue areas depict pockets of nature. The result is something like a patchwork.

FIGURE 4.4   *Nature in the city*

Compared to Figure 4.2, Figure 4.4 does allow for a certain amount of cohabitation of the urban and the natural. However, there are problems. If, for example, we change the scale, and focus on any of the boundaries between the urban and the natural in Figure 4.4, we would effectively be looking at a line which closely resembles the boundary shown in Figure 4.2. In other words, in making a patchwork of city and nature, we have left the ideas of city and nature more or less intact: they are still separate items. The only aspect that we have changed is that we are now saying that the urban and the natural are different at a reduced scale. The next extract helps us to go beyond even this more finely grained separation of the urban and the natural. We are gradually moving towards being able to talk about city-nature formations as more dynamic.

ACTIVITY 4.5   Now read Extract 4.3 by Elizabeth Wilson. When you have read it, think about the different ways in which Wilson uses nature in the city to think about space and the city. Is it possible to redraw the spaces in Figure 4.4 in order to incorporate Wilson's spatial thinking? ◆

EXTRACT 4.3
Elizabeth Wilson: 'Fox in the mews'

Recently a fox was sighted in the mews at the back of our house, which is situated on a London square. It was later suggested that the 'fox' might actually have been only a cat, since the eye witness had been drunk at the time. Yet the sighting, real or imagined, is still significant, for in fact everyone in our neighbourhood confidently expects to see a fox sooner or later. We already have squirrels, an owl, jays, even a hedgehog, and all this in Camden Town, the centre of London. To live in this London square has made me much more aware than I ever was before of the extent of nature in the city. Nature penetrates the city, and it appears that the larger the city, the more plants and animals it will shelter. This rather goes against the way in which, in Western thought, nature and the city (the rural and the urban) are positioned as incompatible opposites. Discussion of the way in which the built environment encroaches on the countryside is not unusual, but the urban seeping into the rural is certainly almost always perceived as destructive and polluting, and movement in the opposite direction is seldom mentioned at all. Perhaps it is also seen as an aberration, a transgression against the clear boundaries which seem necessary to our sense of order.

The trap of binary thinking

This is certainly the way in which Lewis Mumford viewed the relationship between town and country. For Mumford, the city was a container, that is to say it was and had to be a finite space. He believed that the 'sprawling giantism' of the twentieth-century city was leading inexorably to megalopolis and thence to necropolis, the death of the city. The sharp division between country and city no longer exists, he wrote. The original container has entirely disappeared: 'As the eye stretches towards the hazy periphery one can pick out no definite shapes … one beholds rather a continuous shapeless mass … the shapelessness of the whole is reflected in the individual part, and the nearer the centre, the less … can the smaller parts be distinguished' [Mumford, 1960, p.619].

Using, significantly, an organic metaphor, he describes such city growth as cancerous, social chromosomes and cells run riot in 'an overgrowth of formless new tissue', and continues: 'The city has absorbed villages and little towns, reducing them to place names … it has … enveloped those urban areas in its physical organization and built up the open land that once served to ensure their identity and integrity … [Then] as one moves away from the centre, the urban growth becomes ever more aimless and discontinuous, more diffuse and unfocused ... Old neighbourhoods and precincts, the social cells of the city, still maintaining some measure of the village pattern, become vestigial. No human eye can take in this metropolitan mass at a glance' [Mumford, 1960, pp.619–20].

This was written over thirty years ago, and Mumford no doubt did not foresee the radical reassessment made by critical theory in recent years of the way in which, or so it is argued, the whole of Western thought is built on binary oppositions. There has, in particular, been a major challenge to the idea of the binary division of gender, but on the whole this challenge has not been extended to urban/rural space, or at least not in the same prescriptive fashion ...

At any rate, since Lewis Mumford was surely mistaken in believing that a city 'freed from inner contradictions' is either possible or desirable, we must live in this realm of smudged boundaries, of pollution and disorder ... It was always the interstices of the city, the forgotten bits between, the corners of the city that somehow escaped, that constituted its charm, forgotten squares, canals, deserted houses – private, secret angles of the vast public space. The fox – rural intruder in the city ... reminds us that it is the unexpected, unplanned and incongruous that has lent city life so much of its delight, as well as, sometimes, its scariness.

Reference

Mumford, L. (1960) *The City in History*, Harmondsworth, Penguin.

Source: Wilson, 1995, pp.146–7 and 160

For Wilson, the image of the fox is a reminder of the often unseen world that cohabits in the space they call a square in central London. Wilson talks about the fox in the mews in a way that might be construed as allegorical. The story of the fox can be read symbolically as the story of urban theory. In many ways the movement of the fox is rather like the movement of feminist accounts of the city – they both disturb the contained, or territorial, spaces that have been the ideal in so much urban planning and dreaming. For Wilson, territorial space is not the only space in town. To expand on this point Wilson notes, for example, how the famous urban writer Lewis Mumford tended to write of cities in a spatially 'contained' way. The city for Mumford was a container, and ideally it should be contained. In other words, the boundaries should not be indistinct, there should be no sprawl. But Wilson is less convinced by this drive for purity and order. Drawing on feminist work which has continuously sought to disturb or disrupt the solidity of such spatial formations, Wilson introduces the possibility of other kinds of space. Where Mumford saw disorder and wanted to implement a correction programme, with everything in its place, Wilson sees disorder as the order of things. In other words, urban living, for Wilson, is about smudged maps, intrusions and movements. This smudging of the boundaries, embodied in the urban fox's path across the London landscape, might be suggestive of other spaces. Furthermore, it might be said that these spaces are not necessarily disordered, but instead work to other orders. The fox, for example, is not disordered. In many ways the fox's life will follow some regular patterns. The latter may well look alien to the spatial logic of a city plan, but it might be misleading to label the fox, or any other transgressor, as disordered. It might be more useful to think of foxes and other city-nature formations as reordering space. As you may have found out for yourself, this is very difficult to draw on a two-dimensional map.

To return to the final paragraph of the extract, Wilson is right to talk of the city's 'scariness'. It would be hopelessly romantic to suggest that the fox in the mews is a reason to celebrate the end of urban-environmental injustices. Similarly, the dark corners and surprising turns of an urban landscape can be extremely unwelcome and dangerous. Nevertheless, in doing away with some of the more restrictive containers that have inhabited urban theory, and looking for the possibilities that open up in the impure spaces of cities, there might well be opportunities for rethinking and reactivating urban-nature politics.

This said, and before we get carried away, we might want to rephrase our questions a little. For it might well be possible to disturb the settled territories of urban theory. This chapter has started to demonstrate some useful turns that our thinking can take in relation to city-nature formations. But there is a difference between saying that we *can*, if *we* choose, think differently about city-nature formations, and asking what we might call, for simplicity's sake, a more empirical question. In particular, it is worth investigating the extent to which city-nature formations do blur the boundaries of urban space. Can we find examples where a lot of work is going into keeping the spaces pure and the boundaries firm? And can we find examples where it seems as though movement across boundaries is becoming more frequent? These are some of the questions that will concern us in the case studies in the next section. In sum, having started with the question, 'what is natural about cities?', and having demonstrated at least some ways in which we might answer this question, we are now looking into some of the different ways in which city-natures take and change shape.

4 *City-nature formations*

In section 3 we have made an assertion that cities and natures exist in particular mixes. We have partly rejected the idea that cities exist at the expense of nature, and this takes us away from some modes of analysis that can constrain our thinking and action in relation to nature politics in the city. We can, and I believe should, go further. In particular, we have said relatively little about the *types* of city-nature mixes that can be said to exist. We have already mentioned the plant and animal species that are visible and easily imagined types of city-nature formation, but there are others, including city-nature landscapes, flows of matter, energy and information, and city-nature networks, all of which tie together the natural and the social within, through and between cities. We will come across each of these, along with a few more animal and plant species, in the remaining sections of this chapter. The aim will be to consider carefully how these city-nature formations can coexist in the intense spaces of a city. First, it will be useful to revisit briefly the notion of urban rhythms, and apply it to the analysis of city-nature formations.

The term 'urban rhythms', which was introduced through the work of the French writer Henri Lefebvre, is discussed elsewhere in this series (see **Allen, 1999**). The rhythm metaphor may serve to remind us that the city beats with differing degrees of regularity and tempo; this variety may itself change, both as a condition for and an expression of urban differences. As rhythms overlap they may eventually fall into the same beats, and in doing so they might become louder and more successful. Perhaps this is the case for the street pigeons that move with the crowds and throng around any bench where a crumb is likely to be had. Likewise, the rat scurries in sewers and is so in tune with city rhythms that McWhirter (1996, p.108) refers to the species as 'humanity's urban doppelgänger'. Another possibility (and this might also apply to rats) is that there may be silences between beats that allow another rhythm to develop. Perhaps this is where urban foxes make their home, moving across the quieter time-spaces, one step behind the main beat, often unseen, but no less grateful for the detritus of the human rhythms.

Yet another possibility is that there may be more destructive clashes of rhythm, where one beat succeeds in drowning out the other. It may help to think of the ways in which waves and rhythms, when brought together, interfere with one another to produce new, or accentuate old, patterns. So, as **Allen (1999)** has suggested, urban patterns are more complex and multifaceted than a conventional mapping of urban space might indicate. Where there is apparent settlement, there may well be unseen or ignored movements which help to produce the effect of order. Similarly, movements may be so regular that they form a settlement, a routine. Certainly, diversity (of rhythms, life-styles, city-nature formations) doesn't suit everyone and everything. Indeed, city-nature formations may involve a good deal of organizing on behalf of the dominant rhythm. It is to these various possibilities that we now turn.

I want to look at two case studies of city-nature formations and tease out how they run with, or rub up against, other urban rhythms. The first case examines how a city-nature formation and its accompanying and varied rhythms are threatened by a desire for a reordered city-nature formation, whose rhythms are thought to produce a more settled pattern. The second case looks at city-nature formations whose rhythms are said to be already negatively interfering with what some see as desirable city forms.

4.1 CASE I: CONSERVATION IN THE CITY

Some city-nature formations might be regarded more highly than others. Here is some background to a study of a conservation struggle on the edges of London in the early 1990s (Harrison and Burgess, 1994).

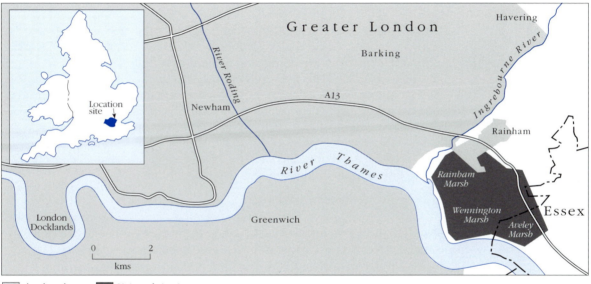

☐ developed area ■ Universal city site

Rainham Marshes, near London (see Figure 4.5), were designated a site of special scientific interest (SSSI) in 1986, on account of their position as part of the inner Thames grazing marshes and their distinctive assemblage of plants and animals. In other words, they were recognized as an important nature conservation site in terms of their resident fauna (mainly bird and insect life, some of international importance) and flora (mainly grasses, some of which were nationally rare). This recognition supposedly protected them from damaging activities and development. In early 1989, a consortium of developers, which included the Music Corporation of America (MCA) as a main player, announced a proposal to create a £2.4 billion entertainment complex on the site. The proposed complex would carry the name 'Universal City' and would look something like the Universal Studios theme parks in the USA. The development would undoubtedly have created large numbers of jobs in the construction and service sectors, and would have brought tourists and a welcome economic boost to this part of London.

FIGURE 4.5
*Location map of
Rainham Marshes*

Following a degree of concern over the conservation status of the site (although this concern was not one shared by the Conservative government of the time), MCA mounted a campaign to assure local and national conservationists that the proposed development would not only help to conserve nature in the city, but would enhance it. It should be noted, before we look in more detail at the ways in which city-natures were being produced at Rainham, that Universal City did not proceed further than the drawing-board. Moreover, this failure to implement the development was not the direct result of a nature conservation argument. Although some concern over nature at the site did result in the consortium's developing a more costly plan, the failure to proceed was more probably attributable to the onset of economic recession in northern Europe in the early 1990s.

The following extract from Harrison and Burgess tells us something about the city-nature formations that existed on the marshes in the late 1980s:

> As with many habitat islands in lowland Britain, Rainham's wildlife status is in part a product of the social relations which have prevailed in the postwar period. The marshes are dominated by land-uses typical of the urban fringes throughout Britain. The site survives neither as an area typical of wilderness beyond the economic margins of productive agriculture and forestry, nor as one typical of 'aestheticised nature' so central to the philosophy and practice of other land designations such as National and Country Parks. Rather, nature at Rainham is 'produced' and dependent upon land-uses which themselves are seldom welcomed by those people who live amongst them – waste disposal, silt dredging and active military training.

(Harrison and Burgess, 1994, p.294)

Compared to the urban fox, these city-nature formations are much harder to romanticize. Mud-flats and marshes have rarely been seen as cherished landscapes in the Western aesthetic tradition. Indeed, when Harrison and Burgess talked to people who lived close by, a good deal of scepticism was expressed as to what exactly was worth conserving on the marshes. One resident called the area a 'rat-infested dump', while others thought of it as a no-go area, especially for women and younger children. Whilst these same residents were impressed by the number of species that existed at Rainham, there were at least two issues that they thought were significant. First, the species they had either seen or had been told about were, to many of the residents, unexciting. One local conservationist captured this mood when she exclaimed, 'What we need is a dolphin' in order to win the conservation battle. Second, in their focus group discussions, the residents suggested that the marshes were poorly managed. They found it incongruous that the designation of the area as an SSSI was accompanied by little or no obvious attempt to conserve nature. The fly-tipping, silt deposits, military practices and scrambling motorbikes continued unabated.

Partly as a result of these kind of sentiments, and partly on account of the groundswell of environmental sympathies in Britain at the time, MCA came up with a cleverly packaged plan. 'Package' is a useful term here. In many ways the developers were able to latch on to the poorly packaged city-nature formations that previously existed at Rainham. With an eye on public opinion and on what sells, the developers effectively sought to repackage these city-natures. So, in place of a rat-infested dump, whose urban rhythms nevertheless provided time-spaces for a wide range of animal and plant species, MCA offered a landscaped development, including a nature and leisure park and a managed wetland area. What looked like a rather haphazard city-nature formation was resold to the public in a more glitzy, media- and marketing-friendly version.

ACTIVITY 4.6 Look closely at Figure 4.6, which shows the plans for the re-landscaping of Rainham. List the differences between this imagined city-nature formation and the one that local residents were describing. Think about the different movements, spatial types and rhythms that characterize the two landscapes. ◆

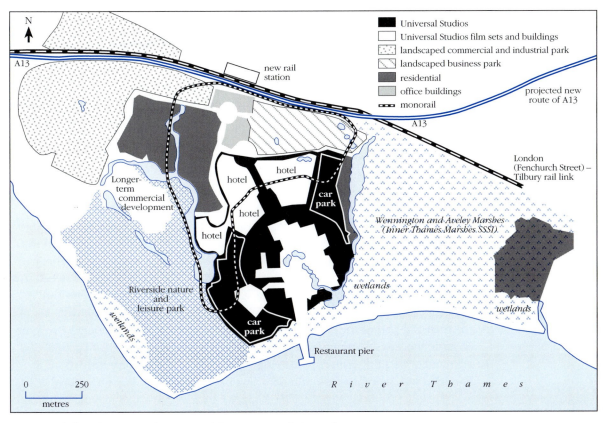

FIGURE 4.6 *The main elements of the Universal City scheme*

When I look at the plans, and try to envisage what the development would look like, I am struck by the kinds of order that would emerge at the site. The scheme is a neat projection of certain human values on to a flat map. It speaks of settlement and containment (and is more Lewis Mumford than Elizabeth Wilson), and of territories for this, and territories for that. It is neatly packaged. Nature is contained. On the ground, you would need to cross a boundary to enter the ecology park. This might well be managerially efficient: when humans cross boundaries they can be asked to pay for the privilege. Thinking in terms of rhythms, this map can be thought of as a projection of specific city rhythms, which fit neatly with those that operate up-river in the financial part of the capital. But the rhythms of non-humans are seldom so easy to map and control in this way. Reducing the extent of the SSSI might irretrievably change the routes and rhythms of the plant and animal species at the site: no amount of landscaping would impress an endangered dragon-fly that had recently been sighted on the marshes.

Another issue that strikes me as I compare the plans with the existing marshes is the question of space and how it is seen in the context of the developers' 'rhetoric' that Harrison and Burgess identify. Here is another short excerpt from their study:

> Hartley Booth, chief executive of British Urban Development [a member of the consortium with MCA], is quoted in an MCA press release as saying 'we shall together be taking a derelict landscape, suffering from pollution and decay in the natural habitat and transforming it into a community where people can live and work in harmony with the environment'. Through this rhetoric, MCA – a multi-million dollar, multi-national entertainments industry – became an organization that cared for nature; it became an environmental steward.
>
> (Harrison and Burgess, 1994, p.300)

We might ask what the term 'community' was doing here. Furthermore, following the previous chapter, we might note how this term was being used to generate a sense of wholeness (and note that the word 'harmony' soon follows). As Linda McDowell pointed out in Chapter 3, community is a construct that can result in the attempt to remove difference from the neighbourhood or street. In a sense this is what was going on in the statement from MCA. The orders and rhythms of the marshes were being usurped by those that suited the rhythms of the financial part of the city. As the site plan suggests, the developers were keen to settle the site down, and turn it into a more productive place for capital accumulation. Furthermore, this reorientation of the city-nature formations at Rainham was being executed under the guise of a duty of care for nature.

The MCA-led consortium portrayed itself and its plans as environmentally exemplary. Indeed, it had partially succeeded in this by employing the idea of sustainable development. In its terms, the consortium would develop the site with due care and attention to the longer-term environmental requirements and conditions on the marshes. Furthermore, in mobilizing a pollution/degradation urban foundational story, which suggested that urbanization had left the marshes in a poor state, the consortium was suggesting that it had the know-how and money to reverse this neglect and dereliction. The consortium, and humanity in general, could now develop sites like Rainham and not repeat the past mistakes of forgetting about nature. In itself this is laudable; it speaks of integrating human and natural activities in meaningful and constructive ways. If we are interested in making city-nature formations more beneficial all round, then MCA seems to be taking us where we want to go. But the concept of urban sustainable development is too easily co-opted by developers to dress up their plans as green. In effect, the term can often simply be a cover for business as usual, with the addition of some aesthetically pleasing landscaping (see Whatmore and Boucher, 1993). Look again at the plans. It strikes me that this is not so much about integration; it's manipulation. The plans mark a city-nature formation that is undoubtedly constructed on certain terms – terms that fit more closely to the rhythms of capital accumulation than they do to those of dragonflies and rats. In contrast, and no doubt through a whole series of accidents and half-baked plans, the 'undeveloped' marshes at Rainham consist of very different city-nature formations. Indeed, in some senses the integration of city-nature rhythms is more pronounced than for the territorial spaces of the theme park. The orders, routes and rhythms of migrating birds, pollen, insects, sewage, silt and detritus and their interrelations, all of which linger to varying degrees on the marshes, provide another form of city-natures.

So how do we adjudicate on this kind of debate? My answer is that it might be sensible to avoid suggesting that one city-nature formation is more natural than another. City-nature formations rely on huge through-flows of information, materials and energies. Furthermore, it seems to me that it might be useful to depart from seeking a return to a settled, mythical, natural landscape, which can only be bounded and fenced in. As we have repeatedly seen, this desire for a return to Eden is part of a conservative logic that oversimplifies city-nature formations and human/non-human relations.

Let's turn to our second case study, where city-nature formations, and the rhythms that they produce and move to, have started to interfere with some other city rhythms.

4.2   CASE II: LIVING TOGETHER OF CITY-NATURES

Street pigeons and Canadian geese are exemplary city-nature formations. Street pigeons are an embodiment of urban nature. They comprise a single species (*columba livia*) descended from domesticated rock doves (Palameta, 1996, p.32). Their range is almost without equal in the bird world – inhabiting most, if not all, of the cities in both hemispheres. Moreover, street pigeons are urban in another sense. They are a product of a long social history of pigeon fancying and breeding. This history has helped to equip the birds for life on, and sometimes under, the streets.

> Because of their affinity for things human, rock doves proved easy to domesticate, and the process of domestication prepared them still further for life among us. People who kept pigeons were fascinated by their ability to 'home' – to return faithfully to favoured roosting sites from great distances. Keeping and racing homing pigeons became a popular pastime. High quality birds were often sold or traded, and breeders selected for qualities that would maximize their profits, such as navigational ability. Generations of selective breeding produced pigeons that were more sociable, less afraid of humans, and more spatially adept than their ancestors … Homing pigeons that escaped their lofts and began to live on the street found themselves armed not only with a naturally evolved sense of direction, but also with an artificially bred ability to remember subtle details about visual landmarks. Thus, pigeons are prepared to solve problems never faced by rock doves. For example, when a rock dove wants to leave a cave, it merely has to move towards the light. A pigeon wishing to leave the inside of a subway station, on the other hand, is faced with an array of synthetic lights and reflective surfaces. City pigeons, it seems, orient themselves according to memorized 'maps' of their surrounding.
>
> (Palameta, 1996, pp.32–3)

It is not only the space of cities that street pigeons successfully negotiate. They also navigate city times. Research suggests that pigeons identify people who habitually feed them, and wait on their arrival at the usual time and the usual place (Palameta, 1996).

Like many bird species, pigeon densities are higher in urban areas than surrounding rural areas. In Paris, for example, bird populations are around three times greater in terms of density than is the case for the adjacent rural landscape (Art Orienté Objet, 1996, p.37). However, like many urban dwellers who prove too successful, pigeon populations are not always welcome guests in the city. Year-round feeding and warm temperatures lead to high breeding rates, with birds sometimes reproducing six or seven times a year. High densities of pigeon can sometimes lead to increased rates of disease and infection, some of which are transmissible to humans. Moreover, as each pigeon produces something like nine kilos of guano per year, it is not only the direct threat to human health that is at stake. A whole architecture of spikes and netting has grown up to prevent pigeons, or rather their excrement, from damaging buildings and reducing the visual appeal of urban settings.

So pigeons that at one moment provide an essential feature of urban place myths (imagine town squares without the hustle and bustle of its pigeon population) can, at the next moment, be labelled as pests. As a result, many cities have public campaigns to discourage people from feeding the birds (see Figure 4.7). Such campaigns are often accompanied by a programme of trapping and killing so that the birds don't starve, in unsightly fashion, on the streets. In Paris, and very likely in many other cities, the killing is done in secret. Indeed, there is a 'secret' department in the Paris Town Hall which organizes the cull. Workers pretend to be doing research, banding or doing a census, when in fact they are removing birds from the streets. The secrecy is an attempt to avoid a public outcry, and is interesting because it says something about the sensibilities of a proportion of urban dwellers to their non-human neighbours. Indeed, secrecy is a feature of human dealings with other non-human neighbours. Here is a brief look at another bird in the city, the Canada goose.

Pigeons are an everyday sight in towns and cities throughout the world. In fact they are accepted as part of urban life. However, the feral pigeon is the cause of nuisance, filth and health problems - they can even cause extensive damage to the fabric of buildings.

Whilst there are a small number of people who feed the pigeons in town centres, most shoppers and visitors positively dislike the nuisance of their scavenging and mess. They would prefer to avoid them and obviously will seek pigeon free areas to visit and shop - which can't be good for business.

Pigeons lead a very simple life so the remedies to the problems they cause are not complex.

Pigeons need food and somewhere to sleep and make more pigeons. They also like a nice vantage point to sit and look for food. If these needs are not met in one location they will go to one where they are.

How can you help reduce the problems of feral pigeons in your area....

Make sure your organisation produces no food waste that can be left available for them.

Encourage your customers to dispose of waste food properly.

Protect any ledges or other features of your building to stop them becoming landing sites.

Ensure any disused parts of your premises are sealed against access.

It really is that simple and if all the businesses in the area take this advice the problems will be reduced and will go away. The result will be happy visitors and shoppers - which just has to be good for business!

CITY OF Stoke ON Trent

The City Council offers advice about protecting your premises and pest control, call...

☎ 01782 232307

FIGURE 4.7
A leaflet produced by Stoke City Council to discourage people from feeding pigeons

In London's St. James's Park in the summer, in the early hours of the morning, hundreds of geese are shot with silenced guns. Meanwhile, 200 goose eggs a year are taken from their nests in Milton Keynes, hard-boiled, and then replaced (Brown, 1995). The unsuspecting parents continue to incubate the dead yolk and thereby fail to produce a new batch. The cull is justified because the birds are regarded as 'bullies' and, if that were not enough, they aren't 'native'. Here again, rather like the Rainham case, we have an example of a preference for a specific kind of nature – one that doesn't move as irregularly as some humans might want. Canada geese have cut through some boundaries.

The idea of a native species is, of course, ecologically suspect and socially offensive. Suggesting that certain animals or plants belong to certain nation states is peculiar, to say the least. It only serves to illustrate the historical development of the life sciences in a period which was marked by the formation of nation states. And the nature–nation–native link is made possible only if we have a static view of ecology. Whilst it is important to be mindful of the suffering incurred by thoughtless introductions of species, it is also important to avoid thinking in ways that fix nature in a mythical, settled past. City-nature formations are likely to involve movements which challenge any notion of cities or of natures that view either as static, settled items.

The living together of city-natures is by no means straightforward, and I have desisted from trying to offer universal rules or general patterns that can be followed. One point that should be emphasized is that city-nature formations, like the bird species I have mentioned and Rainham Marshes, exist through complex histories and geographies. They, like other inhabitants of cities, should not be compared to so-called pure categories like nature and urban. As we have seen, these pure categories were never as pure as some people may have wanted to believe. Instead, city-nature formations are always a result of complex flows and tentative settlements. The next section looks at how this more complex view of city-natures can be utilized.

4.3   CITY-NATURE SPACES AND PRACTICES

I have been arguing that we should be suspicious of any attempt to invoke pure nature in a debate over the characteristics of cities. I want to turn now to the work of a group of academics who are attempting to make some of these ideas operational. Wolch, West and Gaines (1995) seek to develop what they call a 'transspecies urban theory', and, in the reading which you will shortly be asked to study, they are looking for transspecies urban practices. The word 'transspecies' is used by the authors as a means to move urban writing away from its traditional, human-centred focus. For these authors, cities are not simply social products; they also involve non-humans. For Wolch *et al.*, it is animals in particular that need to be acknowledged in our understanding of cities. Their work is based on experiences of cities in the USA.

ACTIVITY 4.7   Now turn to Reading 4A, 'Transspecies urban theory', by Wolch, West and Gaines. When you have read it, consider the following questions:

- How have urban theories ignored city-natures?

- What kinds of spatial arrangements would, in the authors' opinion, favour transspecies urban practices?

- What do you think about the charge that focusing on non-human natures is frivolous and obscures more immediate issues of environmental social justice? ◆

With respect to the first question, my short answer would be: in almost every way possible. This is the case even though ecological metaphors have been instrumental for well over a century in US and European urban theory and planning (from philanthropic improvers of the nineteenth century to the urban sociology of the early twentieth-century Chicago School and the professional planners and sustainable architects of the mid to late twentieth century). This relative silence has been most pronounced for non-human animals. Nevertheless, it seems that, in parts of the USA at least, it is becoming more difficult to ignore non-human neighbours. This may partly be a result of changing sensibilities, although the extent to which this is expressed through different social and cultural affiliations is clearly an important consideration. It may also be a result of the astounding success, in terms of densities, of some urban species.

In thinking about ways of living with this mix of city-natures, it might be useful to borrow and adapt a figure from urban theory. The figure I have in mind is the urban stranger. As you read in Chapter 3, the stranger has been an important figure for many urban writers. Simmel (1921), Wirth (1964), Sennett (1971) and Young (1990), among many others, have all grappled with strangeness and difference in the city. The stranger, remember, is both present and absent in the city. S/he tends to confound simple spatial types. Being an outsider, but also being on the inside in terms of being present, the stranger challenges any simple model of community or cohesion in the city. My question would be: to what extent is it useful to think of non-human city-natures as strangers? By that I am suggesting that non-humans are neither absent from the city, nor truly of the city.

If cities are places where not only all manner of different people live in close proximity (see **Allen, 1999**) but also all manner of animals, plants and eco-social relations, then interspecies negotiation of urban space will be a matter of everyday importance. This may involve the manufacturing of indifference (as people and pigeons traverse a railway-station without too much regard for the other's business), or there may be attempts to segregate the 'urban' and the 'natural' (as rats are poisoned and pigeons are cleared off the streets). Another possibility is that, rather than difference leading to mutual withdrawal, it may also help to produce conditions for mutual recognition (**Allen, 1999**) – a reappraisal, produced through different kinds of interaction between (and within) groups. People feed, move out of the way for and even hold (sometimes furtive) conversations with animals and plants in cities. Furthermore, there is no expectation that either party will fully understand the other, or even that they can share a language.

Perhaps, then, there is a remoteness *and* a proximity to city-nature formations which may well provide moments of mutual recognition, and the formation of characteristically urban human/non-human ethical practices. These practices may well explain the secrecy of bird culling in some cities. Perhaps the recognition of city-nature communities lies behind the call for transspecies urban practices. As I read Reading 4A, I understood that the aim is not to incarcerate and domesticate more of non-human nature, or to make it 'culture'. The spatial logic of animals in the city has changed (something zoological gardens are increasingly trying to come to terms with). So, in addition to the conventional container-speak of zoning and conservation sites, Wolch *et al.* talk of other kinds of space. They refer to corridors, networks, buffers, flows, connectivity, and by doing so convey the attempts that are being made to find alternative urban spatial logics – logics that speak of movement as well as settlement. Whilst none of these spatial arrangements is beyond criticism, and certainly not immune to co-option (as the authors note when they mention the questionable use of habitat conservation plans by developers to circumvent the Endangered Species Act), they do express a willingness to rethink cities in ways that resonate with city-natures. They speak at least of a will to live together with the comings and goings, the movements, of city-natures.

Having thought about transspecies urban theory, and the spaces of city-nature formations, we can now turn to the thorny issue of 'relevance'. Clearly, there are social and cultural differences in human attitudes to non-human city-natures (although these are far from static and I would be wary of predicting and projecting these on to specific groups in specific cities). Nevertheless, and as we have already partly seen in the cases I have traced above, urban differences can produce conflicts of opinion and priority over city-natures. The task that remains in this chapter is to work through ways in which the approaches that we have formulated so far can have a bearing on issues that surround socio-environmental justice and city-natures. In order that we may do this, it will be necessary to revise and possibly extend what we mean by city-nature formations.

5 *City-nature movements*

In the previous sections we have claimed that cities are far from being unnatural. In order to stake such a claim, I have relied almost entirely on a handful of visible and possibly familiar city companions – the pigeon, the rat and a dragon-fly in Europe, which were joined in the last reading by urban deer, coyotes and cougars (among others) in the USA. These figures have helped us to develop some arguments about city-natures, their movements, rhythms, spatial forms and the changes they are undergoing. Yet there is more to city-nature formations than a selection of animal and bird species, and there is more variety to city-nature formations than the examples used so far in this chapter might suggest. In a sense, this acknowledgement of the breadth of what we can mean by city-nature formations is an acknowledgement that there is more at stake. Putting broadly imagined city-natures on the move has a number of significant consequences. We will deal with these in two ways. First, we will remain with the easily imagined city-natures within cities, but we will suggest the ways in which city-nature movements, embodied in plant and animal species, are vital resources for many different kinds of urban living and settlement. The second way of underlining the significance of talking about city-nature formations will involve taking a slightly different view of cities. Broadly speaking, this will involve considering the movements of city-nature formations not only in the worlds within cities, but also in city worlds, as they move between and through cities, and from cities to countryside and back again. The latter will highlight the importance of uneven distributions of city-natures.

5.1   ASSERTIVE CITY-NATURES

To what extent is urban nature simply a luxury that can be enjoyed and possibly campaigned for by those people who are most comfortable, or at ease, in cities? Is city-nature simply for the park *flaneurs*, people with time and urban space on their hands, like W.H. Hudson? Is it only for the wealthy who can afford gardens, and is the celebration of urban foxes and calls for their protection simply an expression of a pastoral nostalgia – a cowardly response to more pressing human-urban problems? In these supposedly rapid and intense times, does urban nature mark a yearning for a more settled community, for a slower pace? If so, city-nature politics might not develop in the ways we might have hoped. After all, we set up this chapter to talk about how we can put city-natures on the move and, by doing so, provide some resources for reworking city-nature politics. In this brief section I want to flag something that has been implicit throughout the chapter so far. That is, city-nature formations can take assertive as well as more palliative shapes; and assertiveness can be thought of as a form of defiance.

Betty Smith's bestseller, *A Tree Grows in Brooklyn*, is an example of how human aspirations can be linked to city-natures. The tree of the title, *ailanthus altissima* (the 'tree of heaven'), like the novel's heroine, survives city life despite the odds. Another example is provided by 'defiant urban gardens', a term used by Helphand (1997), a landscape historian writing in the USA, who reviews the diaries and accounts of Eastern European urban ghettos during the Nazi occupation. Helphand notes the prevalence of passages which refer to the cultivation of flowers and the production of vegetables and fruits in the ghettos. For example, the teenage journal of Mary Berg describes how bombed out sites in Warsaw were appropriated by the Toporal Society (The Society to Encourage Agriculture among Jews). They distributed seeds, planted vegetables and medicinal herbs, nurtured trees and even created a park (Helphand, 1997, p.114). From the diaries, it is clear that this cultivation served as a means to survival and as a way of coping with the despair of the situation.

In many contemporary cities, agricultural food production is on the rise. In the Sparkbrook area of inner-city Birmingham, where there were high levels of unemployment in the 1980s and 1990s, hundreds of people became involved in the Ashram Acres project. Derelict gardens were transformed, and valuable urban crops including okra, karella and coriander were cultivated by the residents (Girardet, 1992). This growth of urban cultivation requires that some notions of what it means to be urban have to be changed. The changing levels of tolerance towards urban agriculture in many African cities is a good example. What follows is a description of the growing of crops and rearing of animals in Harare, the capital city of Zimbabwe. The account is drawn from the research of Beacon Mbiba (1994) and Geoffrey Mudimu (1996), both of whom work at the University of Zimbabwe.

In 1992, Harare had an official population of just under one million people (Mbiba, 1994). In terms of the built-up area, of the 56,000 or so hectares that are designated urban, some 10,000 are open land. This proportion of 'undeveloped' land has been partly provided for by the 1976 Regional Town and Country Planning Act. The open land is varied. There are stream banks and wetlands, which are regulated by the Natural Resources Act of 1975, outdoor recreational areas, road and rail reserves, and land reserved for future residential and industrial development. It is estimated that 50 to 60 per cent of this open land was under cultivation in 1995. Between 1990 and 1994, the area under cultivation increased by 93 per cent. According to Mudimu (1996), this rapid increase in urban farming has resulted from the large growth in the urban population since independence in 1980, the worsening economic conditions since the mid 1980s, the rise in commodity prices, especially food, and the financial hardships brought about by the economic structural adjustment programme initiated in 1991. The latter 'resulted in many families in the city turning to urban cultivation as an alternative source of food in order to save on food expenditure and raise cash income' (Mudimu, 1996, p.181). Although urban farming is not restricted to the urban poor, it is the most economically vulnerable residents of the city who are most likely to engage in cultivating open land. For economic policy and cultural reasons, it is also women who are most likely to be engaged in this activity.

ACTIVITY 4.8   Now turn to Reading 6A, 'Gender, poverty and economic adjustment in Harare, Zimbabwe' by Nazneen Kanji, which you will find at the end of Chapter 6. Since the reading is mainly intended for use on Chapter 6, you can afford just to read it through at this stage. For the moment, use the reading simply to give you more of a flavour of the economic, gender and poverty issues that underpin this discussion of urban agriculture. ◆

The attitude of the authorities in Harare to urban agriculture has largely been negative. Some of this is based on reasonable concerns. Land erosion, loss of habitats, reduction in tree cover and the loss of recreational space are all cited as negative impacts of city farming (Mudimu, 1996). Other concerns are the result of a general distaste towards 'rural' land uses in an 'urban' context. For example, the 1976 Regional Town and County Planning Act excludes urban agriculture from the definition of land development. Whilst city farming is not banned in cities, it can be controlled through statutory instruments. Likewise, the Urban Council Act empowers the authorities to clear crops where there is a likelihood of a fire risk, where there may be health problems, or if they are unsightly (Mudimu, 1996). In Zimbabwe, the intolerance towards urban agriculture leads to the sporadic harassment of urban cultivators by the authorities (Mbiba, 1994). The following extract from Mudimu demonstrates the strength of feeling over this matter of urban land use. It also suggests that a more accommodative course of action is needed from the authorities if food production and protection of the urban environment are to be secured:

> In 1990 and 1991, the City Council of Harare mounted a campaign against urban agriculture, slashing maturing crops in an attempt to stop agricultural activities. The response of the women cultivators was to fight what they considered to be colonial and male attitudes to city planning with regard to alternative land uses … The women mobilised themselves to lobby local city councillors and constituent Members of Parliament for a general change in attitude towards urban agriculture. In their view, the open and undeveloped urban lands were under-utilized. The women also contended that they needed the land to grow crops to supplement incomes and food supplies. They also made their resistance felt by continuing their agricultural practices in the face of the Council's opposition. In response to this persistence and the accompanying political pressure, the City Council conceded ground in 1992 by allocating some designated land to women operating through organized groups which formed farming co-operatives … This option gave some residents the right to grow crops and at the same time allowed the City Council some measure of control to allocate land for cultivation and also monitor the impacts of cultivation. Despite this move, unsanctioned or 'illegal' cultivation by both members of the co-operatives who have plots outside the identified co-operative lands as well as other individuals who are not members of co-operatives has persisted.

(Mudimu, 1996, p.183)

This partial accommodation still has a long way to go. In Mbiba's (1994) view, something of a cultural shift is needed in Zimbabwe's cities in order to develop a more open definition of urban land use and urban governance. In a similar vein to the women cultivators, Mbiba views the prohibition of urban farming as based on an inappropriate urban model. He also notes how urban agriculture often produces higher yields than rural production, and that urban agriculture was accepted in pre-colonial African cities as part of the urban way of life. Some colonial governments also promoted this activity, although Mbiba suggests that in Harare, the pre-1980 colonial government tolerated urban agriculture as a means to excuse the paying of low wages to urban Africans.

In terms of the more recent conflicts that have emerged in Harare over the use of land for agriculture, the solution, according to Mbiba, is to move to a more accommodative style of urban living and governance. Mbiba suggests that the experience of Maseru, the capital of Lesotho, provides a useful lesson. In Maseru, the keeping of dairy cows, maize cultivation, sheep and pig rearing, and vegetable and fruit production are all prominent and conspicuous activities. Moreover, the authorities in Maseru take an enabling approach to urban agriculture: for example, the town provides special veterinary services through the Ministry of Agriculture's Livestock Division (Mbiba, 1994, p.192). Clearly, there are fewer anxieties here about what it means to be urban.

Here are some important points to draw from this discussion:

- Focusing on city-nature formations need not fuel a nostalgic, backward-looking, anti-urban attitude.

- In Harare, Maseru and Birmingham, there are signs that city-nature formations can play a role in some assertive and progressive forms of urban-environmental politics.

- Redefinitions of what it means to be urban that include recognition of city-natures may help to produce more sustainable forms of urban living. The role of urban agriculture in relieving, though by no means solving, urban poverty in Harare may be an important step in the effort to build better cities in which to live.

5.2   EXTENDING CITY-NATURE FORMATIONS

Our examples of Mexico City, bird life in London, the species and time-spaces of Rainham Marshes and the women farmers in Harare have alerted us to the complex geographies and histories of city-nature formations. These formations don't simply exist in contained spaces within cities; they also move across and between cities, and incorporate places and processes that we normally associate with the non-urban. Another way of stating this is to employ the idea of the *ecological footprint* (see also **Massey, Allen and Pile, 1999**; Hague *et al.*, 1996). The footprint idea is an attempt to imagine the total area of land required to sustain a settlement. Clearly, specifying the extent and boundaries of a footprint is a difficult task, largely because of the complexity of the connections and disconnections that have already been discussed in previous chapters. However, the idea does at least provide a useful way of starting to picture the movements that go to make up a settlement like a city. We can imagine the footprint as an extended web or network, a city-nature formation which, in terms of the intensity of connections, is focused on cities. This is similar to the idea of cities as a geographic plexus (see **Massey, Allen and Pile, 1999**). You may remember that it was Lewis Mumford who coined the term 'geographical plexus' and that Mumford was for the most part talking about the bringing together of people in cities and the resulting complexities of social interactions (**Pile, 1999; Massey, 1999b**). In this chapter we have extended this idea to include not only the people but also the animals, plants, materials, energies and information that move into, out of, between and within cities. There are at least two ways of thinking about this set of movements and the city-nature formations that they produce. We can use the idea of the footprint to convey both of these approaches, but the spatial imagery is different to the extent that the foot starts to look more like a network in our second version. Here's the first way of thinking about a footprint.

Guha and Martinez-Allier (1997), writing about Barcelona in the 1990s, provide a neat image of a city's ecological relations. They view the city as a large organism, and describe the web of electricity power lines that spread over the Catalonian landscape, seen from the air, as a 'huge spider's web' (p.140). The convergence of cables on the consumption capital of the region prompts the comment that one soon sees 'where the spider is lurking'.

My reservation with this imagery is that the traffic is pictured as moving in one direction only. We are back to viewing the city as a parasite, a monster that is sucking in resources and producing waste. Again, there is an anti-urbanism here, and it is based upon a questionable distinction between country and city, and between rural and urban environments. Once this spatial division is unsettled (something we have done numerous times in this chapter), we can start to see other movements.

This is where our second approach to the geographical plexus/footprint comes in. The British geographer David Harvey adopts this approach in his writing on cities and natures. He does so by first making what might seem like an extraordinary statement; 'there is nothing *unnatural* about New York city' (Harvey, 1996, p.186; emphasis in original). Now, the skyline of New York is familiar to most of us, but I'm sure that it would take some imaginative thinking for many people (and I include myself) to be able to see New York in the same terms as a favourite wilderness scene. Harvey's point, though, and it is one of the points of this chapter, is to reject the distinction between rural and urban. In the terms we have used here, both are city-nature formations.

ACTIVITY 4.9   Now read the following short extract from David Harvey's book, *Justice, Nature and the Geography of Difference*. When you have read it, try to answer the following question:

● In what ways does Harvey's notion of city ecosystems extend the idea of the ecological footprint? ◆

It is fundamentally mistaken … to speak of the impact of society *on* the ecosystem as if these are two separate systems in interaction with each other. The typical manner of depicting the world around us in terms of a box labelled 'society' in interaction with a box labelled 'environment' not only makes little intuitive sense (try drawing the boundary between the boxes in your own daily life) but it also has just as little fundamental theoretical or historical justification. Money flows and commodity movements, for example, have to be regarded as fundamental to contemporary ecosystems (particularly given urbanization), not only because of the past geographical transfer of plant and animal species from one environment to another … but also because these flows formed and continue to form a coordinating network that keeps contemporary ecological habits reproducing and changing in the particular way they do. If those money flows ceased tomorrow, then the disruption within the world's ecosystems would be enormous. And as the flows shift and change their character, as is always the case given the uneven geographical development of capitalism, so the creative impulses embedded in any socio-ecological system will also shift and change in ways that may be stressful, contradictory or salutary as the case may be.

(Harvey, 1996, pp.186-7)

Harvey's first point in this extract is that the distinction between the natural and the social is an arbitrary division. He then explains that cities are as natural as they are social. Notice how this does not simply involve picturing cities as consumption centres. Cities are also conceived of as parts of complex networks. These networks, which are forever forming, collapsing and reforming, are important devices for understanding the ordering of ecosystems both within and outside the conventional city boundaries (boundaries which start to make less sense when we think of the ways in which places are woven together through networking processes (the 'work' of the net*work*). Harvey's point is that even the remotest of landscapes – and Antarctica, deep ocean floors and rainforests might spring to mind – are touched by the threads of these networks. In city-nature terms, the landscape that we label a city may simply be the place where these movements of capital, materials and energies are more intensely controlled. It is in the financial districts of cities, for example, where movements are more likely to be speeded up, slowed down, stopped and restarted largely on account of connections and threads combining and rubbing up against other threads and rhythms (see Chapter 2). Nevertheless, we shouldn't get carried away with the notion of connection, as if this word leads us to imagine everything, everyone and everywhere being attached by a universal, global membrane. As Chapter 1 made clear, and as Harvey (1996) suggests when he talks about shifts and uneven development, there are disconnections where people, places and natures are cut off for a duration. Likewise, there are different kinds of connection. Certain threads become more active than others, and we are left to think of differential power, its production and its effects. Such matters are picked up in the following chapters of this book.

Here are some of the important points that you should take from this extension of city-natures:

- By putting city-natures on the move, we have collapsed any obvious distinction between the urban and the rural. Aspects of shanty settlements, rice crops, oceanic crust and merchant banks are city-nature formations.

- In unsettling distinctions made on the basis of 'naturalness', we have used the idea of networking practices which serve to generate threads that link the rhythms of vascular plants in the Amazon with the cadences of corporate capital in Brasilia, New York and elsewhere.

- Cities, by virtue of all these threads, are sites where a good deal of ordering work is produced. This work extends and maintains threads, and co-ordinates them to fit in with particular rhythms. It is true to say that, like all stitching work, networks can sometimes be undone, and people in cities may not always be in control. Networks are a useful metaphor because they, and the threads that sew them together, do not necessarily swamp city-nature formations.

- All this talk of ordering, networking and intensities, and in particular the privileging of certain sites as more-or-less ordering, and more-or-less ordered, suggests that the power over city-nature formations, and the power to effect change, is expressed unevenly within and between cities (a matter that is developed in Chapter 5).

6 *Conclusion*

Let me conclude by returning to Mexico City. Remember how the author of 'Mexico City trashed' (Extract 4.1) seemed to be blaming the Spanish colonizers for the socio-environmental ills that are currently faced by the city's inhabitants? Recall too the suggestion that a major contribution to the city's problems was caused by the destruction of nature in the city. My question would now be: how can the concept of city-nature formations help us to understand the socio-environmental problems that millions of people live with in parts of Mexico City? Clearly, we can reject the idea that the problems are caused by the city's unnaturalness. The city is as natural as the rest of Mexico. But that is not to say that everywhere is the same, or that all cities are the same. We need to pay attention to the times and spaces of city-nature formations, and to the uneven characteristics that develop and sometimes hold within cities and between cities. In order to account for and develop action upon environmental injustices in Mexico City, it is not sufficient to view the city as the accumulation of elements in one place over time. To think in this foundational way can be constraining. Instead, Mexico City's city-nature formations can be conceived of as involving movements between places as well as settlement within places. Moreover, following **Massey (1999a)**, cities can be thought of as meeting places. The coming together of city-nature formations, the resulting interferences as rhythms clash or combine, the threads that trace connections and disconnections and produce desirable and less desirable effects – all of these ideas can be mobilized to generate new insights into the problems and opportunities that arise from a living together of city-natures.

References

Allen, J. (1999) 'Worlds within cities', in Massey, D. *et al.* (eds).

Art Orienté Objet (1996) 'Savage Paris', in Dion, M. and Rockman, A. (eds).

Brown, P. (1995) 'Feathers fly over calls to cull alien birds', *The Guardian*, 1 July, p.8.

Dion, M. and Rockman, A. (eds) (1996) *Concrete Jungle*, New York, Juno Books.

Gilroy, P. (1993) *The Black Atlantic: Modernity and Double Consciousness*, London, Verso.

Girardet, H. (1992) *The Gaia Atlas of Cities: New Directions for Sustainable Living,* London, Gaia Books.

Guha, R. and Martinez-Allier, J. (1997) V*arieties of Environmentalism – Essays North and South*, London, Earthscan.

Hague, C., Raemakers, J. and Prior, A. (1996) 'Urban environments: past, present and future', in Sarre, P. and Blunden, J. (eds) *Environment, Population and Development*, 2nd edn, London, Hodder and Stoughton in association with The Open University.

Harrison, C. and Burgess, J. (1994) 'Social constructions of nature: a case study of conflicts over the development of Rainham Marshes', *Transactions of the Institute of British Geographers*, vol.19, no.3, pp.291–310.

Harvey, D. (1996) *Justice, Nature and the Geography of Difference*, Oxford, Blackwell.

Helphand, K. (1997) 'Defiant gardens', *Journal of Garden History*, vol.17, no.2, pp.101–21.

Hudson, W.H. (1898/1969) *Birds in London*, Newton Abbott, David and Charles.

McWhirter, C. (1996) 'They scurry among us', in Dion, M. and Rockman, A. (eds).

Massey, D. (1999a) 'Cities in the world', in Massey, D. *et al.* (eds).

Massey, D. (1999b) 'On space and the city', in Massey, D. *et al.* (eds).

Massey, D., Allen, J. and Pile, S. (eds) (1999) *City Worlds*, London, Routledge/The Open University (Book 1 in this series).

Mbiba, B. (1994) 'Institutional responses to uncontrolled urban cultivation in Harare: prohibitive or accommodative?', *Environment and Urbanization*, vol.6, no.1, pp.188–202.

Mudimu, G.D. (1996) 'Urban agricultural activities and women's strategies in sustaining family livelihoods in Harare, Zimbabwe', *Singapore Journal of Tropical Geography*, vol.17, no.2, pp.179–94.

Palameta, B. (1996) 'Pigeons, the smart bird', in Dion, M. and Rockman, A. (eds).

Pile, S. (1999) 'What is a city?', in Massey, D. *et al.* (eds).

Pysek, P. (1993) 'Factors affecting the diversity of flora and vegetation in central European settlements', *Vegetatio*, vol.106, no.1, pp.89–100.

Ross, A. (1994) *The Chicago Gangster Theory of Life*, London, Verso.

Sennett, R. (1971) *The Uses of Disorder: Personal Identity and City Life*, London, Allen Lane.

Simmel, G. (1921) 'The sociological significance of the stranger', in Parks, R.E. and Burgess, E.W. (eds) *Introduction to the Science of Sociology*, Chicago, University of Chicago Press.

Simon, J. (1996) 'Mexico City trashed', in Dion, M. and Rockman, A. (eds).

Smith, B. (1943) *A Tree Grows in Brooklyn*, New York, HarperCollins.

Tuan, Y-F. (1978) 'The city: its distance from nature', *The Geographical Review*, vol.68, no.1, pp.1–12.

Tuan, Y-F. (1984) *Dominance and Affection – The Making of Pets*, New Haven, CT, Yale University Press.

Whatmore, S. and Boucher, S. (1993) 'Bargaining with nature: the discourse and practice of environmental planning gain', *Transactions of the Institute of British Geographers*, vol.18, no.2, pp.166–78.

Wilson, E. (1995) 'The rhetoric of urban space', *New Left Review,* no.209, pp.146–60.

Wirth, L. (1964) *On Cities and Social Life*, Chicago, University of Chicago Press.

Wolch, J., West, K. and Gaines, T. (1995) 'Transspecies urban theory', *Environment and Planning D: Society and Space*, vol.13, no.6, pp.735–60.

Young, I.M. (1990) *Justice and the Politics of Difference*, Princeton, NJ, Princeton University Press.

READING 4A
Jennifer R. Wolch, Kathleen West and Thomas E. Gaines: 'Transspecies urban theory'

Redesigning nature's metropolis

Efforts are increasing to alter the nature of interactions between people and animals in the city, to change everyday practices of urban planners and environmental designers, and to defend more forcefully the interests of urban wildlife. Ultimately these efforts constitute a nascent *transspecies urban practice*, as yet poorly documented and undertheorized. In this section, we describe this emergent set of urban practices which are beginning to influence urbanization, human–animal interactions with wildlife, and urban wildlife ecology.

Learning to coexist

Urban populations increasingly demand alternatives to traditional, extermination-oriented animal-control policies. But wildlife specialists argue that larger scale management is also required to insure that wildlife populations are kept within carrying capacity of available habitat and to remove individual animals that threaten human safety. Such methods are controversial. For example, bitter disputes over the most appropriate and humane way to protect people and deer are common. This has initiated the search for 'technofixes' (for example, conception inhibition for deer and other urban wild-animal populations) which are seen as the preferred way to resolve such conflicts (Jones and Witham, 1990).

Education of urban residents is currently the most commonly used tactic to increase knowledge and understanding of, and respect for, wild-animal neighbors. Education about appropriate building, landscaping, and maintenance of urban development is used by many jurisdictions to reduce the attractiveness of the built environment for specific 'nuisance' animals in order to minimize costly damage to homes and to preclude the need for exterminations. Similarly, educational campaigns seek to convey the risks associated with wild animals and the range of precautionary behaviour needed to avert such risks (such as Lyme disease, or what to do if confronted by a cougar). Moreover, education is used to try to limit the many human activities that are intended to benefit wild animals but have the opposite effect, such as improper feeding, which can result in injury, death, or disease for animals and/or people. Some educational programs stress public participation, in the belief that it will result in better wildlife management, gained through residents' active involvement in management efforts and the transfer of their on-site information to professional managers (Canter, 1977). Other educational programs promote the development of small wildlife areas for children to play in and for them to make contact with and gain understanding of wild animals (Schicker, 1987).

There are limits to education and/or participation approaches. Certain human practices are detrimental to wild animals, stimulating some jurisdictions to enact *regulatory controls*. For example, conventional landscaping produces biologically sterile, resource-intensive environments, leading some cities to pass landscaping regulations which emphasize native species and which are designed to reduce resource dependence and create habitat for wildlife. Other human practices not typically geared to the needs of wildlife include many common residential architectures and approaches to building maintenance, garbage storage, fencing, and companion-animal keeping. In areas where conflict has emerged around such practices, jurisdictions are increasingly turning to regulatory power to alter them.

Planning minimum-impact urbanization

Wild animals were never a focus of urban and regional planning. Since the passage of the US Endangered Species Act (ESA) in 1973, however, planners have been forced to grapple with minimizing the impact of human activities (urban and rural) on threatened or endangered species, if not on all forms of wildlife. In response to the ESA, Leedy *et al.* (1978) identified five now familiar types of *land-use tools* for reducing the impact of urbanization on wild animals: zoning (including urban limit lines and wildlife overlay zones); public or non-profit land acquisition; transfer of development rights; environmental impact statements; and wildlife impact and habitat conservation linkage fees (Nelson *et al.*, 1992). None of these tools is without severe and well-known technical, political, and economic problems.

The ESA has also spawned *habitat conservation plans* (HCPs), regional landscape-scale planning efforts to avoid fragmentation inherent in project-by-project planning and local zoning control. Some argue that HCPs allow developers a way around the ESA (as HCPs allow some 'takings' of the endangered wildlife species in question) and thus reduce political threats to the Act's integrity (Saldana, 1994). Only a small number of HCPs have been developed or are in progress. Beatley's (1994) evaluation of these efforts suggests that, despite some benefits, there are serious challenges inherent in HCP efforts. These center on whether sufficient habitat will be preserved, on whether reserves created will be too isolated from one another and thus threaten genetic viability, on problems of securing funds for technical analyses, planning, and compensation required to offset losses to developers and landowners, on lack of interim conservation measures during multiyear planning processes, and on the ESA-driven single-species (versus a multispecies) approach.

Despite the pressures for habitat conservation from the ESA, minimum-impact planning for urban wildlife has not been a priority either for planning academics or for practitioners. Concern for wildlife is largely absent from the progressive planning agenda, for example. Despite the rise of the so-called 'new urbanism' in architecture and design, and the fashionability of the more widespread *sustainable cities movement*, planners rarely define sustainability in relation to wild animals. The 'new urbanism' emphasizes sustainability through high-density, mixed-use urban development, and promotes the maintenance of certain natural habitat features as necessary for humane city life. But it remains strictly anthropocentric in perspective except in those instances in which public protest has required developers to maintain or restore wildlife habitat. Although more explicitly environmental, the sustainable cities movement aims to reduce human impacts on the natural environment through environmentally sound systems of solid-waste treatment, energy production, transportation, housing, and so on, and the development of urban agriculture capable of supporting local residents (Platt *et al.*, 1994; Stren *et al.*, 1992; Van der Ryn and Calthorpe, 1991). But, although such approaches have long-term benefits for all living things, the sustainable cities literature pays little attention to questions of wildlife per se. An interesting exception is *A Green City Program for San Francisco Bay Area Cities and Towns* (Berg *et al.*, 1989), which recommends riparian setback requirements, review of toxic releases for their impacts on wildlife, habit restoration, a Department of Natural Life to work on behalf of urban wildness, citizen education, and mechanisms to fund habitat maintenance and the 'creation' of 'new wild places' (pp.48–9).

Linking town and country

The environmental design professions have become increasingly engaged in the implications of urban design for wildlife ecology and biodiversity. Ironically, many of the basic approaches heralded by today's landscape ecologists were promoted by the garden cities and regional plan advocates of the late 19th and early 20th centuries (such as Patrick Geddes, Ebenezer Howard, Lewis Mumford, Benton MacKay, and Frederick Law Olmsted and his sons), and, during the 1970s, the urban design approach of McHarg (1969). Most of the early plans did not attempt specifically to preserve wildlife, but nevertheless look astonishingly like metropolitan habitat plans being developed to link town and country now, almost a century later.

Environmental designers draw on understandings from conservation biology and landscape ecology to propose new metropolitan landscapes fit for occupation by wildlife as well as people (Forman and Godron, 1986). At the regional scale, *wildlife reserve networks* and *wildlife corridor plans* are currently in vogue, and examples can be found in many metropolitan areas (Little, 1990; Smith and Hellmund, 1993). Wildlife reserve networks are systems of one or more core reserves (that is, undisturbed ecosystems) linked through movement corridors that provide connectivity in the landscape matrix, allowing (in theory) even animals with large home ranges to survive and travel through metropolitan landscapes. Reserves and corridors must be surrounded by buffer zones which permit only gradually more intense human land use. Wildlife corridors serve not only as movement paths between 'mainland' habitats beyond the urban fringe but as methods to connect gene pools and achieve overall landscape connectivity. For animals with small home ranges, corridors can themselves become habitats if they are wide enough and support native vegetation.

Neither the reserve network models nor the wildlife corridors have escaped criticism. One key question is how big core reserves have to be in order to prevent extinction (Soule and Simberloff, 1986). For large predators the answer is 'very big' and certainly many times larger than any metropolitan area reserve. The effectiveness of corridors has been questioned because of the deleterious effects of edge interactions; for instance, Simberloff and Cox (1987) argue that corridors are costly, and may help transmit diseases, facilitate the spread of exotics, decrease genetic variation or disrupt local adaptations and coadapted gene complexes, spread fire or other contagious catastrophes, and increase exposure to hunters or poachers and other predators. Noss (1987, p.162) however, maintains that 'perhaps the best argument for corridors is that the original landscape was interconnected'.

One excellent example of metropolitan-scale reserve network and corridor planning comes from Soule *et al.*'s work in San Diego County's chaparral canyons. From their ecological findings, Soule (1991) provided a set of land-use guidelines emphasizing critical features of network design. These include maximum size and minimal fragmentation to reduce edge effects, maximum linkage between reserve nodes of the network to minimize genetic isolation, and maintenance of coyote populations in canyons where development endangers native bird species (because coyotes keep bird predators in check). More generally, cougar ecologist Beier recommends that planners identify wildlife corridors through the application of geographic information systems and gap analysis, and then circulate maps which mark the location of corridors, and identify the species for which they are designed. Such acts of political cartography – what Beier terms '*putting* [the animals] *on the map*' (1993, p.106, emphasis in the original) – are effective in mobilizing public support for wildlife and corridor protection and signal the prospect of conflict to potential developers.

Wildlife-oriented residential landscape architecture remains relatively uncommon and driven mostly by endangered species laws (Lane, 1991). Despite McHarg's legacy in landscape architecture and environmental design, the number of wildlife-oriented urban developments is minuscule. Most examples are new developments (as opposed to retrofits), sited at the urban fringe, planned for low densities and thus oriented for upper-income residents only.

In general, planned unit development (PUD) designs allow more extensive and less fragmented open space. Comparative studies indicate that PUD designs reduce impacts on wildlife compared with traditional subdivision layouts (Goldstein *et al.*, 1983). For example, Franklin and Wilkinson (1986) assessed the impacts of PUD versus conventional subdivision designs on avian biodiversity, finding that the PUD was superior in its ability to retain bird species. Some subdivision designs have been driven by urban real-estate marketing considerations as much as concern for biodiversity. For example, the Tuscon-area development of La Reserve is a subdivision located directly adjacent to the Pusch Ridge Wilderness, a vast public wildlands set aside as mountain sheep habitat. Home sites bordering the sheep reserve are situated on large lots. Prospective buyers had to agree to 'donate' the back portion of their property adjacent to the reserve to a nonprofit wilderness management organization, pay a monthly 'urban wildlife research' fee, and follow a variety of landscaping, architectural, and behavioural guidelines affecting the visual and noise-related impacts of their dwelling units on the sheep. In this instance, proximity to wildlife was deemed by developer and homebuyers alike to be an amenity worth paying for; finished homes were extremely expensive by local standards, not only because of construction quality, site and unit size, etc.,

but because of proximity to wildlife. The extent to which La Reserve's provisions have protected the reclusive mountain sheep remains to be seen. The estimated population of sheep in the Pusch Ridge Wilderness continues to fall, but it is unclear how much of this population loss can be attributed to the La Reserve development versus Wilderness area users [for example, hikers with dogs, campers, etc. (W.W. Shaw, personal communication, 1994)].

Defending the interests of urban wildlife

In addition to traditional educational or management, planning, and design strategies for creating more space for wildlife in cities, a growing number of *urban environmental organizations* struggle for the preservation of urban wetlands, forests, and other wildlife habitat, because of the ecological value, amenity benefits, and importance to wildlife of these areas. Such groups also engage in restoration efforts, initiating stream daylighting projects, coastal cleanup campaigns, and helping communities and individual landowners restore habitat for wildlife. In addition, *community-based organizations* are increasingly involved in battles to protect urban deer herds, coyotes, cougars, birds, and other wild animals. These efforts may arise out of mainstream environmental organizations or animal rights groups, but they often ignite spontaneously in the face of threats from developers or unpopular wildlife-management practices [such as culling or trapping (see LaGanga, 1993; McAninch and Parker, 1991)].

Virtually all information about these efforts is restricted either to media reports or to scattered case studies, so there are few definitive conclusions to be drawn. However, anecdotal evidence suggests that political action around urban wildlife has exposed deep division within the environmental movement. The rapid growth in urban environmental activism among communities of color, especially the environmental justice movement, has sensitized mainstream (largely white and middle-class) organizations to problems of environmental racism and has forced recognition of the linkages between social and environmental justice. But many groups pay only lip service to social justice issues. Many activists of color continue to consider traditional environmental priorities such as concern for the survival of wildlands and wildlife – especially in cities – as at best a frivolous obsession of affluent white suburban environmentalists and at worst reflective of an unwillingness, rooted in class position and racism, to acknowledge the environmental problems faced by people of color.

References

Beatley, T. (1994) *Habitat Conservation Planning: Endangered Species and Urban Growth*, Austin, University of Texas Press.

Beier, P. (1993) 'Determining minimum habitat areas and habitat corridors for cougars', *Conservation Biology*, 7, pp.94–108.

Berg, P. *et al.* (eds) (1989) *A Green City Program for San Francisco Bay Area Cities and Towns,* San Francisco, Planet Forum Books.

Canter, L.W. (1977) 'Public participation in environmental decision making', in Canter, L.W. (ed.) *Environmental Impact Assessment*, New York, McGraw-Hill.

Forman, R.T.T. and Godron, M. (1986) *Landscape Ecology*, New York, Wiley.

Franklin, T.M. and Wilkinson, J.R. (1986) 'Predicted effects on bird life of two development plans for an urban estate', in Stenberg, K. and Shaw, W.W., *Wildlife Conservation and New Residential Development*, Tucson, University of Arizona.

Goldstein, E.L. *et al.* (1983) 'Wildlife and greenspace planning in medium-scale residential developments', *Urban Ecology*, 7, pp.201–14.

Jones, J.M. and Witham, J.H. (1990) 'Post-translocation survival and movements of metropolitan white-tailed deer', *Wildlife Society Bulletin*, 18, pp.434–41.

LaGanga, M.L. (1993) 'Officials to kill Venice ducks to halt virus', *Los Angeles Times*, 22 May.

Lane, K.F. (1991) 'Landscape planning and wildlife: methods and motives', in Adams, L.W. and Leedy, D.L, *Wildlife Conservation in Metropolitan Environments*, Columbia, National Institute for Urban Wildlife.

Leedy, D. *et al.* (1978) *Planning for Wildlife in Cities and Suburbs*, Washington, US Government Printing Office.

Little, C.E. (1990) *Greenways for America*, Baltimore, Johns Hopkins University Press.

McAninch, J.B. and Parker, J.M. (1991) 'Urban deer management programs: a facilitated approach', *Transactions of the 56th North American Wildlife and Natural Resources Conference*, 56, pp.428–36.

McHarg, I. (1969) *Design with Nature,* New York, Doubleday/Natural History Press.

Nelson, A.C. *et al.* (1992) 'New fangled impact fees: both the environment and new development benefit from environmental linkage fees', *Planning*, 58, pp.20–4.

Noss, R.F. (1987) 'Corridors in real landscapes: a reply to Simberloff and Cox', *Conservation Biology*, 1, pp.159–64.

Platt, R.H. *et al.* (eds) (1994) *The Ecological City: Preserving and Restoring Urban Biodiversity*, Amherst, University of Massachusetts Press.

Saldana, L. (1994) 'MSCP plans the future of conservation in San Diego', *Earth Times*, February/March, pp.4–5.

Schicker, L. (1987) 'Design criteria for children and wildlife in residential developments', in Adams, L.W. and Leedy, D.L., *Integrating Man and Nature in the Metropolitan Environment*, Columbia, National Institute for Urban Wildlife.

Simberloff, D. and Cox, J. (1987) 'Consequences and costs of conservation corridors', *Conservation Biology*, 1, pp.63–71.

Smith, D.S. and Hellmund, P.C. (1993) *Ecology of Greenways: Design and Function of Linear Conservation Areas*, Minneapolis, University of Minnesota Press.

Soule, M.E. (1991) 'Land use planning and wildlife maintenance: guidelines for conserving wildlife in an urban landscape', *Journal of the American Planning Association*, 57, pp.313–23.

Soule, M.E. and Simberloff, D. (1986) 'What do genetics and ecology tell us about the design of nature reserves?', *Biological Conservation*, 35, pp.19–40.

Stren, R. *et al.* (1992) *Sustainable Cities: Urbanization and the Environment in International Perspective*, Boulder, Westview Press.

Van der Ryn, S. and Calthorpe, P. (1991) *Sustainable Cities: A New Design Synthesis for Cities, Suburbs and Towns*, San Francisco, Sierra Club Books.

Source: Wolch, West and Gaines, 1995, pp.745–9

CHAPTER 5

Cities of power and influence: settled formations

by John Allen

1 *Introduction*

In previous chapters, the webs of cross-cutting relationships which make up city life have been highlighted through a variety of connections which combine and settle in particular ways. Whether it be the comings and goings of different cultures within cities, or nature turning up on the doorstep in often surprising and unsettling ways, the concern has largely been one and the same: to explore the resultant intensity and mix as things come together in one place. And so it is with power and cities.

If nature is regarded by many as a stranger to the city, power is rarely so. The governance of cities, a concern to trace who gets to exercise power over whom, and whose interests are realized at the expense of others, is a familiar feature of the urban landscape. The ability of certain groups to bend or influence the will of others – for example in relation to transport choices – has long been a characteristic feature of city life and politics. In this chapter we shall remain with this feature, but shift the focus from that of interest groups and decision-making practices to the wider networks of power within which cities are embedded (see **Massey, 1999**). In line with debates in earlier chapters, the focus will be upon the different strands and connections which are thought to lie behind the power and influence of cities. As such, a good part of the inquiry will focus upon the movement and dispersal of resources, practices, and materials of all kinds; it will also look at the ways in which today, for example, the likes of New York or Tokyo are said to intensify their power through an ability to settle the flows of economic power and fix their location, however provisionally. In a world that is apparently becoming ever more mobile and rapid in pace, the need for such cities to direct and influence the significant global networks has fast become an economic convention.

From the outset it may be apparent that some cities are more powerful than others and that over time cities may grow visibly in power or see their powers fade and decline. The 'global' reach of many of today's most powerful cities stands in contrast to those cities whose moment of influence has passed or to those whose fortunes have shifted precariously as the networks of which they are a part no longer sustain them. Whether it is the changing fortunes of Mexico City since its long entanglement in Imperial Spanish connections centuries ago, or the more recent transformations in the influence of cities such as Moscow and Hong Kong who find the basis of their power shifted almost overnight, the settlement of power is not something that is achieved simply. It is for this reason that the chapter is as much concerned with questions of connection and disconnection as it is with questions of power and influence. Above all, we will be seeking to understand why

● power and influence across the networks appears to concentrate in so few cities

and also seeking to show

● the complex of networks through which cities sustain, enhance or lose their power over time.

These are the prime issues that interest us in this chapter.

1.1   FORMATIONS OF POWER AND INFLUENCE

Much of the time we will be talking about cities *of* power, but we need to be clear in our minds what we mean by this expression. There are two aspects to consider, one around the formation of city power, and a second around the plural nature of that power: that is, the many different sources of power to be found in cities and their different sites of intensity.

On the first point, we know from Chapter 1 that cities are best not seen as a kind of stored warehouse of resources where, for example, economic or cultural assets are routinely called upon to exert a dominating influence or to intervene decisively in global affairs. Cities may in one sense 'contain' power, but that is rather different from saying that power necessarily 'belongs' to some cities and not others; or from saying that cities represent some kind of awesome force which is simply waiting to be dusted down and passed on intact from one urban generation to the next. In thinking, for instance, about the cultural influence of Paris as the centre of fashion for the best part of the nineteenth and twentieth centuries – across much of the western world at least – what is it exactly that has made it a beacon of style? What was it that produced Paris as a city of cultural authority, as a site of imitation whose themes of elegance and style spread in influence from city to city and beyond? Can we trace its influence back to a particular historical turn of events – say, to the presence of a fashion-conscious royal court in the nineteenth century? Or should we be looking at the skills of the couturiers in Paris at that time as the basis of Paris' authority? Or perhaps we should turn our attention to the more recent economic power of the French high fashion houses such as Dior, Givenchy or Cardin?

FIGURE 5.1
Christian Dior – a powerhouse of fashion?

ACTIVITY 5.1   Extract 5.1 is about Paris as an influential centre of fashion and provides a brief glimpse of what lay behind the formation of the *haute couture* industry of Paris and how it has reproduced itself over time. Read it carefully and make up your own mind as to the basis of Paris' settled cultural power and authority. ◆

EXTRACT 5.1
Michael Storper and Robert Salais: 'Paris as a capital of fashion'

Paris was a center of clothing fashion in the nineteenth century because it was the home of Europe's most centralized court: men and women from England to the west and Poland and Russia to the east looked to Paris to know how to dress (but not the Italians, who had their own skills in this domain). The imperial couple in Paris had no competition as fashion leaders at the time: the Prussian court was austere, and in London there were few *mondanités* after the death of Victoria's husband, Prince Albert. Because of its commitment to luxury – not so far from the qualities demanded by Colbert's manufactories more than 150 years earlier – we might call this an economy of splendor; the whole system was propelled by the centralized fashion authority of the aristocratic hierarchy. The system grew progressively more splendid throughout the nineteenth century, in the bourgeois-imperial society that succeeded the Revolution, the fetish of fashion grew, and clothing design was considered one of the *beaux arts*. Bourgeois men, too, engaged in dandyism in the nineteenth century. For women, the situation was summed up in the slogan of the *Magasin des Modes* (the most important of the fashion stores) that 'boredom was born one day out of uniformity'. Fashion, with its obsession with change, was supposed to fight boredom.

Women's clothing was extremely complicated and cumbersome until 1885, when a major transformation began to take shape. Redfern, couturier to the Duchess of Wales, invented the modern women's outfit – a jacket and skirt in cloth, without the customary ornaments or constraints. In 1907, Poiret (who eliminated the corset) took advantage of the revolution in women's clothing to reshape the division of labor, designing the garment and turning it over to the *modeliste* to execute. The fashion authority of the designer was now effectively sundered from customized production, and made more independent of the client.

In the years after World War I, this separation was extended even further. Thus the opening of the first real *maison de haute couture* by Gabrielle Chanel, in 1924, did not at all represent the triumph of customization; rather, it marked an attempt to rationalize high fashion production under the authority of the couturiers, for Chanel abandoned continuous creation in favor of the system of two annual collections each with about 150 models, with two minor half-season collections in between. This enabled her to assemble a specialized staff for design, cutting, sewing, and completion. Patou, Lanvin, and Schiaparelli did the same in the 1920s, and all rose to become world-famous exporters before the beginning of the depression of 1929.

The Paris high fashion system continued its upward ascent in the middle of the twentieth century: the magazine *Marie Claire* appeared in 1937 as its mouthpiece, and other famous fashion houses, such as Balmain, Dior, Givenchy, and Cardin, opened immediately after World War II. The high fashion system was nourished by the continuing image of Paris as a world fashion capital. It is also important to realize that as late as 1955, haute couture served as much as 35% of the Parisian market, compared to less than 10% for U.S. buyers in 1940 or 20% for Germany. In the 1950s, a high fashion house might make 10,000 dresses per year (as compared to a few hundred today), and employ, full time, 350 models and several hundred production workers. With about twenty of these houses located in Paris, an important production complex was in place …

At its apogee, the *haute couture* industry of Paris shared many of the characteristics Marshall identified in describing nineteenth-century English industrial districts, in particular the tight linkages between producers, the circulation of specific information among them, and ongoing endogenous innovation within the district's product specialization. In this particular version, the reputation and authority of the couturiers, and their tight linkages with a restricted clientèle, were the ties that bound the system together.

Source: Storper and Salais, 1997, pp.121–3

Aside from the many different actors involved and the peculiarities of history, what struck me was the absence of a single cause. Instead of seeking that decisive historical moment, the extract draws our attention to a combination of past and present qualities, a mix of symbolic and economic resources drawn through different sources and connections, which enabled the fashion houses of Paris to stabilize and to reproduce their dominant influence. More to the point, there is nothing to suggest that such cultural power or authority 'belongs' to Paris on an indefinite basis. On the contrary, the élite tastes of *haute couture*, once the embodiment of fashion, no longer appear to hold the straightforward appeal and influence that they once did in the face of a more diverse, individualized set of tastes in western fashion (Lipovetsky, 1994). In that sense,

● the particular power *of* cities such as Paris is best understood as something which is continuously produced in various *settled formations* through its networked relations, and not as some locatable 'reserve'.

And, following on from this, it is important to note that fashion is not the only source of influence to which Paris can lay claim. In terms of world music, for instance, Paris occupies a rather different position in the networks of influence which oscillate between places as far away as Senegal and Mali. If we were to pursue this line of inquiry outside of the domain of cultural influence, no doubt it would be possible to identify economic and political groupings with powerful attachments beyond the city to transnational élites elsewhere, such as those connected to France's colonial past and its territories or to Paris' current global economic ambitions. The point is that

●   cities hold a *diversity* of power arrangements, with different parts of a city locked into different sets of rhythms and connections, with different degrees of influence and authority.

Just as we find it difficult to talk about the city as a singular entity, so also is it unrealistic to think about cities as containers for a single source of power. An individual city such as Paris, or Madrid, or Miami, can be powerful in many respects, and different parts of those cities may represent particular concentrations of influence and control.

Bearing these two points in mind, the rest of the chapter is devoted to unravelling what is it that makes certain cities a dominant influence within particular networks. The basis of an answer to this question is set out in section 2 and is progressively advanced through the two main sections. In section 3 we consider two different yet related accounts of *where* power settles and *why*. A key issue addressed here is why power appears to concentrate in some cities and not others. Following that, in section 4, we ask whether cities 'run' the networks of connections through their concentration of resources and power *or* whether the networks 'generate' cities as sites of power through their interconnections. In doing so, we extend the analysis of power further through a consideration of collaborative as well as instrumental modes of power. Power does not always crush and the power of cities, regardless of source, can work in various modes, bringing people together as well as dividing and separating them.

2 *What makes a city powerful?*

At one level, we have already started to answer this question. Powerful cities are those which exert some kind of influence or control over what happens around them. It is as if the further away that influence and involvement extends, the wider the ambit of its concerns, the more powerful a city looms in our imagination. Earlier, mention was made of New York and Tokyo and their like in this respect, with the importance of both cities stemming from the fact that the world of banking and finance seems to have little choice other than to go through the financial districts of those cities. They stand at the *intersections*, so to speak, of all that matters in global economic terms. In that sense, both cities are locations for something far greater, far more significant in scale, than their actual size would have led us to believe. They may rank among the mega-cities in terms of population numbers, but it is not their size which makes them appear powerful (see **Massey, 1999**). And if such an impression holds for the economic power of New York and Tokyo, then cities such as Mecca convey much the same thing culturally in the wider Islamic world, as does Paris in the world of style and fashion. In their own inimitable way, each of these cities is expressive of something more than itself. They are the points at which the lines of authority and influence meet.

As valid as such an answer may seem, however, it only just scratches the surface of what enables these cities to exercise a degree of power. In truth, such an answer does not address what actually makes a city powerful. To understand that,

● we need to consider what it is that comes together to produce certain forms of power and influence in cities. More to the point, we need to know how the different strands and connections come together and what is made from them.

The mere fact that all manner of information and resources and lines of communication interact in places like Taipei or Toronto does not automatically produce something which we can recognizably call power. Networks of power and influence are not like electrical 'circuits' that are simply switched on by fortunate cities in the right locations. The strands and connections which comprise the numerous networks have to be constructed; forged through the resources of cities and combined in ways that settle and fix their influence. It is in this sense that power is produced.

But produced out of what? There are at least three aspects to consider.

The first is that of *people* and their social interaction. If we think of Shanghai as a city of economic dynamism about to make its mark on the world-stage as a powerful actor, for instance, or of Paris as a city of cultural authority in more ways than one, then it could just be that the right kinds of people, with the right kinds of skills and connections came together in these cities and the intensity is, in a very real sense, powerful. Similarly, if we take the example of London or

New York's financial community, it could be argued that their acknowledged economic power is currently the product of the right blend of 'knowledge élites' specializing in all manner of financial innovations and monetary risks, together with the right kind of legal and political connections, as well as the appropriate cultural capital. We find all the high-powered lawyers and other high-flyers attracted to these intense concentrations precisely because these are the places where things happen. Without the combination of home-grown banks, overseas finance houses, law firms and the like, and more especially the vitality and creativity of their staff, the economic power of such cities today may well have slipped away to other locations. Much the same, perhaps, could be said of Washington, DC, which contains people with the right political connections and competencies, or indeed of Brussels, whose political influence is felt throughout the European Union.

On this view, power practically condenses into an audit of a city's people, skills and connections. Over time, then, it would appear that such cities have been able to use or 'work' their human resources to maintain a competitive advantage over other potential rivals. Their advantage is cumulative, in so far as they are aided by the inertia of established connections and accumulated expertise which gives these particular cities a head-start over others. If it were the case that any city could become powerful overnight, such historical legacies would count for nothing. But they do count. For they make it that much more difficult for other, less established cities to rival them. As such, there is the possibility of various established financial, cultural and political élites enjoying power as an enduring settlement, rather than as a fleeting moment.

But this is only part of the story. The power of cities is often more than an effect produced through the mobilization of human resources, even if it has become settled over time. Powerful cities are made up of more than people and connections, and resources are not simply human.

So a second aspect to consider is the *mix of resources* – not just those relating to people, but also to the kind of environment, the type of institutions, the flows of information, the symbolic assets, and so forth.

Take for example the notion of the tele-mediated city discussed in Chapter 1, where a premium on electronic communications and information technologies has arguably enabled a handful of cities on the world-stage (or more often, a handful of streets in a handful of cities) to stabilize the throughflow of information and knowledge to their advantage. The mix of resources involved – technological, social and economic – backed up by a strong institutional presence can, however, give many a capital or core metropolitan city a dominant position in the different networks of influence (see Amin and Thrift, 1995, on the institutional 'holding down' of global processes). Likewise, the symbolic power of the Los Angeles' film industry, which is often claimed to exert an excessive cultural influence – almost tantamount to cultural domination by the US – over audiences across the globe, could be linked directly to the peculiar mix of its assets. The type of investment involved, the range of technologies, the channels of distribution, the combination of cultures, the spread of creative bodies, the nature of the landscape, and so on, may each, in their own way, have helped to produce Los Angeles and Hollywood as a powerful cultural force. If it were Bombay (Mumbai) and the Indian film industry under consideration, we would probably be looking for a similar, although by no means identical, cluster of resources to account for its extensive cultural reach and location.

Again, then, it is the peculiar combination, the different ways in which the connections are drawn, which gives us an insight into how cities are able to marshal their resources to significant effect. And from this we can draw out a third, critical characteristic which, as you may have been aware, has been present all along: namely, *the practices of power* which highlight that it is *what you put together* with the resources that matters, not simply the amount of human and material resources at your disposal.

So, on this account if a city aspires to join the ranks of the powerful, it should adopt the ways of doing things that have proven to be successful in influential cities or, even better, attempt to steal a march on them. Singapore may well be a case in point. In the 1990s the government and press of Singapore started to refer to themselves as 'an intelligent island'.

FIGURE 5.2   *Singapore's skyline: an intelligent island in the making?*

ACTIVITY 5.2   To obtain a sense of what that might involve, read Extract 5.2, 'Towards an "intelligent island"'. As you progress through it, make a note of the kinds of skills, connections, and mix of resources referred to and, more importantly, what is supposed to arise from their combination. ◆

EXTRACT 5.2
Ranald Richardson: 'Towards an "intelligent island"'

The IT2000 masterplan seeks to develop Singapore into an Intelligent Island, where information technology (IT) is pervasive in every aspect of its society – at work, home and play. An integrated and extensively connected National Information Infrastructure (NII) will be put in place to link computers and other information appliances in homes, offices, schools and factories across Singapore. Singapore, the Intelligent Island, will be a global centre for excellence for science and technology, a high value-added location for production and a critical strategic node in global networks for commerce, communications and information (National Computer Board, 1997).

The 'intelligent island' masterplan has five goals:

1   developing a global hub;

2   transforming the economy;

3   enhancing the potential of individuals';

4   linking communities locally and globally;

5   improving quality of life.

A key element in the plan is *Singapore ONE* (One Network for Everyone), a national high-capacity network that will link businesses, schools and homes. The infrastructure will have a core broadband network based on ATM switching and optical fibre technologies connecting several local access networks. At the local access level to reach the homes, offices and the public, existing or new networks will be used. The plan envisages Singapore ONE as a means of linking together the whole of society, co-ordinating disparate existing networks. It is hoped that the network will allow increased homeworking, homeshopping, learning from home and access to government services …

… Initially Singapore ONE will have government as an anchor tenant, using the bandwidth to make government services available to the public. It will then aim to gain a critical mass of non-government services and thus become a commercial network. There are a number of incentives to encourage this. The 'Pioneer Club' is meant for companies or organisations who are committed to taking an early initiative to transform Singapore into an Intelligent Island through the development of multimedia applications and services in the Singapore ONE project. Members are eligible for financial and fiscal incentives and for preferential tariffs.

A range of initiatives are planned around Singapore ONE including:

●   an Electronic Commerce Hotbed to 'jump-start' electronic commerce in Singapore;

●   infrastructure for enabling electronic ID to improve security in trading;

●   a national contacts database; and,

● I-HUB, a national project bringing about a nation-wide information infrastructure to link together network centres of all the Internet Access Providers, commercial network service providers, government networks and other access networks.

The IT2000 plan also envisages a number of 'technology corridors', with physical and electronic connections between strategic high tech sites such as science parks, the universities and the airport.

The plan goes beyond infrastructure to consider education, skills training, and developing an IT culture. The NCB claims that IT education begins as young as four years old and that multimedia computer-aided instruction courseware are already being used in some kindergartens.

Source: Richardson, 1997, pp.7–8

The first point to note perhaps about IT2000 is that it has the status of a plan. Clearly Singapore wishes to be at the forefront of the information revolution, directing and influencing the growing networks of money, information and knowledge that circulate much of the globe. It wants to be a global hub and, as such, it has designs on being a 'smart city', with all the power that is said to entail. But to achieve this aim, it has officially set itself the task of becoming creative: creative in the manipulation of signs and symbols, creative in managing the rhythms and flows of information, creative in all the ways that are thought to matter in a changing global economy. Singapore's city economy is itself rather small in terms of stocks and shares, but that is not the measure of economic power adopted. Power in this context, when the right mix of resources is in place (in this case IT resources, including an IT culture), increases in proportion to the imaginative and creative practices employed in a city. Something *extra* is said to arise out of the *combination*, something which delivers the intensity of relations which will enable Singapore to fix and direct the flows of money, information and knowledge across the region. Storper and Salais (1997) were saying much the same in relation to Paris' high fashion system (Extract 5.1), where the tight connections, the circulation of skills and know-how, produced something innovative which made the reputation and authority of the couturiers appear unassailable.

The point, however, is a broader one than that of the fate of either Paris as a centre of fashion or Singapore as a 'smart city'. It is, to return to our initial formulation, about how the different strengths and connections come together in such a way as to produce an *intense* set of interactions we might recognize as a powerful city.

● All kinds of things come together in cities – people, practices, ideas, resources among them – some from far away, others from close by, but it is what is made out of that interaction which has the potential to make a city powerful.

Perhaps that gives us a clue as to why power and influence of whatever kind appears to be concentrated in few rather than many cities.

3 *Where does power settle?*

We have already seen that different kinds of power flow through different kinds of networks, with different cities, invariably only parts of them, at their centre. In an age of global decentralization and transnational connections, however, only one kind of power has been elevated to the world-stage as a major force within which cities occupy the pivotal role: that of economic power. This is not to say that cultural or symbolic resources are not mobilized in the exercise of economic power, nor that political connections are absent from it; it is merely to register that it is the practices of finance, business and the economy more widely which have largely captured the attention of contemporary theorists of power and the city. In this section, therefore, we are going to consider two of the most significant approaches to *where* economic power settles and *why*. The first, which draws extensively but not exclusively on the work of Saskia Sassen, has tended to emphasize the role of certain 'powerful' cities in running the global networks of economic production and finance. The second approach, based primarily on the work of Manuel Castells, takes a less settled approach to power and favours a more relational understanding of cities and power. In both approaches, the ever-present powers of nation-states tend to be sidelined and this is an issue to which we will return. For the moment, we will focus on the major claims of each approach to see what it is that they have to say about cities, networks and power.

3.1   … IN GLOBAL CITIES

If we pick up again on the idea that the power of cities is expressed through their dominance of the networks which flow through them, then their location at the intersection of such flows, the nodal points, has frequently been portrayed as the 'commanding heights'. Among those who initially conceived of economic power as centred in this way, such as Friedmann and Wolff (1982), the cities of London, Paris, Tokyo, New York, Los Angeles and Chicago were placed at the top of a hierarchy of world or global cities. These were considered to be the core primary cities which lay at the intersection of the most important linkages. Below them, ranked according to the degree of power they commanded on the world-stage, were placed such cities as Madrid, Milan, Miami, Singapore and Sydney. Figure 5.3 gives you a clear sense of the hierarchy and its ranking. The higher the rank, the greater the assumed power, with a whole string of cities in relentless competition with one another to secure a higher placing through the acquisition of ever more command-and-control functions (Friedmann, 1995).

Much of the thinking behind this rank order of cities rested squarely on the assumption that economic power is something that is lodged in the headquarters of transnational organizations. On this view, the resources and decision-making powers of the big corporations were marshalled at the centre and used to secure

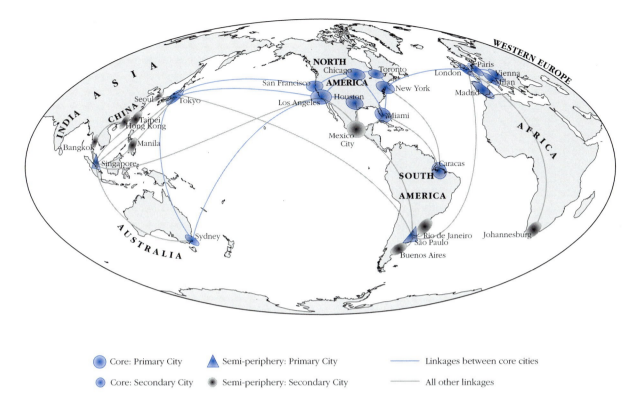

FIGURE 5.3 *The ranked hierarchy of world cities (from Friedmann, 1986)*

organizational and institutional goals in far-flung locations, notwithstanding of course an element of resistance. A glance at the location of transnational corporate headquarters quickly revealed their concentration in a few cities worldwide. These cities stood at the apex of corporate and financial hierarchies and, without too much reflection, a hierarchical list of world cities soon fell into place.

But why did the big corporations choose to locate in so few cities? And how does the obvious power that they wield actually bring about something as complex as the co-ordination and control of the global economy? Powerful as they may be, are the big corporations really that much in command? These are the questions which troubled a number of observers, especially Saskia Sassen.

The starting-point for her analysis of why the cities of New York, London and Tokyo appear to occupy such dominant positions in the working of the global economy is the scale and complexity of the latter today, rather than the specificity of the cities themselves. The intensity of economic connections, the increased pace of cross-border transactions, and the magnitude of economic flows, combined with the global dispersal of economic activities, prompted Sassen (1991) to see global cities as the *necessary* outcome of a more complex economic geography. Contrary to the expectations of many, as Chapter 1 noted, the rapid developments in information and communications technologies actually strengthened the strategic role of cities as sites through which the integration of today's complex economies could be achieved. The fast and footloose nature of industries, exaggerated by the growth of telematics (that is, the transmission of computerized information over long distances), generated a need for places where the work of globalization – the management and manipulation of intense interactions – could effectively be performed. Those places, according to Sassen, are indeed those listed in earlier accounts of world cities, but for a different set of reasons from those initially given.

If New York, London and Tokyo, among others, are strategic locations capable of exercising control over an increasingly globalized world, then this is because such cities represent an agglomeration of specialist activities. It is certainly not because such cities simply have a concentration of multinationals headquarters. Importantly for Sassen, such a capability for global control cannot be 'read off' from a city directory of multinationals, finance houses and the like; it is something that is *produced* through an amalgam of critical service activities. We can obtain a better sense of what this production complex entails by turning directly to the reading by Sassen (Reading 5A).

ACTIVITY 5.3 You should now turn to Reading 5A by Saskia Sassen (1995) entitled 'On concentration and centrality in the global city', which you will find at the end of the chapter. As you read through it, you will find reference early on to the combined processes – of the global dispersal of economic activity and of global integration – which Sassen argues lie behind the strategic role of global cities today. Following that, I would like you to focus upon two key aspects of her argument:

● What is meant by the *production* of a global capability and the *practice* of global control?

● Which practices are brought together in global cities to *stabilize* the global flows of economic activity? ◆

It is a fairly compact read and some of the points represent a summary of Sassen's arguments in *The Global City* (1991). Nonetheless, it is possible to glean the line of her thinking. Remember the point we made earlier about the power of cities being produced through their networked relations and mobilized resources? For Sassen, the logic is much the same:

● global cities produce a capability for global control through bringing together in one place the right kinds of people, skills and expertise, as well as playing host to the latest in communications technologies.

In drawing so many diverse specialists together – financial analysts and investment bankers, lawyers and accountants, management consultants and programmers, many of whom will have honed their expert knowledges in other central locations – places like New York and London, or Sydney and Frankfurt effectively produce what it takes to orchestrate and run the global economy. Such a combination is not put in place overnight, as indicated earlier, which is why the likes of Mombasa or Calcutta would struggle to achieve such a capability. History plays its part, advantaging some cities over others, laying down the preconditions which may serve to concentrate future possibilities.

Above all, then, it is a combination of past and present practices that enables certain cities to produce the capacity to fix the intense cross-cutting nature of economic flows and transactions which comprise what we understand today as the global economy.

So, to pick up on the practice of global control, according to Sassen, it is precisely what all the different kinds of high-level service professionals *produce together* which constitutes the economic power of global cities. Working in close proximity to one another, the dense clustering of people, knowledge and information enables such cities to respond to the immediacy of the markets and the general speed-up and acceleration of global economic activities. Figure 5.4 gives you an impression of what this looks like on the ground in the three strategic cities of New York, London and Tokyo. Note the proximity and overlap of many of the activities and indeed the relatively small areas of the three cities in which they are settled. As the movement of business, money, workers and ideas notches up a pace, only a handful of cities are in a position to truly practise the skills of global control and to do, as it were, the negotiation work of globalization. It is in this sense that Sassen is able to stress that the global economy is not so much 'out there', as it is *embedded* in the institutional processes of particular cities and practised by a new class of urban professionals.

It is also important to register the fact that the processes involved extend far beyond those performed by the headquarters of multinationals. As Sassen argues, whilst many of the largest corporations choose to locate their headquarters in the big cities, it is not their power alone which makes a city economically powerful. Rather, many of the large corporations locate their headquarters in such cities precisely because that is where they can tap into the integrated complex of specialist services necessary for the practice of global control. And, in so doing, they are seen to enhance the power and influence of the dominant cities, and to add to their concentration.

At this point it is probably useful to pause for a moment and to contemplate what kind of global city, in terms of its shape and geographical layout, is actually being evoked by Sassen. With so many things coming together to form what appears to be an intense cluster of interaction, it is tempting to imagine a global city as comprising a powerful central core of the kind represented in Figure 5.4 and a rather disempowered set of outlying districts or suburbs. Not so, asserts Sassen, pointing to a more differentiated geography of centrality. If anything, she wishes to argue that there is no one pattern of centrality but many. Included among them, as you will have read, is the idea of a metropolitan grid of nodes: that is, a decentred collection of commercial and business districts which has been effectively recentred through state-of-the art telematics. This web of activity stretched across a region and linked together by advanced telecommunications is put forward by Sassen as a new way of thinking about the centrality of powerful cities. And if the 'centre' is gridded or striated in this way, then those parts of a city which fall in the spaces in-between – the poorer, disconnected districts – really do constitute the disempowered. In London, for

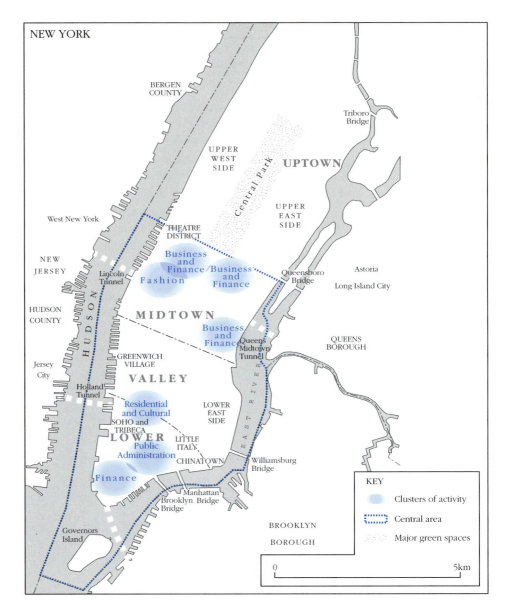

NEW YORK

BERGEN
COUNTY

Triboro
Bridge

UPPER
WEST
SIDE

Central Park

UPTOWN

West New York

UPPER
EAST
SIDE

THEATRE
DISTRICT

NEW
JERSEY

Business
and
Finance

Business
and
Finance

Queensboro
Bridge

Astoria

Lincoln
Tunnel

Fashion

Long Island City

HUDSON
COUNTY

MIDTOWN

Business
and
Finance

QUEENS
BOROUGH

Jersey
City

GREENWICH
VILLAGE

Queens
Midtown
Tunnel

Holland
Tunnel

VALLEY

LOWER
EAST
SIDE

Residential
and Cultural

SOHO and
TRIBECA

LOWER

Public
Administration

LITTLE
ITALY

CHINATOWN

Williamsburg
Bridge

Finance

Manhattan
Brooklyn Bridge
Bridge

BROOKLYN

BOROUGH

Governors
Island

HUDSON

EAST RIVER

KEY

Clusters of activity

Central area

Major green spaces

0   5km

FIGURE 5.4 *Dynamic service clusters in New York, London and Tokyo*

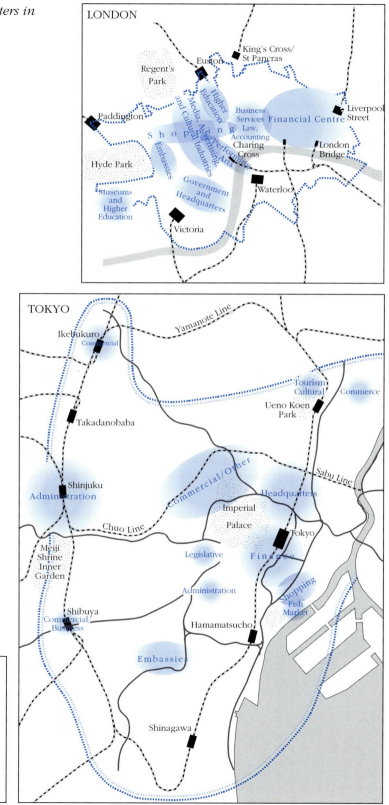

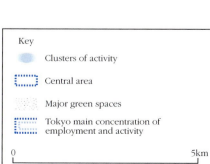

instance, it would be the overlooked spaces of Hackney or Tower Hamlets which sit alongside the new Docklands developments, yet are by-passed entirely by the new communications infrastructures. In Paris it would be the likes of Aubervilliers on the edge of the city (see Chapter 3 section 6.1) which would fall through the communications grid which stretches out to La Défense and beyond. And so on.

The one constant to Sassen's argument, however, is that no matter what form the spatial economy of the 'centre' takes, whether the singular core of an old established city or a decentred complex of nodes, the 'centres' of the most powerful cities remain connected to one another via the digital highways and electronic corridors. It is this global network of cities, with New York, London and Tokyo dominant within it, that encompasses the range of strategic sites through which the global economy is considered to be controlled and co-ordinated. Other cities besides the 'big three' occupy a functional niche within this transnational hierarchy – Miami, Sydney, Toronto, Hong Kong, Singapore, São Paulo, Paris, Frankfurt, and Madrid among them – and together they make up one global system of interconnected 'centres'.

What holds the system together is more than a series of electronic impulses, however. Above all, it is the various global command-and-control functions embedded in the centres of the different cities which are said to comprise the network, although this in itself tells us more about the attributes of cities and what goes on in them, than it does about the ties that bind them (see Taylor, 1997). On the latter, it is frequently the cultural ties between a highly mobile, professional class of workers which is thought to provide the connections between cities (see Hannerz, 1996) and, more importantly, the contact points which facilitate the transfer of creative assets, skills and knowledges. Indeed, the earlier example of Singapore and its aspirations to become a 'smart city' over and above that of its immediate rivals in South East Asia is a case in point. The energetic policy of the Singapore government to increase its rank status in the hierarchy of cities (including extending an invitation to Friedmann to speak on world cities: see Friedmann, 1995) and thereby enhance its economic power and influence, rests heavily on the notion that success is to be found in putting together *in situ* the right combination of imaginative knowledges and creative practices. The production of a global capability in this context

● stems from the possibility of attracting from elsewhere best practice in managing and settling a greater proportion of the economic flows and transactions worldwide. In short, the greater the concentration of flows stabilized, the more *intense* the power that is said to radiate from it.

In all such accounts, however, it is never quite clear how the economically powerful cities in the system – whether established centres such as New York or emergent nodes such as Singapore – actually *extend* their influence across the far-flung networks of economic activity. If such cities are 'running' the global networks through their ability to concentrate resources or to centralize key practices, then it should be possible to trace the relationships of control *through* the networks once they are extended or stretched beyond their central command points. Yet we know a lot about the city locations at which power is produced, but far less about how it flows or how it is effective at locations distant in space and time. For that, we need to turn to a different viewpoint on power and cities.

3.2 ... IN CITY NETWORKS

Rather than view global cities as those which contain or trap power, another view takes us towards a more mobile conception of power. The work of Manuel Castells, in particular the first volume of his trilogy on the information age entitled *The Rise of the Network Society* (1996), stresses the role of global cities as a *process*, rather than a series of discrete places. By this, he wishes to stress the point that the kind of critical service activities brought together in places such as New York, Tokyo, Hong Kong, Singapore, Paris, Madrid or London cannot be divorced from the networks of which they are a part. In stronger terms, global cities as a phenomenon would not exist but for the connections which constitute them. As such, any talk of cities as the 'commanding heights' of the global economy is to invite misunderstanding. There are no 'heights' in the sense that power is amassed in a few locations worldwide and then extended across distant economic landscapes to the far reaches of the global economy. On the contrary,

● city powers are mobilized *through* networks; it is what flows *through* the networks which empowers particular groups and generates certain cities as sites of power.

As Castells puts it:

> ... while the actual location of high-level centres in each period is critical for the distribution of wealth and power in the world, from the perspective of the spatial logic of the new system what matters is the versatility of its networks. The global city is not a place, but a process. A process by which centres of production and consumption of advanced services, and their ancillary local societies, are connected in a global network, while simultaneously downplaying the linkages with their hinterlands on the basis of information flows.

> (Castells, 1996, p.386)

So let's take a closer look at the versatility of networks and what it is that connects some cities to the network and not others.

At the core of Castells' argument is the assumption that the global economy today, and indeed society at large, is constructed around movement and flows: most obviously, those of information, ideas, symbols, money, technology, and the forms of social interaction which lead to their exchange and assimilation. The ability to interact with others distant in space and time, a form of co-presence brought about by the developments in communications technologies, is critical to all this. The very fact, argues Castells, that certain élite groups in different parts of the globe can be networked to overcome the barriers of distance, lifts them out, so to speak, from the particularities of place and locates them in the 'space of flows'. The term 'space of flows' has a rather ethereal quality to it, but Castells grounds it in the following ways.

The first and most straightforward meaning of the term refers to the networks of communication – the various public and private electronic networks, including the internet and company intranets – which allow for continuous interaction to take place at the global scale in real time. Instantaneous electronic communications, whilst obviously facilitating the expert knowledges clustered in particular cities, as Sassen argues, also has the potential to bring together absent others in simultaneous time. The networks in that sense almost appear to have a life of their own, the logic and architecture of which has the capacity to define 'new space, very much like railways defined "economic regions" and "national markets" in the industrial economy' (Castells, 1996, p.412).

It is this process of definition which provides a second material basis to the 'space of flows'. For such networks construct the connections between cities and, according to Castells, assign each one a role and a weight in the hierarchy of wealth and power-making which shapes their relative prospects and economic fortunes. In the industrial economy, so the argument runs, the powerful cities were those located at the critical points of material networks – the railway hubs, the deep sea-ports or major road intersections, which facilitated the flow of material goods. Cities such as Chicago, for example, made their reputation and their fortune on such a basis (see **Pile, 1999**). In an information age, however, where knowledge networks are valued over material ones, a different set of cities is defined by the flows of signs, symbols and creative practices. New York, London, Tokyo, Paris and indeed Chicago are still among them, albeit on the basis of their new-found ability to strategically direct such flows, and they have been joined, according to Castells, by the likes of Madrid, São Paulo, Buenos Aires, Mexico City, Taipei, Milan, Amsterdam, Frankfurt and Zurich. Singapore's IT2000 plan, you may recall, is an attempt by its government to be up there among these cities.

Indeed, today, argues Castells, the position of such cities is best understood as either co-ordinating *hubs* in world markets, switching-points for the smooth transmission of information and knowledge, or as strategic *nodes* in the networks, centres which orchestrate and direct the many flows and interactions which comprise the networks. As the global economic networks which make up 'the global economy' expand, alter in nature, or contract, so the different kinds of knowledges and services required lead to the incorporation of new cities whilst simultaneously shedding links to others. In the context of ever-changing linkages, connections and disconnections shape the fortunes of particular cities over time, leaving some cities and groups high-and-dry whilst providing a life-line to those whose services and skills are in demand.

This brings us to the third and best grounded dimension to the 'space of flows', namely the city-based élites which comprise the power-holders. Located in the great cities, the professional élites are at one and the same time part of these cities, yet isolated from the majority of those around them. They are part of places such as New York or Tokyo in so far as their cultural codes are grounded in the exclusive circles of such cities, but detached from such locations by virtue of their privileged position in the networks. In that sense, they are distant from many of those close by and near to those in far-off locations who share similar cultural and symbolic traits. What Castells is trying to convey here is that the points of attachment are more to the networks than to the cities involved. The cultural and symbolic codes are embedded in the flows of interaction, not in the peculiarities of a particular urban context, and it is this which enables the professional élites to gain access to the networks and to circulate freely. As such, the cultural capital of the transnational networks is said to transcend the historical or local style of any one city in the global network.

ACTIVITY 5.4   It is perhaps harder to imagine how this network of cities holds together in comparison to that of Sassen's account or even that of Friedmann's. Figure 5.5, for instance, is the same projection as Figure 5.3 with the flows between various hubs and nodes emphasized at the expense of the cities. But this inevitably is a snapshot, which makes it difficult to think about those switched off, so to speak, from the network as they alter in form and shape. It is also difficult to represent in this form the idea of transnational élites which owe more to the connections between cities than to any one particular city. How would you begin to represent global cities as a *process*, rather than a series of discrete places? This, I should add, is something for you to *think* about, not necessarily to resolve or map out. ◆

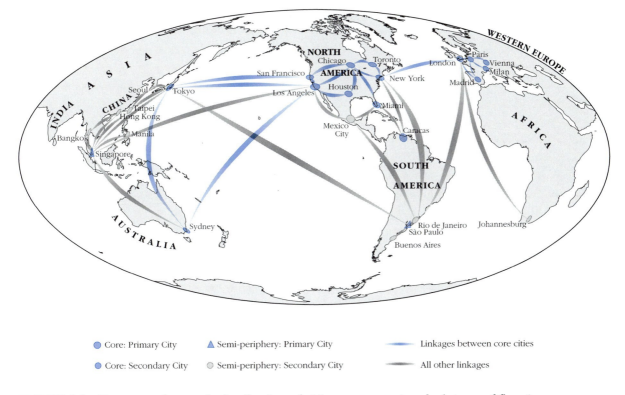

FIGURE 5.5   *Not so much a ranked collection of cities, more a networked space of flows?*

Whatever you may think of Castells' argument, of the assumptions he makes around technology, for example, it should be quite clear that in contrast to Sassen's account

- it is the forms of interaction and exchange which take place *through* the networks which are thought to be constitutive of city power. Élite groupings in cities such as New York or Tokyo mobilize their power through the networks, through the co-present interaction which enables them to shrink the space and time between each other and to construct closer, integrated ties and connections.

In common with Sassen, Castells recognizes that only a certain number of cities are at any one time in a position to stabilize and direct the flows of economic transactions, by virtue of their information-based production complexes. Where the two accounts radically diverge, however, is over the source of the economic power and influence of global cities. For Castells, power is concentrated in the networked space of flows, whereas for Sassen it is concentrated in those groups who exercise the command-and-control functions embedded in global cities.

In other words, it comes back to the question of whether the networks themselves 'generate' cities as sites of power through their interconnections *or* whether cities 'run' the networks through their concentration of resources and expertise.

4 *City networks or networks of cities?*

The formulation of the question in this way probably overstates the differences between the two accounts, although the usefulness of the distinction is that it does draw attention to two of the major ways in which cities, networks and power have been conceived. If you recall our initial concern to understand why power and influence appear to concentrate in relatively few cities worldwide, then it matters whether or not we think about power as something which 'belongs' to particular cities and not others. To be fair to Sassen, she is aware of the settled formation of power in global cities, but it is easy to slip into the idea that 'powerful' cities do things or are influential in certain ways simply because they possess the necessary solid capacity. To think in this way, where certain cities and their élite groupings are the 'centres' of power, runs the risk of playing down the network of connections which, Castells argues, sustains them. More to the point, it can lead to the impression that networks are simply 'things' which are 'out there' waiting to be tapped into by élite groupings from their city locations. Once a system of cities is in place of, say, the kind represented in Figure 5.3, on this view all that remains is for cities to tap into it and to extend their power and influence across it.

This is not the way it works, however. For one thing, networks change over time and with them the connections drawn between particular cities. As **Massey (1999)** has demonstrated in relation to Tenochtitlán/Mexico City, cities can become disconnected from the networks which previously sustained them and inserted into new arrangements and sets of relationships. If we think of networks as constitutive of city power, then

- any change to those networks will affect the make-up of cities and their prospects. As networks change, some cities may find themselves by-passed by a new pattern of flows to the extent that their power fades and with it their reach and influence. Others may be faced with the task of reconstructing their links to particular networks or of attracting the resources necessary to construct new sets of relationships and connections.

In this section, we will look first at the *dynamic* nature of networks which, because of their frequently shifting pattern of relationships, have the potential to 'make' some cities more or less powerful than others. The city-ness of power in this context, it should be noted, owes as much to the connections mobilized by cities, as it does to their ability to stabilize the resources which (may) flow through them. Following that, we press home the point that cities are frequently sites of more than one kind of power, despite the singular logic implied in much of the global cities thinking. Different kinds of networks, with different kinds of power relations, co-exist in cities, some with considerable cross-over, others with little, but together they are a testament to the *plurality* of networks and rhythms which characterize cities of power.

4.1   INSIDE/OUTSIDE NETWORKS

In the discussion of Castells' ideas, we drew attention to his argument that global cities – or, rather, powerful cities – would not exist but for the connections which constitute them. On this view, when certain connections are not sustained or when they are severed or played down, perhaps because of a pooling of different kinds of resources elsewhere in a network, a once-'powerful' city may find itself sidelined or displaced. In considering this point, however, we do not have to restrict our thinking to big, economically 'powerful' cities. The fluctuating or rather unsettled fortunes of cities such as Moscow, Berlin or Prague at the present time, for example, come to mind.

FIGURE 5.6   *Moscow's changing regimes: from Lenin to the money markets*

All manner of people, practices and resources came together in a place like Moscow under the Soviet regime, for instance, but once that empire of relationships collapsed in the late 1980s, the city and its diverse groupings had to re-negotiate a different set of futures for themselves in rather different kinds of networks and surroundings. In this context, the ability to make new connections, to construct new sets of relationships, has to be understood alongside an appreciation of those networks already constructed, both within Russia and beyond its boundaries. Attempts by various kinds of economic groupings in post-socialist cities such as Moscow to enhance their power and influence by forging alliances through western networks requires the connections to be made, not simply tapped into (see Gritsui, 1997).

And this is no simple matter. In the first place it involves a re-assessment of the skills and resources pooled under a previous regime which drew its legitimacy from the state plan rather than the market. In addition, it involves the negotiation and manoeuvre within a different regime of interests and practices to persuade others that particular economic resources – money, capital, information, goods and the like – should be channelled through their cities and settled there. There are no guarantees in such processes and the now 'weak' power of those on the outside looking in to established, yet changing, networks is an indication of the ways in which hierarchies may themselves be rendered unstable over time.

Much the same can perhaps be said of the changing fortunes of Hong Kong as a global city. Its strategic location in East Asia as a node within the global networks of money, property and information has, until recently, ensured that its dense clustering of communications, expertise and people with the right connections has been able to respond to the immediacy of the markets and to direct the regional flows of business and finance through Hong Kong. Almost overnight, however, the city's position within such global networks has been unsettled by its re-connection in July 1997 to the Chinese mainland and its new-found status as a Special Administrative Region of the People's Republic of China. It is perhaps premature to draw any firm conclusions as to the long-term significance of this change in political status for Hong Kong, but if Castells is accurate in his observation that the points of attachment of city-based élites are more to the networks than to the cities themselves, then Ronald Skeldon may well be correct to point to the forthcoming provincial status of the city and the subsequent loss of economic power.

ACTIVITY 5.5   You may find it easier to judge for yourselves as to whether Hong Kong's power and influence is on the wane following its shift in political status by casting a quick eye over Skeldon's essay, 'Hong Kong: colonial city to global city to provincial city?' – Reading 5B at the end of the chapter.

The title of the essay is clearly indicative of Skeldon's thinking but, on the basis of what you have gleaned from section 3 about connections and disconnections between cities, make up your own mind about whether or not Hong Kong has, in Castells' words, been 'switched off' the global network.

For this, you will have to consider the evolving policies of the Chinese government in relation to the outside world, as well as the potential mobility of the city-based élites to effect change. ◆

It is, as Skeldon acknowledges, obviously impossible to predict the future trajectory of Hong Kong, with much depending on the settled nature of the city's people, expertise and resources. Clearly there is a sense that the power of Hong Kong is related to its openness to wider, global networks and if this is curtailed then the status of a provincial city may beckon. As such, the likes of Hong Kong, or Moscow for that matter, will have known what it was like to be one part of a powerful network of cities. In the latter case, it was the evaporation of power from the socialist network of cities which, to a large degree, prompted the re-assessment of its mix of resources and practices. For a number of other cities, however, this assessment has taken place against a quite different backdrop where the provision of the requisite infrastructure is regarded as the starting-point for entry into the economically powerful networks. In other words, if a city wants the 'space of flows' to go through its hub, then it has to be open to the outside and to the mix of relations that attracts such flows.

FIGURE 5.7   *Downtown Kuala Lumpur – the basis of a modern city in the making?*

Kuala Lumpur is a case in point. Disentangled from the declining networks of British Imperial rule, Malaysia, largely through the ambitious programme of its government and Prime Minister, Mahathir Mohamed, has set itself the task of establishing Kuala Lumpur as the 'information hub' of East Asia in the twenty-first century. The initial aim is to construct the city as a vibrant 'centre', with all the communications capabilities necessary to interact with distant others in simultaneous time and so, in turn, place the city on the regional map as *the* point at which all the lines of information and knowledge intersect. The programme involves an explicit vision of the future of Malaysia, in which the trajectory of Kuala Lumpur is firmly rooted in the information age. In common with Singapore, it shares an aspiration to be a 'smart city', but in this case the initiative is starting from scratch with the construction of a *Multimedia Super Corridor* (MSC) that stretches 50 kilometres south from the twin Petronas Towers in Kuala Lumpur. Extract 5.3 is taken from a Malaysian government publication on the MSC and gives an idea of the scope of vision involved. Figure 5.8 provides an indication of the physical scale of Phase 1 of the project.

As you will have noted, the 'greenfield' corridor will play host to two purpose-built cities, Putrajaya and Cyberjaya. The former is designed as a paperless

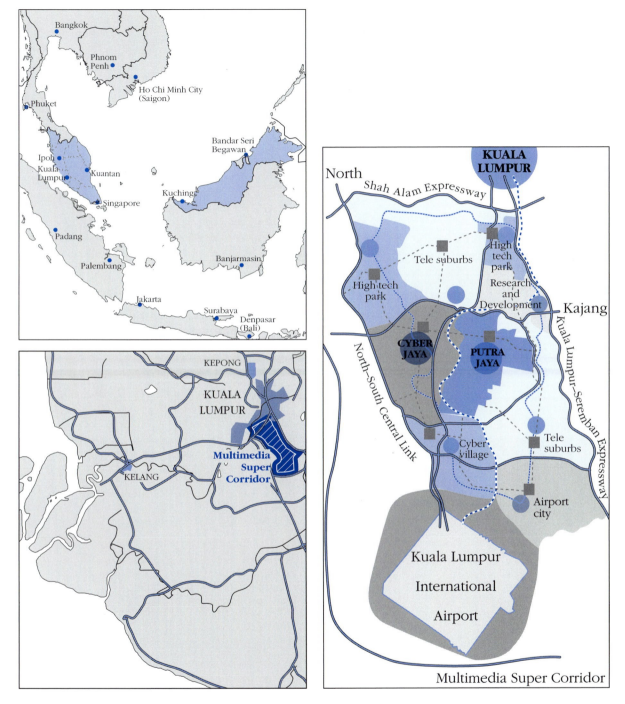

FIGURE 5.8   *The Multimedia Super Corridor planned for Kuala Lumpur*

EXTRACT 5.3
Multimedia Development Corporation: 'Unlocking the full potential of the information age'

Welcoming the Information Age

Malaysia welcomes the advent of the Information Age with its promise of a new world order where information, ideas, people, goods and services move across borders in the most cost-effective and liberal ways.

As traditional boundaries disappear, and as companies, capital, consumers, communications and cultures become truly global, new approaches and attitudes to business are required.

Malaysia upholds the virtues of the new world order, believing that the globe is collectively moving towards a 'century of the world', a century of worldwide peace and shared prosperity among nations.

Malaysia has chosen to be open and pragmatic in dealing with change, and is committed to working with other world citizens to encourage creativity, innovation and entrepreneurship.

Creating the Multimedia Super Corridor

As a first step, Malaysia has created the Multimedia Super Corridor – a world-first, world-class act – to help companies of the world test the limits of technology and prepare themselves for the future. The MSC will also accelerate Malaysia's entry into the Information Age, and through it, help actualise Vision 2020.

The MSC will bring together, for the first time ever, an integrated environment with all the unique elements and attributes necessary to create the perfect global multimedia climate. It is a length of greenfield 'corridor', 15 kilometres wide and 50 kilometres long, that starts from the Kuala Lumpur City Centre (KLCC), itself an intelligent precinct, which houses the world's tallest buildings – down south to the site of the region's largest international airport, the Kuala Lumpur International Airport (KLIA), which will be commissioned in 1998.

Two of the world's first Smart Cities are being developed in the Corridor: *Putrajaya*, the new seat of government and administrative capital of Malaysia where the concept of electronic government will be introduced; and *Cyberjaya*, an intelligent city with multimedia industries, R&D centres, a Multimedia University and operational headquarters for multinationals wishing to direct their worldwide manufacturing and trading activities using multimedia technology.

Set to deliver a number of sophisticated investment, business, R&D and lifestyle options, the MSC will be:

Unlocking the Full potential of the Information Age

• *A vehicle for attracting world-class technology-led companies to Malaysia, and developing local industries*
• *A Multimedia Utopia offering a productive intelligent environment within which a multimedia value chain of goods and service will be produced and delivered across the globe*
• *An island of excellence with multimedia-specific capabilities, technologies, infrastructure, legislation, policies, and systems for competitive advantage*
• *A test bed for invention, research, and other ground-breaking multimedia developments spearheaded by seven multimedia applications*
• *A global community living on the leading-edge of the Information Society*
• *A world of Smart Homes, Smart Cities, Smart Schools, Smart Cards and Smart Partnerships*

The Multimedia Development Corporation envisions a 20-year time-frame for the full implementation and execution of the MSC, when Malaysia will have achieved leadership in the Information Age.

There will be three phases of activity:

Phase I: Under this phase, the MDC will successfully create the Multimedia Super Corridor, attract a core group of world-class companies, launch seven Flagship Applications, put in place a world-leading framework of cyberlaws, and establish Cyberjaya and Putrajaya as world-first intelligent cities.

Phase II: The MDC envisages that during this period, it will link the MSC to other cybercities in Malaysia and the world. It will create a web of corridors and establish a second cluster of world-class companies. It will also set global standards in flagship applications, champion cyberlaws within the global society, and establish a number of intelligent globally-linked cities.

Phase II: During this final phase, it is expected that Malaysia will be transformed into a knowledge-based society – being a true global test bed for new multimedia and IT applications and a cradle for a record number of multimedia companies. It would have a cluster of intelligent cities linked to the global information super highway, and become the platform for the International Cybercourt of Justice.

Source: Multimedia Development Corporation, 1998, pp.10–11

administrative capital and future seat of the government, whereas the latter is intended to operate as a site of innovation and creativity around multimedia applications. In particular, Cyberjaya is expected to act as some kind of catalyst, attracting world-class talent from overseas and putting in place the mix of resources necessary to act as an intense hub of information networks. Linking all the sites together, from Kuala Lumpur City Centre in the north southward, is a high-speed, fibre-optic backbone which, in many ways, makes the whole project not dissimilar to the decentred collection of nodes that Sassen spoke about in relation to the 'new centrality' of powerful cities.

At the time of writing (1998), much of the fibre-optic infrastructure is in place, a framework of 'cyberlaws' established, and an optimism present which can lead one to believe that the completion of Phase 1 is just a matter of time and convenience. In practice, Kuala Lumpur's entry into the economically powerful network of cities is far from assured and the construction of the links fraught with cultural, political and economic uncertainty, not least because Mahathir has openly rejected a simple, 'western-style' notion of economic success based on a neo-liberal 'network of ideas' which currently underpins global city formation (see Khoo Boo Teik, 1996). In the following chapters, we will consider the significance of this network of neo-liberal ideas for cities across the globe, but for the moment I would like to draw from the Malaysian example two general points.

The first point that you should bear in mind is that the mere concentration of resources, as noted in section 2, is not sufficient for cities to be able to mobilize power through the networks, nor to become the centre of powerful regional geographies. In fact, the mere presence of a state-of-the-art infrastructure guarantees nothing. Do you remember the earlier point about the significance of cumulative advantages built up by global cities over time? Well, such legacies of power broadly ensure that the 'space of flows' does not alter its pattern or direction simply in response to the lure of modern infrastructure. It is not an accident or a whim of history that cities such as New York have enjoyed power on a settled basis for a considerable period of time, accumulating advantages from an *existing* position of strength and influence.

Equally, and this is the second point, that does not mean to say that the network of powerful cities is fixed in its formation. Cities do grow (as well as shrink) in power. Post-war Tokyo is a good illustration of a city whose economic fortunes grew rapidly in recent history, and so too – up to the present – is Hong Kong. But neither of them, it should be stressed, started out from the disadvantaged position of a Bamako or a Calcutta or a Gaberones prior to their shift in fortune. Kuala Lumpur operates under a specific set of historical preconditions and economic legacies which leave open the possibility of powerful city formation as a process – as an active strategy. Whilst a state-of-the art infrastructure is part of that process, it is, however, the wider – predominantly western – network of connections and the ties forged through cities which will circumscribe its 'success'.

FIGURE 5.9   *Gaberones: at a historical disadvantage in terms of economic power and influence*

This relates to a further observation which can be drawn from Mahathir's stance in response to 'the West's' equation of neoliberal power with success. It is the recognition that there are other kinds of thinking, other kinds of values, and indeed other sources of power within cities besides those which have fuelled recent market-driven success. It is to these juxtapositions of power and their effects that we now turn.

4.2 JUXTAPOSITIONS OF POWER

So far in this chapter we have made little reference to the diversity of power arrangements within cities, signalled in section 1.1, or to their different rhythms and connections. In part, this simplification was merely pragmatic, so that we could consider the constitutive power of city networks without losing ourselves in a maze of detail. But the cost of this was to gloss over the possibility that,

● within cities, different networks of power may cross over one another with consequent effects, or that they may simply co-exist with little or no effect.

Mahathir's embrace of the economic networks of a coming information age, yet rejection of a 'western' notion of success based on individualism, speculation and 'vague' ethical standards, is symptomatic of the former possibility.

Leaving the significance of the Malaysian Prime Minister to one side, if that is indeed possible, the co-existence of modern economic networks and Islamic networks within Kuala Lumpur has produced what Joel Kahn (1997) has called an Islamic trajectory to modernization. Alongside an interesting mix of temporalities and varied rhythms in the daily life of Kuala Lumpur, the cross-cutting élites of the city are the foci of quite different geographies beyond the city. On the one side, there is the familiar western-dominated economic network of cities and, on the other, there is the Islamic network of connections which stretches as far as Jeddah and the Gulf States and takes in parts of the Asian Muslim world. More importantly, the difference between the two networks does not amount to a simple economic–cultural divide, nor a straightforward modern–traditional split. Kuala Lumpur has its share of Islamic banks, for instance, which operate along Islamic lines and combine principles of modern money management in so far as they do not conflict with their religious codes. The extension of credit facilities, for example, whether equity financing, leasing arrangements, sale and purchase, or trade financing, is entirely feasible within the bounds of an Islamic profit-orientated system which, above all, prohibits interest (Khoo Boo Teik, 1996).

The co-existence of the two sets of relationships and conventions, therefore, and their potential cross-over and subsequent effects, has set Kuala Lumpur on a rather different trajectory from that of simply another 'information city' in the western mould. To what extent the city and its multimedia corridor will fuse elements of the two networks remains, for now, an open question, and one that may at a later date bring into sharp relief the relative power embedded within the two contrasting networks. But even an answer to this question would not fully convey the plurality of Kuala Lumpur.

For another set of networked relationships also runs through Kuala Lumpur, complete with its own powerful nodes and influential associations. Yet one whose beneficiaries are far less attached to the location than they are to the network itself. The network in question is that forged over time by the overseas Chinese, some 56 million who are spread across the cities of the Pacific rim and beyond. Far less obvious in the demonstration of their

FIGURE 5.10 *Kuala Lumpur as the focus of both 'modern' and Islamic influences: (from top clockwise) the Jamik Mosque, the railway station, the National Mosque and the Malayan Bank*

economic power compared with other groupings with similar assets and influence, the overseas Chinese, as you will see from Chapter 7, remain a potent, yet understated, city-based force.

In fact, it is not uncommon in the singular images presented of powerful cities to gloss over or to miss the presence of alternative networks of influence and authority embedded within them. If we think back to Sassen's account of the three major global cities discussed earlier – New York, London and Tokyo – it is the economic power of the high-level service professionals and their connections which receives the bulk of her attention. In her book, *The Global City* (1991), however, she also draws attention to the large concentrations of immigrant workers, many of whom are drawn from the less developed economies, who appear to occupy at best a support role in relation to the dynamism of global cities. Seen from the vantage point of global financial power, many of these migrant workers – the Latinos, the Chinese, and the Koreans for example – are perceived as merely 'powerless'. But this is largely to caricature their presence. Whilst for Sassen such migrant groups are relegated to the 'downside' of growth in global cities, the co-existence of different networks and trajectories speaks to a rather different 'power geometry'. This was apparent, for example, in the mix of international groupings outlined by Skeldon which gave Hong Kong much of its global vibrancy and, indeed, intensity. Viewed from a different angle, the lines of connection and power embedded in a variety of ethnic diasporas tell a story of networks stretching from city to city across national borders transmitting remittances, people, ideas and the like along their length and breadth. This, then, is a different kind of dynamism and a rather different 'space of flows' from that considered earlier, which although potentially less hierarchical, is no less constitutive of power relationships than other networks considered thus far.

Present-day transnational ethnic communities may, through ties of association rather than lines of authority, pool resources and ideas to produce the kinds of innovative designs or tastes which may ultimately lead, for example, to forms of cultural power similar to those mentioned in section 1. At a minimum, such communities would add to the distinctive international mix of relations and cosmopolitan blends which are almost a prerequisite for global cities. The 'something extra' that we spoke about in relation to 'what makes a city powerful', where the combination and juxtaposition of ideas, people, and information delivers the intensity of relations which underlies power, does not arise solely among existing groups of professional influence and authority. If that were so, there would be little that would surprise us in global terms or little to consider outside the power and influence of money and markets. The question of powerful cities would, then, largely be closed, or at least answered in a rather predictable fashion. It is, however, the mix and diversity of power arrangements in a city, as Chapter 1 in particular emphasized, which frequently provides for an openness to change and the possibility of different outcomes. Although, as we have observed, these are not always openly registered.

ACTIVITY 5.6   Bearing these points in mind, it might be a worthwhile exercise to turn back to the clusters of interaction outlined in Figure 5.4 for the cities of New York, London and Tokyo. I would like you now to give some thought to the possibility that there may be other sites of activity, other groupings, with their own extensive networks of influence across these cities which are not marked on the maps. That is, different groups, different skills, different cultural influences, and so forth, which do not register in the singular images of these cities, yet may well form part of their dynamism. In London, for instance, the centres of fashion, art and design clustered in parts of the east end of the city would not register on such maps. Yet their role as centres of influence and their involvement, especially across mainland European cities, in the circulation of ideas and aesthetic forms, should not be understated, either culturally or economically.

If you are uncertain about those particular cities, think back to Chapter 3, to the account of Birmingham developed by Linda McDowell and to the possible transnational ethnic groupings involved in the city at large or, for that matter, to any major city with which you have some degree of familiarity. What plurality of power arrangements comes readily to mind, beyond that of simply the dominant economic clusters? ◆

Don't worry if you found it rather difficult to move beyond the singular image of powerful cities. Such representations are, after all, powerful in themselves. If we turn back to section 1.1, we can see from the example of Paris the plurality of networks which intersect within that city, not all of which readily spring to mind in the urban imagination. But the point here is more than simply to draw attention to the proximity and juxtaposition of different sites of influence across cities. It is also to recognize that these different networks exert their influence in different ways. The claim that part of some global cities act as 'command centres' in the global economy, for instance, is a rather different claim from that which states that cities of cultural power draw their influence through imitation and copy. Equally, the tendency to see hierarchical rather than associational relations as the hallmark of these networks limits our understanding of what power can do. People do not always lose out if they are on the receiving end of relationships of power (see **Allen, 1999**). Those groups within cities, including Castells' transnational élites, who exert some kind of influence or control over what happens elsewhere may do so through collaboration as much as through competition.

In the next and penultimate chapters, you will be introduced to each of these possibilities and to what is implied by the term 'global reach and influence'.

5 *Conclusion*

In this chapter, we have been concerned to show how the power of cities is something that is produced over time in various settled formations. Far from being the 'property' of certain cities, power may be built-up over time yet lost overnight. Any claim to permanency, which suggests that power 'belongs' to some cities and not to others on a long-term basis is merely the impression gained through successive mobilizations. As such, the city-ness of power that we have spent much time exploring is best understood as the outcome of the connections mobilized by groups within cities, whether they be based on cultural, political or economic networks of relations. Moreover, if connections and disconnections are the stuff of the networks, then an ability to settle the resources which flow through them is a prerequisite of power and influence. We have seen that this is so from the outset, with Paris' role as an established, influential centre of fashion outlined, right through to the final section, where the strategic attempts of élites in Kuala Lumpur to enable the city to become a powerful, 'information hub' in East Asia, was considered.

We have also seen that power and influence across the globe tend to concentrate; that the intense mix of relationships that has the potential to make a city powerful is remarkably uneven geographically. This is the case whether the issue under consideration is the apparent singularity of economic power, concentrated as it is in so few global cities, or the proximity and juxtaposition of different kinds of power across cities. To put it succinctly, the points at which the networks intersect and the lines of influence meet are few and far between.

But, as we know, this is not solely because key cities 'run' the networks of connections through their concentration of resources. Rather, it is also because the networks are themselves *constitutive* of power that any shift in the make-up of globally networked relationships alters the connections drawn between cities. There is a two-way process involved, where cities at one time on the inside of powerful networks may find themselves by-passed by an altogether new set of connections and relationships. As we have had cause to note, city powers may be mobilized through networked relationships, but few cities have the capacity to attract and to settle the flows which circulate around and across them.

References

Allen, J. (1999) 'Spatial assemblages of power: from domination to empowerment' in Massey, D., Allen, J. and Sarre, P. (eds) *Human Geography Today*, Cambridge, Polity Press.

Amin, A. and Thrift, N. (1995) *Globalization, Institutions, and Regional Development*, Oxford, Oxford University Press.

Castells, M. (1996) *The Information Age: Economy, Society and Culture,* Volume 1: *The Rise of the Network Society,* Oxford, Basil Blackwell.

Friedmann, J. (1986) 'The world city hypothesis', *Development and Change*, vol.17, no.1, pp.69–84.

Friedmann, J. (1995) 'Where we stand: a decade of world city research' in Knox, P.L. and Taylor, P.J. (eds).

Friedmann, J. and Wolff, G. (1982) 'World city formation: an agenda for research and action', *International Journal of Urban and Regional Research*, vol.6, pp.309–43.

Gritsui, O. (1997) 'The economic restructuring of Moscow in the international context', *GeoJournal*, vol.42, no.4, pp.341–7.

Hannerz, U. (1996) *Transnational Connections: Culture, People, Places,* London and New York, Routledge.

Kahn, J. (1997) 'Demons, commodities and the history of anthropology' in Carrier, J.G. (ed.) *Meanings of the Market*, Oxford and New York, Berg.

Khoo Boo Teik (1996) *Paradoxes of Mahathirism*, Oxford, Oxford University Press.

Knox, P.L. and Taylor, P.J. (eds) (1995) *World Cities in a World-System*, Cambridge, Cambridge University Press.

Lipovetsky, G. (1994) *The Empire of Fashion: Dressing Modern Democracy,* Princeton, NJ, Princeton University Press.

Marshall, D. 1973) *Industrial England, 1776–1851*, London, Routledge.

Massey, D. (1999) 'Cities in the world' in Massey, D., *et al.* **(eds).**

Massey, D., Allen, J. and Pile, S. (eds) (1999) *City Worlds*, **London, Routledge/The Open University (Book 1 in this series).**

Multimedia Development Corporation (1998) *Unlocking the Full Potential of the Information Age*, Kuala Lumpur, Malaysia, MDC.

National Computer Board (1997) *Information Technology 2000, Singapore Unlimited*, Singapore.

Pile, S. (1999) 'What is a city?' in Massey, D., Allen, J. and Pile, S. (eds).

Richardson, R. (1997) *Towards the Information Society in Southeast Asia? Experiences in Singapore and Malaysia*, A Briefing Report to Northern Informatics, Newcastle Upon Tyne, Centre for Urban and Regional Development Studies, University of Newcastle Upon Tyne.

Sassen, S. (1991) *The Global City: New York, London and Tokyo*, Princeton, NJ, Princeton University Press.

Sassen, S. (1995) 'On concentration and centrality in the global city' in Knox, P.L. and Taylor, P.J. (eds).

Storper, M. and Salais, R. (1997) *Worlds of Production: The Action Frameworks of the Economy*, Cambridge, MA, Harvard University Press.

Skeldon, R. (1997) 'Hong Kong: colonial city to global city to provincial city?', *Cities*, vol.14, no.5, pp.265–71.

Taylor, P.J. (1997) 'Hierarchical tendencies amongst world cities: a global research proposal', *Cities*, vol.14, no. 6, pp.323–32.

READING 5A
Saskia Sassen: 'On concentration and centrality in the global city'

At the global level, a key dynamic explaining the place of major cities in the world economy is that they concentrate the infrastructure and the servicing that produce a capability for global control. The latter is essential if geographic dispersal of economic activity – whether factories, offices, or financial markets – is to take place under continued concentration of ownership and profit appropriation. This capability for global control cannot simply be subsumed under the structural aspects of the globalization of economic activity. It needs to be produced. It is insufficient to posit, or take for granted, the awesome power of large corporations.

By focusing on the production of this capability we add a neglected dimension to the familiar issue of the power of large corporations. The emphasis shifts to the practice of global control: the work of producing and reproducing the organization and management of a global production system and a global market-place for finance, both under conditions of economic concentration. Power is essential in the organization of the world economy, but so is production: including the production of those inputs that constitute the capability for global control and the infrastructure of jobs involved in this production. This allows us to focus on cities and on the urban social order associated with these activities.

Much analysis and general commentary on the global economy and the new growth sectors does not incorporate these multiple dimensions. Elsewhere I have argued that what we could think of as the dominant narrative or mainstream account of economic globalization is a narrative of eviction (Sassen, 1993, 1994a). Key concepts in the dominant account – globalization, information economy, and telematics – all suggest that place no longer matters and that the only type of worker that matters is the highly educated professional. This account privileges the capability for global transmission over the concentrations of built infrastructure that make transmission possible; information outputs over the workers producing those outputs, from specialists to secretaries; and the new transnational corporate culture over the multiplicity of cultural environments, including reterritorialized immigrant cultures, within which many of the 'other' jobs of the global information economy take place. In brief, the dominant narrative concerns itself with the upper circuits of capital, not the lower ones, and with the global capacities of major economic actors, not the infrastructure of facilities and jobs underlying those capacities. This narrow focus has the effect of evicting from the account the place-boundedness of significant components of the global information economy.

The intersection of globalization and the shift to services

To understand the new or sharply expanded role of a particular kind of city in the world economy since early 1980s, we need to focus on the intersection of two major processes. The first is the sharp growth in the globalization of economic activity; this has raised the scale and the complexity of transactions thereby feeding the growth of top-level multinational headquarter functions and the growth of advanced corporate services. It is important to note that even though globalization raises the scale and complexity of these operations, they are also evident at smaller geographic scales and lower orders of complexity, as is the case with firms that operate regionally. Thus while regionally oriented firms need not negotiate the complexities of international borders and the regulations of different countries, they are still faced with a regionally dispersed network of operations that requires centralized control and servicing.

The second process we need to consider is the growing *service intensity* in the organization of all industries (Sassen, 1991, pp.166–8). This has contributed to a massive growth in the demand for services by firms in all industries, from mining and manufacturing to finance and consumer services. Cities are key sites for the production of services for firms. Hence the increase in service intensity in the organization of all industries has had a significant growth effect on cities in the 1980s. It is important to recognize that this growth in services for firms is evident in cities at different levels of a nation's urban system. Some of these cities cater to regional or subnational markets; others cater to national markets, and yet others cater to global markets. In this context, globalization becomes a question of scale and added complexity.

The key process from the perspective of the urban economy is the growing demand for services by firms in all industries, and the fact that cities are preferred production sites for such services, whether at the global, national, or regional level. As a result we see in cities the formation of a new *urban economic core* of banking and service activities that comes to replace the older, typically manufacturing oriented, core.

In the case of cities that are major international business centres, the scale, power, and profit levels of this new core suggest that we are seeing the formation of a new urban economy. This is so in at least two regards. First, even though these cities have long been centres for business and finance, since the late 1970s there have been dramatic changes in the structure of the business and financial sectors, as well as sharp increases in the overall magnitude of these sectors and their weight in the urban economy. Second, the ascendance of the new finance and services complex, particularly international finance, engenders what may be regarded as a new economic regime, that is, although this sector may account for only a fraction of the economy of a city, it imposes itself on that larger economy. Most notably, the possibility for superprofits in finance has the effect of devalorizing manufacturing insofar as the latter cannot generate the superprofits typical in much financial activity.

This is not to say that everything in the economy of these cities has changed. On the contrary, they still show a great deal of continuity and many similarities with cities that are not global nodes. Rather, the implantation of global processes and markets has meant that the internationalized sector of the economy has expanded sharply and has imposed a new valorization dynamic: that is, a new set of criteria for valuing or pricing various economic activities and outcomes. This has had devastating effects on large sectors of the urban economy. High prices and profit levels in the internationalized sector and its ancillary activities, such as top-of-the-line restaurants and hotels, have made it increasingly difficult for other sectors to compete for space and investments. Many of these other sectors have experienced considerable downgrading and/or displacement, as, for example, neighbourhood shops tailored to local needs are replaced by upmarket boutiques and restaurants catering to new high-income urban élites.

Though at a different order of magnitude, these trends also became evident during the late 1980s in a number of major cities in the developing world that have become integrated into various world markets: São Paulo, Buenos Aires, Bangkok, Taipei, and Mexico City are only a few examples. Also, here the new urban core was fed by the deregulation of financial markets, ascendance of finance and specialized services, and integration into the world markets. The opening of stock markets to foreign investors and the privatization of what were once public sector firms have been crucial institutional arenas for this articulation. Given the vast size of some of these cities, the impact of this new core on the broader city is not always as evident as in central London or Frankfurt, but the transformation is still very real.

...

The formation of a new production complex

According to standard conceptions about information industries, the rapid growth and disproportionate concentration of producer services in central cities should not have happened. Because they are thoroughly embedded in the most advanced information technologies, producer services could be expected to have locational options that bypass the high costs and congestion typical of major cities. But cities offer agglomeration economies and highly innovative environments. The growing complexity, diversity, and specialization of the services required has contributed to the economic viability of a free-standing specialized service sector.

The production process in these services benefits from proximity to other specialized services. This is especially the case in the leading and most innovative sectors of these industries. Complexity and innovation often require multiple highly specialized inputs from several industries. The production of a financial instrument, for example, requires inputs from accounting, advertising, legal expertise, economic consulting, public relations, designers, and printers. The particular characteristics of production of these services, especially those involved in complex and innovative operations, explain their pronounced concentration in major cities. The commonly heard explanation that high-level professionals require face-to-face interactions, needs to be refined in several ways. Producer services, unlike other types of services, are not necessarily dependent on spatial proximity to the consumers, i.e. firms served. Rather, economies occur in such specialized firms when they locate close to others that produce key inputs or whose proximity makes possible joint production of certain service offerings. The accounting firm can service its clients at a distance, but the nature of its service depends on proximity to specialists, lawyers, and programmers. Moreover, concentration arises out of the needs and expectations of the people likely to be employed in these new high-skill jobs, who tend to be attracted to the amenities and lifestyles that large urban centres can offer. Frequently, what is thought of as face-to-face communication is actually a production process that requires multiple simultaneous inputs and feedbacks. At the current stage of technical development, immediate and simultaneous access to the pertinent experts is still the most effective way, especially when dealing with a highly complex product. The concentration of the most advanced telecommunications and computer network facilities in major cities is a key factor in what I refer to as the production process of these industries.[1]

Further, time replaces weight in these sectors as a force for agglomeration. In the past, the pressure of the weight of inputs from iron ore compared with unprocessed agricultural products was a major constraint pushing toward agglomeration in sites where the heaviest inputs were located. Today, the acceleration of economic transactions and the premium put on time have created new forces for agglomeration. Increasingly this is not the case in routine operations. But where time is of the essence, as it is today in many of the leading sectors of these industries, the benefits of agglomeration are still extremely high – to the point where it is not simply a cost advantage, but an indispensable arrangement.

This combination of constraints suggests that the agglomeration of producer services in major cities actually constitutes a production complex. This producer services complex is intimately connected to the world of corporate headquarters; they are often thought of as forming a joint headquarters–corporate services complex. But in my reading, we need to distinguish the two. Although it is true that headquarters still tend to be disproportionately concentrated in cities, over the last two decades many have moved out. Headquarters can indeed locate outside cities, but they need a producer services complex somewhere in order to buy or contract for the needed specialized services and financing. Further, headquarters of firms with very high overseas activity or in highly innovative and complex lines of business tend to locate in major cities. In brief, firms in more routinized lines of activity, with predominantly regional or national markets, appear to be increasingly free to move or install their headquarters outside cities. Firms in highly competitive and innovative lines of activity and/or with a strong world market orientation appear to benefit from being located at the centre of major international business centres, no matter how high the costs.

Both types of firms, however, need a corporate services complex to be located somewhere. Where this complex is located is probably increasingly unimportant from the perspective of many, though not all, headquarters. From the perspective of producer services firms, such a specialized complex is most likely to be in a city rather than, for example, a suburban office park. The latter will be the site for producer services firms but not for a services complex. And only such a complex is capable of handling the most advanced and complicated corporate demands.

Elsewhere (Sassen, 1994a), a somewhat detailed empirical examination of several cities served to explore different aspects of this trend towards concentration. Here there is space only for a few observations. The case of Miami, for instance, allows us to see, almost in laboratory-like fashion, how a new international corporate sector can become implanted in a site. It allows us to understand something about the dynamic of globalization in the current period and how it is embedded in place. Miami has emerged as a significant hemispheric site for global city functions, although it lacks a long history as an international banking and business centre as is typical of such global cities as New York or London.

The case of Toronto, a city whose financial district was built up only in recent years, allows us to see to what extent the pressure towards physical concentration is embedded in an economic dynamic rather than simply being the consequence of having inherited a built infrastructure from the past, as one would think was the case in older centres such as London or New York. But Toronto also shows that it is certain industries in particular that are subject to the pressure towards spatial concentration, notably finance and its sister industries (Gad, 1991).

The case of Sydney illuminates the interaction of a vast continental economic scale and pressures towards spatial concentration. Rather than strengthening the multipolarity of the Australian urban system, the developments of the 1980s – increased internationalization of the Australian economy, sharp increases in foreign investment, a strong shift towards finance, real estate, and producer services – contributed to a greater concentration of major economic activities and actors in Sydney. This included a reduction in such activities and actors in Melbourne, for a long time the centre of commercial activity and wealth in Australia (Daly and Stimson, 1992).

Finally, the case of the leading financial centres in the world today is of continued interest, since one might have expected that the growing number of financial centres now integrated into the global market would have reduced the extent of concentration of financial activity in the top centres.[2] One would further expect this outcome given the immense increases in the global volume of transactions. Yet the levels of concentration remain unchanged in the face of massive transformations in the financial industry and in the technological infrastructure upon which this industry depends.

For example, international bank lending grew from US $1.89 trillion in 1980 to US $6.24 trillion in 1991 – a fivefold increase in a mere ten years. Three cities (New York, London, and Tokyo) accounted for 42 per cent of all such international lending in 1980 and for 41 per cent in 1991, according to data from the Bank of International Settlements, the leading institution worldwide in charge of overseeing banking activity. There were compositional changes: Japan's share rose from 6.2 per cent to 15.1 per cent; the UK's share fell from 26.2 per cent to 16.3 per cent; while the US share remained constant. All increased in absolute terms. Beyond these three, Switzerland, France, Germany, and Luxembourg bring the total share of the top centres to 64 per cent in 1991, which is just about the same share these countries had in 1980. One city, Chicago, dominates the world's trading in futures, accounting for 60 per cent of worldwide contracts in options and futures in 1991.

The space economy of the centre

Today there is no longer a simple, straightforward relation between centrality and such geographic entities as the downtown, or the central business district. In the past, and up to quite recently in fact, the centre was synonymous with the downtown or the central business district. Today, the spatial correlate of the centre can assume several geographic forms. It can be the central business district, as it largely still is in New York City, or it can extend into a metropolitan area in the form of a grid of nodes of intense business activity, as we see in Frankfurt (Keil and Ronneberg, forthcoming).

Elsewhere (1991) I have argued that we are also seeing the formation of a transterritorial 'centre' constituted via digital highways and intense economic transactions; I argued then that New York, London, and Tokyo could be seen as constituting such a transterritorial terrain of centrality with regard to a specific complex of industries and activities. And at the limit we may see terrains of centrality that are disembodied, that lack any territorial correlate, that are in the electronically generated space we call cyberspace. Certain components of the financial industry, particularly the foreign currency markets, can be seen as operating partly in cyberspace.[3]

What is the urban form that accommodates this new economic core of activities? Three distinct patterns are emerging in major cities and their regions in the developed countries. First, in the 1980s there was a growing density of workplaces in the traditional urban centre associated with growth in leading sectors and ancillary industries. This type of growth also took place in some of the most dynamic cities in developing countries, such as Bangkok, Taipei, São Paulo, Mexico City, and, toward the end of the decade, Buenos Aires. Second, alongside this central city growth came the formation of dense nodes of commercial development and business activity in a broader urban region, a pattern not evident in developing countries. These nodes assume different forms: suburban office complexes, *edge cities*, and *exopoles*. Though in periperhal areas, these nodes are completely connected to central locations via state-of-the-art electronic means. Thus far, these forms are only rarely evident in developing countries, where

vast urban sprawl with a seemingly endless metropolitanization of the region around cities has been the norm. In developed countries, the revitalized urban centre and the new regional nodes together constitute the spatial base for cities at the top of transnational hierarchies. Third is the growing intensity in the local-ness or marginality of areas and sectors that operate outside that world market-oriented subsystem, and this includes an increase in poverty and disadvantage. This same general dynamic operates in cities with very diverse economic, political, social, and cultural arrangements.

A few questions spring to mind. One question here is whether the type of spatial organization characterized by dense strategic nodes over the broader region does or does not constitute a new form of organizing the territory of the 'centre', rather than, as in the more conventional view, an instance of suburbanization or geographic dispersal. Insofar as these various nodes are articulated through cyber-routes or digital highways they represent the new geographic correlate of the most advanced type of 'centre'. The places that fall outside this new grid of digital highways are peripheralized; one question here is whether it is so to a much higher degree than in earlier periods, when the suburban or non-central economic terrain was integrated into the centre because it was primarily geared *to* the centre.

Another question is whether this new terrain of centrality is differentiated. Basically, is the old central city, still the largest and densest of all the nodes, the most strategic and powerful node? Does it have a sort of gravitational power over the region that makes the new grid of nodes and digital highways cohere as a complex spatial agglomeration? From a larger transnational perspective these are vastly expanded central regions. This reconstitution of the centre is different from agglomeration patterns still prevalent in most cities that have not seen a massive expansion in their role as sites for global city functions and the new regime of accumulation thereby entailed. We are seeing a reorganization of space–time dimensions in the urban economy.

It is under these conditions that the traditional perimeter of the city, a kind of periphery, unfolds its full industrial and structural growth potential. Commercial and office space development lead to a distinct form of decentralized reconcentration of economic activity on the urban periphery. This geographic shift has much to do with the locational decisions of transnational and national firms that make the urban peripheries of the growth centres of the most dynamic industries. It is distinctly not the same as largely residential suburbanization or metropolitanization.

We may be seeing a difference in the pattern of global city formation in parts of the United States and in parts of Western Europe (Keil and Ronneberg, forthcoming; Sassen, 1994). In the United States major cities such as New York and Chicago have large centres that have been rebuilt many times, given the brutal neglect suffered by much urban infrastructure and the imposed obsolescence so characteristic of US cities. This neglect and accelerated obsolescence produce vast spaces for rebuilding the centre according to the requirements of whatever regime of urban accumulation or pattern of spatial organization of the urban economy prevails at a given time.

In Europe urban centres are far more protected and they rarely contain significant stretches of abandoned space; the expansion of workplaces and the need for intelligent buildings necessarily will have to take place partly outside the old centres. One of the most extreme cases is the complex of La Defense, the massive, state-of-the-art office complex developed right outside Paris to avoid harming the built environment inside the city. This is an explicit instance of government policy and planning aimed at addressing the growing demand for central office space of prime quality. Yet another variant of this expansion of the 'centre' on to hitherto peripheral land can be seen in London's Docklands. Similar projects for recentralizing peripheral areas were launched in several major cities in Europe, North America, and Japan during the 1980s.

Conclusion: concentration and the redefinition of the centre

The central concern in this chapter has been to examine the fact of locational concentration of leading sectors in urban centres. This concentration has occurred in the face of the globalization of much economic activity, massive increases in the volume of international transactions, and revolutionary changes in technology that neutralize distance.

The specific issues discussed provide insights into the dynamics of contemporary globalization processes as they materialize in specific places. Concentration remains a critical dimension particularly in the leading sectors such as financial services. The production process in these industries is one reason for this, and the continuing concentration of the most advanced communications facilities in major cities is a second important reason. There are, however, a multiplicity of spatial correlates for this concentration, and in this sense we see emerging a new geography of the centre, one that can involve a metropolitan grid of nodes connected through advanced telematics. These are not suburbs in the way we conceived them twenty years ago, but a new form or space of centrality.

Notes

1   The telecommunications infrastructure also contributes to concentration of leading sectors in major cities. Long-distance communications systems increasingly use fibre optic wires. These have several advantages over traditional copper wire: large carrying capacity, high speed, more security, and higher signal strength. Fibre systems tend to connect major communications hubs because they are not easily spliced and hence not suitable for connecting multiple lateral sites. Fibre systems tend to be installed along existing rights of way, whether rail, water, or highways (Moss, 1986). The growing use of fibre optic systems thus tends to strengthen the major existing telecommunication concentrations and therefore the existing hierarchies.

2   Furthermore, this unchanged level of concentration has happened at a time when financial services are more mobile than ever before: globalization, deregulation (an essential ingredient for globalization), and securitization have been the key to this mobility, in the context of massive advances in telecommunications and electronic networks. One result is growing competition among centres for hypermobile financial activity. In my view there has been an overemphasis on competition in general and in specialized accounts on this subject. As I have argued elsewhere (Sassen, 1991, Ch.7), there is also a functional division of labour among various major financial centres. In that sense we can think of a transnational system with multiple locations.

3   This also tells us something about cyberspace – often read as a purely technological event and in that sense a space of innocence. The cyberspaces of finance are spaces where profits are produced and power is thereby constituted. Insofar as these technologies strengthen the profit-making capability of finance and make possible the hypermobility of finance capital, they also contribute to the often devastating impacts of the ascendance of finance on other industries, on particular sectors of the population, and on whole economies. Cyberspace, like any other space, can be inscribed in a multiplicity of ways: some benevolent or enlightening; others, not (see Sassen, 1993).

References

Daly, M.T. and Stimson, R. (1992) 'Sydney: Australia's gateway and financial capital' in Blakely, E. and Stimpson, T.J. (eds) *New Cities of the Pacific Rim*, Ch.18, Berkeley, CA, University of California, Institute for Urban and Regional Development.

Gad, G. (1991) 'Toronto's financial district', *Canadian Urban Landscape*, vol.1, pp.203–7.

Keil, R. and Ronneberg, K. (forthcoming) 'The city turned inside out: spatial strategies and local politics' in Hitz, H. (ed.) *Financial Metropoles in Restructuring: Zurich and Frankfurt En Route to Postfordism.*

Moss, M. (1986) 'Telecommunications and the future of cities', *Land Development Studies*, vol.3, pp.3–44.

Sassen, S. (1991) *The Global City: New York, London, Tokyo,* Princeton, NJ, Princeton University Press.

Sassen, S. (co-curator) (1993) *Trade Routes*, Catalogue for an exhibition at the New Museum of Contemporary Art, New York City.

Sassen, S. (1994a) *Cities in a World Economy*, Thousand Oaks, CA, Pine Forge/Sage.

Sassen, S. (1994b) 'Analytic borderlands: race, gender and nationality in the new city' in King, A. (ed.) *Representing the City*, London, Macmillan.

Source: Sassen, 1995, pp.63–75

READING 5B
Ronald Skeldon: 'Hong Kong: colonial city to global city to provincial city?'

Hong Kong: the transition to a global city

While there were village populations and a viable economy in the territory then encompassed by the Crown Colony of Hong Kong before the British arrived in January 1841, there were certainly no urban centres or urban-based economic activities. Under the British, Hong Kong became the quintessential colonial city, an enclave economy dependent upon trade carried out through compradors, an entrepôt supplying distant markets, but with little intrinsic value except for its location and superb port. The transformation of this colonial backwater into one of the major industrial cities of the world in the decades following the Korean War has been the opiate of development economists' analyses for some considerable time and is an integral part of the Asian economic 'miracle'. Generally still regarded as an NIE (newly industrialized economy), Hong Kong has seen a marked reduction in the numbers employed in manufacturing, from almost 920,000 in 1985 to 570,000 in 1994. Hence, in terms of its labour force, Hong Kong is actually a 'newly deindustrializing economy' as it moves towards service-based activities, a pattern that it shares with Japan, South Korea and Taiwan.

The above scenario is somewhat deceptive as Hong Kong remains a major centre of manufacturing. Its industrial base has been shifted across the border into Shenzhen and beyond, with a conservative estimate of some three million workers employed in enterprises and joint ventures located in China but controlled by Hong Kong. What has emerged since the Deng reforms, implemented from 1979, is a vast mega-urban region spreading from Hong Kong around the Pearl River delta to incorporate not only the Special Administrative Region of Shenzhen, but also the southern Chinese provincial capital of Guangzhou and its satellite towns, and as far as Zhuhai and Macau. This is a functionally integrated economic system, as well as being a rapidly expanding urban region, with Hong Kong constituting the service and financial centre linking this region to the international economy and society. Hong Kong and its region have thus emerged as a significant node in the regional and global economy, with Hong Kong essentially becoming a global city. Hong Kong has emerged as the world's eighth largest trading entity in terms of value of merchandise trade, and it operates the world's busiest container port. Whether it can maintain this position after it is returned to the sovereignty of China is not only a critical question for the future of the city itself but also a fascinating test-case for the relative roles of economic and political factors in development. The economic fundamentals appear strong but the principal unknowns lie in the degree of future political interference in an economy that is currently relatively free from direct government intervention.

Patterns of immigration: settlers and workers

While they are of fundamental importance, size and manufacturing on their own do not a global city make, as Friedmann and his discussants have so clearly demonstrated (Friedmann, 1986; Knox and Taylor, 1995). The way in which global cities integrate into the world economy and their position as major sites of concentration of international capital and as nodes of global communication are also critical in assessing whether a city has attained such a rank. So, too, are the patterns of population migration. Global cities are the 'points of destination of both domestic and/or international migrants' (Friedmann, 1986, p.75). There are, however, other more specific characteristics of the migration to global cities. Two main flows of migration tend to emerge: an élite flow of transnational workers and a majority of poorer immigrants who serve the former (Sassen, 1991). Clear class polarization results. The classic case to date is that of New York, where poor Puerto Rican, Haitian or other Caribbean or Latin American immigrants serve the 'new aristocracy of labour' (Waldinger, 1992), those highly paid stockbrokers, bankers or senior company executives in the central urban area. The former provide the cleaning staff, the security guards and the legions in the catering industry from delivery boys to chefs, waiters and dishwashers who serve the needs of the élite group. Many of the latter themselves are migrants sent by their companies from overseas and from other areas within the US.

New York may be the archetypal world city, but Hong Kong shows interesting variations on the theme. Hong Kong was a creation not only of British colonialism but, more practically, of migration from China. This occurred in several waves, the most recent of which was in the late 1970s when some 400,000 people entered Hong Kong within five years. These migrants, like those in previous waves, were dominated by young men who were absorbed primarily into the manufacturing sector and who helped to relieve upward pressure on wages in that key sector. Since then, immigration has been tightly controlled: from 1983 to the end of 1993 numbers were set at 75 a day, then the intake was raised to 105 a day and, subsequently, from 1 July 1995, to 150 a day. The vast majority of these migrants from China, who in 1995 numbered 45,986 persons, were dependants of Hong Kong residents and they were not likely to be direct entrants into the labour force. For example, between 44 and 50 per cent of the annual intake from 1987 to 1995 was made up of children of Hong Kong residents, with a further 35–44 per cent composed of wives. The increase in numbers in the lead-up to 1997 is to allow a more gradual inflow of those who will be entitled to residence in the city under the terms of the Basic Law after 1 July 1997 …

…

In the years leading up to the transition to Chinese sovereignty, one of the leading public concerns has been rising levels of emigration: the brain drain from Hong Kong. Unquestionably, there has been an increase in the number of those leaving Hong Kong and, equally unquestionably, part of the reason why these Hong Kongers have left has been fears over the transition (Skeldon, 1994, 1995). However, during the period of rising emigration, there was increasing immigration, not just of dependants and contract workers, … but of the highly skilled, including professionals and managers and their families. The annual inflow of professionals and middle managers with employment visas rose from around 2000 in the mid-1980s to over 7000 in the peak year of 1994 … These, as in the case of the other global cities, are representatives of the 'new aristocracy of labour', the élite group which the foreign domestic contract workers also serve. Hence, like the migration to the other global cities, the flows are polarized into two quite different groups: élites and workers. However, unlike the other global cities, the latter are all contract workers, not immigrants, and neither group has any right to permanent residence in Hong Kong.

…

Internationalization and the global city

The critical features of the global city, by definition, are its international character and its international linkages, yet the very fear that the Chinese authorities have about Hong Kong is its 'internationalization'. By this, they imply the interference of foreigners in China's internal affairs. This sensitive point regarding sovereignty is well taken, but so too is the guarantee that the Special Administrative Region of Hong Kong is to be given a high degree of autonomy. Hong Kong's system is supposed to remain unchanged for 50 years after the return to Chinese sovereignty, although quite what this means for a city where constant and rapid change is a way of life has never been clearly specified. The right to travel overseas for Hong Kong residents is as much an integral part of its economic success as is the ease with which the élite migrants can enter to carry out their business. As has been made clear above, apart from the movement of dependants of Hong Kong residents, there is no immigration, as such, into the city but, currently with the exception of the British, all entrants come in under contract or with a valid work visa. However, if justification can be made, rarely is entry denied. What is also clear from recent research in the expatriate community is that there is considerable job mobility within the labour market after skilled or professional migrants have gained initial entry. 'The existence of a pool of skilled foreign labour is an essential element of a global city such as Hong Kong, and may be of significance to the future economic growth of such cities' (Findlay *et al.,* 1996, p.59).

The extent to which this relatively free access and relatively free labour market for foreigners will continue after 1 July 1997 is, of course, unknown. It is, however, difficult to disagree with one assessment that concludes that China will attempt to limit the dependence of sectors of Hong Kong's economy on foreign labour (Menski, 1995). The latest data available on immigration to Hong Kong show that, with the exception of legal migration from China, there has been a sharp downturn in the numbers entering the territory both as skilled and professional migrants and as domestic contract workers, the first decrease in well over a decade. Even though the numbers of entrants during 1995 were still high by recent historical standards, the drop was about twenty per cent compared with the previous year. Certainly, not too much should be read into this decline as the results of a single year do not necessarily indicate the start of a trend. Furthermore, the Hong Kong labour market reacts very quickly to changes in the international economy: there was indeed a marked slowdown in the growth of consumer demand during 1995 and in the growth of GDP per capita from 5.4 per cent in 1994 to 4.6 per cent the following year. This slowing was not unique to Hong Kong but characterized all of the more advanced rapidly developing economies of East Asia, a decline that need not suggest an end to 'the Asian miracle' (see *The Economist,* 1–7 March 1997, pp.23–5). Thus, as in the consideration of patterns of emigration, it is extremely difficult to separate regional and global trends from those caused by local circumstance.

Nevertheless, immigration is one of the clearest indicators of the internationalization of a city. It creates communities whose loyalties, even if neutral, may be perceived to be directed back to the areas of origin rather than forwards towards the destination which, in Hong Kong's case, can be translated into a 'love of China'. Clearly, these sentiments could be attributed not only to the expatriate community but also to returning Hong Kong migrants who hold options of residence elsewhere. While there are expatriate business communities within China itself, about which we know very little, there is no city in China where at least ten per cent of the population has close linkages overseas and whose primary lines of communication are outwards rather than inwards. The absorption of this expatriate group, together with the returned Hong Kong Chinese, will be a supreme test for a leadership paranoid about foreign intervention on Chinese soil. The migrant communities in Hong Kong, unlike those in the other global cities, have no right of abode and could rapidly fade away.

Conclusions through a glass darkly

Hong Kong's position as a leading economic centre might look secure. Its role as broker between China and the Overseas Chinese communities should maintain its advantage over Shanghai, for example. Nevertheless, the situation in which the expatriate communities find themselves seems precarious. These, after all, represent the fundamental linkages to the international economy and sustain Hong Kong's position as a global city. Should China seek to limit the dependence of sectors of the economy on foreign labour, and should the process of localization diffuse deeply beyond the civil service, then there is a real danger that Hong Kong will retreat from its position as a global city to become but another, if extremely large, provincial city in China. The history of decolonization is not exactly reassuring on this point as, when politics holds sway over economics in the context of rising nationalism, the trend is towards more homogeneous communities. This is the antithesis of the polyethnicity required for the global city.

In Friedmann's original formulation of a system of world cities, Singapore was seen as the only world city in Asia. More multicultural than any other centre in Asia, Singapore draws its labour force from an increasingly diverse range of sources and presently perhaps one fifth of its labour, skilled and unskilled, is foreign. A key urban node in the booming Asia-Pacific Rim, it also has access to the emerging Indian Ocean Rim economies. Hong Kong and Singapore as twin, but very different, dynamic growth centres in eastern Asia theoretically should be able to maintain their positions if we simply examine their economic potential. The future of both, however, is likely to depend more upon unpredictable political factors. The expansion of Singapore's economic activity is into two independent, and much larger and culturally very different, nation states. From 1 July 1997 Hong Kong will be part of a single nation state. Hong Kong is also expanding into an immediate hinterland that is racially homogeneous. Its position as the critical link between the international community, including the Overseas Chinese communities, and potentially the largest market in the world in China, should be able to maintain its position as a global city. However, all will depend upon the policies towards Hong Kong followed by China over the immediate future.

While it is impossible to predict accurately the direction of the policies to be taken by the Chinese government, the omens are hardly auspicious. Within China, the foreign community is still closely monitored and one can only wonder at the future function of the massive Chinese foreign ministry that is nearing completion in the heart of Hong Kong: will the China watchers become the China-watched? Will tensions in the international political arena that involve China be reflected in restrictions on foreign nationals? The issue is compounded by the large number of Hong Kong Chinese who have been educated overseas and/or who hold foreign nationality. The Chinese authorities will never permit Hong Kong to become a 'Trojan horse', bringing within their gates foreigners and those with dual loyalties who carry values that may undermine the Chinese system. And the free movement of Hong Kong residents overseas will only maintain and strengthen ties to foreign areas. It is thus difficult to be sanguine about Hong Kong's future position as a node in such an international migration system without a major shift in attitude and policy in Beijing. The movements of population both into and out of Hong Kong will be an important barometer to assess any shift in that policy. Ultimately, however, the future status of Hong Kong, as well as the status of other global cities in Asia, will depend upon how China interacts with the international and regional communities of nations. Hong Kong's current position (as well as that of Singapore, if for very different reasons) emphasizes the fragility of the global system, and it is but a short step for Hong Kong to move from its current status of global city to that of provincial city.

References

Findlay, A.M., Li, F.L.N., Jowett, A.J. and Skeldon, R. (1996) 'Skilled international migration and the global city, a study of expatriates in Hong Kong', *Transactions, Institute of British Geographers*, vol.21, pp.49–61.

Friedmann, J. (1986) 'The world city hypothesis', *Development and Change,* vol.17, pp.69–83.

Knox, P. and Taylor, P. (eds) (1995) *World Cities in a World System*, Cambridge, Cambridge University Press.

Menski, W. (ed.) (1995) *Coping with 1997: The Reaction of Hong Kong People to the Transfer of Power,* London, Trentham Books

Sassen, S. (1991) *The Global City: New York, London, Tokyo,* Princeton, NJ, Princeton University Press.

Skeldon, R. (ed.) (1994) *Reluctant Exiles? Migration from Hong Kong and the New Overseas Chinese,* M.E. Sharpe and Hong Kong University Press.

Waldinger, R. (1992) 'Taking care of the guests: the impact of immigration on services – an industry case study', *International Journal of Urban and Regional Research,* vol.16, pp.96–113.

Source: Skeldon, 1997, pp.265–71

CHAPTER 6

City rhythms: neo-liberalism and the developing world

by Michael Pryke

1 *Introduction*

If the previous chapter was largely concerned with cities as sites or juxtapositions of power, this chapter is concerned with the effects of such concentrations. Our focus is the current wave of urbanization that **Massey (1999a)** and others have spoken about, which is bringing with it momentous changes in the nature and shape of city life, especially in so-called developing countries. Many of the emerging mega-cities, and some of those not so large, are increasingly subject to the widespread influence of forces that emanate from outside of them. You will recall the significance, outlined in Chapter 1, of the networks of connections which bind together the different fortunes of cities. Well, in this chapter we want to explore those connections which, in a predominantly economic and political context, are presently shaping the lives of many of those who live in the growing cities of the less developed world. In places as far apart as La Paz, Nairobi and Calcutta, a similar set of forces, which may generally be referred to as 'liberalized' global economic processes, are having a profound effect on their diverse fortunes.

Liberalization is perhaps best understood as a particular kind of dynamic which works through the widespread flows of money and private finance and is orchestrated by the likes of such powerful, global institutions as the IMF, the World Bank and the World Trade Organization. Liberalization can be thought of as a persuasive set of ideas, an ideology, and as a set of practices and conventions governed by these dominant institutions. In both senses it can be thought to set up a particular ground-beat, a dominant rhythm so to speak, which conveys a neo-liberal dynamic through city networks. The origin of this ground–beat, this dynamism, is institutionalized and mobilized through the major global cities – New York, London and Tokyo. As centres of control, in the sense outlined by Saskia Sassen in the previous chapter, they direct and orchestrate a series of economic flows which *reach* across the networks of cities, pulling in the likes of Harare, as much as São Paulo, Buenos Aires and Moscow.

The generation of neo-liberal ideas, the construction of deregulation initiatives, the promotion of management practices through private finance packages – in other words, the 'work' of 'liberalization' – is performed in these dominant global cities. But the effects of such measures are felt as far afield as Lima and Dar es Salaam, and with varying intensity. Thus, while developing countries are enmeshed in a post-colonial phase when the form of connection between these countries and their cities should at least have been reformulated in their favour, the economic context of developing countries has been undergoing profound changes, determined in important ways by 'the West'. It is a context shaped increasingly by a particular form of economic connection, one informed not only by the conventions of neo-liberalism but also by the private-sector flows of money and finance that have come to play a dominant role. Empowered by sophisticated information technology, these flows roam the world, often serving to constrain developing countries in so far as they influence how national governments approach the management of their national economies, and in these ways they establish particular conventions for urbanization.

In this chapter, then, we will address the *intensity* and *reach* of neo-liberal influences upon urban transformation in those cities which are frequently on the receiving end of global institutional power. Sections 2 and 3 set up the broad context in which neo-liberal ideas and institutional practices have been formed and played out. Following that, section 4 outlines the usefulness of thinking about the dynamic and influence of such practices as possessing a rhythmic component, a series of regularized beats, which project an organizing and constraining series of polices and influences. Sections 5 and 6 then look explicitly at the impacts that such rhythms and dynamics have had upon everyday urban life in three cities of the developing world – Harare, Dar es Salaam and Bogotá.

It is to the broad context of liberalization and a Latin American scene that we turn first, however.

2 *Scene from the window: recognizing a change in economic context*

In the introduction to a text on Latin American cities the British geographer, Alan Gilbert, introduces his discussion of the urban landscape by noting that, although it might be easy at first to think that all Latin American cities are alike in the way that they are all '... highly unequal and contain wide extremes of poverty and affluence', there is quite a high degree of diversity: Oaxaca, Salvador, Buenos Aires and Lima, for example, 'reflect major differences in culture, climate, poverty and economic function' (Gilbert,1994, p.1).

Gilbert notes, for instance, how Bogotá is a city now full of migrants from all areas of Colombia and has grown rapidly from a population of around 350,000 in 1938 to its present size of around 5.2 million. The city is a mix of centrally located office-blocks and skyscrapers, while to the north are 'sanitized' suburbs and beyond these and also to the south of the central area are much poorer areas. As well as noting the changes in the city's built form, Gilbert draws our attention to something he calls the *regional debt crisis*. This, as we will soon see, is part of the change to the wider economic context of urbanization in Latin America – and elsewhere. But it is helpful to note its existence here, for it is something that is not only reflected in the built form but has changed the social fabric of this and other cities. For instance, in La Paz, the capital of Bolivia, with a population of only 800,000,

> A series of devaluations beginning in February 1982 caused household income to deteriorate in dollar terms. By April 1983, average monthly household income was worth only US$40. Artisans faced increasing hardships as the prices of imported raw materials, tools and machinery skyrocketed while the demand for their services decreased.
>
> (Gilbert, 1994, p.18)

Clearly something about how this city connected into the wider global economy had altered during the early 1980s. In one of the new office-blocks in La Paz, high above the street life, lies a possible clue to the causes of this mix of poverty and wealth.

ACTIVITY 6.1   When reading Extract 6.1, taken from a book written by a writer and journalist who is a specialist on Latin America, you may find it useful to underline those phrases which to you suggest the changed nature of interconnections between La Paz, Bolivia, and the global economy, and which may have set in motion an altered economic context for urbanization. ◆

EXTRACT 6.1
Duncan Green: 'Silent revolution'

From the ninth floor of the World Bank building in La Paz, the marchers look more like ants than people. Over the hum of the air-conditioner and the computers, the bangs of firecrackers and chanted slogans can barely be heard. 'Bolivia will never belong to the gringos' drifts up, tiny and weak through the megaphones.

The sun is setting, catching the snowy peak of Mount Illimani high above the city, glowing on skyscraper and slum alike. 'These people can't be allowed to win', mutters the Bolivian yuppie in the striped shirt, and goes back to talking up the economy.

Much more than nine floors separate the World Bank's Juan Carlos Aguilar from the protesters far below. Juan Carlos is fired with crusading zeal. This sleek, impossibly young technocrat thinks that only his free market panacea can enable Bolivia to leave behind 500 years of poverty and inequality. To his eyes the demonstrators are a relic from Bolivia's revolutionary past, an obstacle to future prosperity that must be side-stepped if they are not to divert politicians from the path of economic righteousness.

FIGURE 6.1 *La Paz*

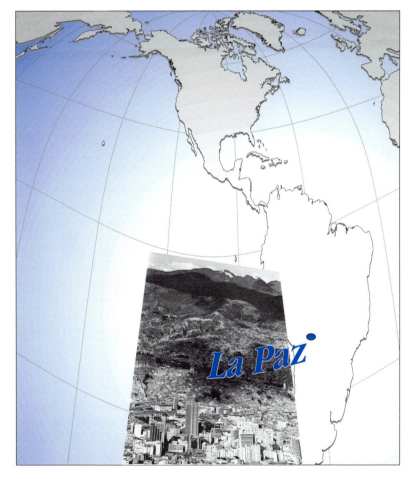

Success is in the eye of the beholder. The silver tongued Juan Carlos spins a seductive web of statistics, speaking purely in abstract economic variables – growth, trade, inflation. Bolivia's hungry masses – unemployed tin-miners and factory workers, street sellers, vagrants and hungry peasant farmers – talk instead about trying to feed their families, grappling with the crumbling health service and their desperate search for some kind of paid job. All are supposedly discussing the well-being of the same nation, yet often there seems almost no connection between their different worlds.

There are Juan Carlos clones throughout Latin America, young, smartly dressed men and women speaking perfect English thanks to their post-graduate degrees from US universities. They run the booming stock exchanges, subsidiaries of transnational corporations, economy ministries and offices of multinational organizations like the World Bank. They, and their friends in Washington, London and Frankfurt, share a common vision and awesome self-belief. They know what is wrong with Latin America, and they know how to put it right …

Latin America's techno-Yuppies are the avant-garde of a 'silent revolution' which has been going on in the region since the early 1980s. It is a revolution which has gone largely unreported outside Latin America, except in the pages of the *Financial Times* and other media catering to foreign investors … It is enshrining market forces at the heart of the region's economy.

Source: Green, 1995, pp.1–2

Extract 6.1 raises a number of issues, admittedly abstract but nonetheless relevant to the changed economic context of urbanization. At this stage it is helpful perhaps to do no more than note them simply for the questions they provoke. For one thing, I found the emphasis on the *market* – on 'talking up the economy', the 'free market panacea', 'enshrining market forces at the heart of the region's economy' – to be quite dramatic. Clearly this is a very particular way of thinking about economic organization, with its appeal to economic variables of growth and inflation, of stock exchanges and so on. It is an approach that would seem to be quite different to the past, and one that has stirred confrontations between the demonstrators on the street and the advocates of 'market' thinking, such as Juan Carlos, those with a 'common vision and awesome self-belief' about how to correct what is wrong with Latin America.

All this began to make me wonder how such a move in thinking – this set of beliefs about economic organization – came to have a grip on La Paz and other cities and countries in Latin America. I also wanted to know why particular emphasis was placed on certain economic processes, on statistics reporting growth rates and inflation, for example, and what the significance was of noting how this 'silent revolution' was largely unreported except in the pages of the financial press catering for foreign investors. For clearly the impact on life within cities has been quite stark over the same period that readers of the *Financial Times* have shown growing interest in Latin America. As Gilbert notes, by 1990 around 36 per cent of Latin America's urban population – 104 million people – were living in poverty. This type of poverty was a new phenomenon (Gilbert, 1994, p.168). It is to the possible causes of this new phenomenon and to the emergence of a free-market panacea that we now turn.

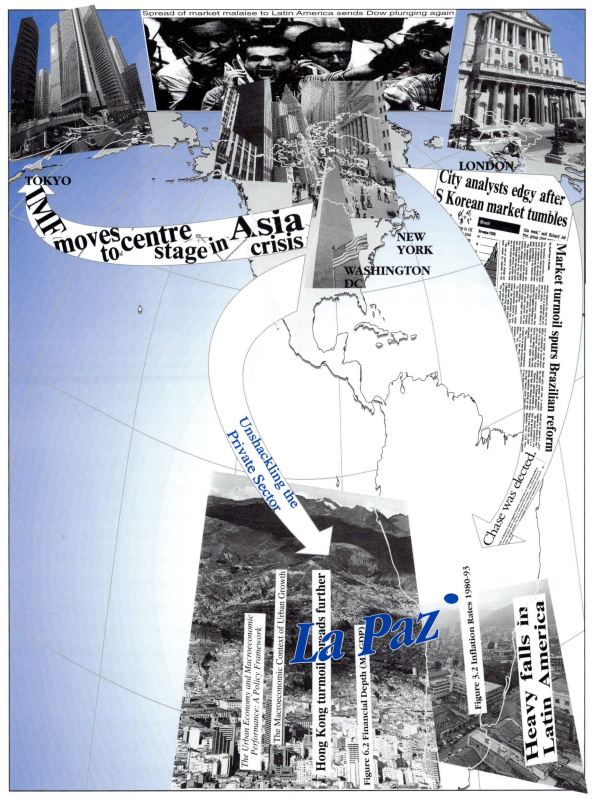

Spread of market malaise to Latin America sends Dow plunging again

TOKYO

LONDON

City analysts edgy after S Korean market tumbles

IMF moves to centre stage in Asia crisis

NEW YORK

WASHINGTON DC

Market turmoil spurs Brazilian reform

Unshackling the Private Sector

Chase was elected.

La Paz

The Urban Economy and Macroeconomic Performance: A Policy Framework

The Macroeconomic Context of Urban Growth

Hong Kong turmoil spreads further

Figure 6.2 Financial Depth (M₃:GDP)

Figure 3.2 Inflation Rates 1980-93

Heavy falls in Latin America

FIGURE 6.2 *Market malaise and urban rhythms: La Paz in a liberalized global economy*

3 The context of neo-liberal urbanization

In the introduction we talked about the 'liberalization of the global economy' and suggested how this has become the dominant force shaping urbanization in developing countries. However, before we move from a liberalized global economy to the processes of urbanization in developing countries in the following section we need to add a little detail. This detail is very much about noting the ideas passing through the interconnections between developing countries and the global economy and how these came to transmit effectively, particularly during the 1980s, the 'global' agenda of economic liberalization and how in turn this established the economic context for urbanization.

To make sense of these interconnections we therefore need to be aware of the basic and very particular ideas – the ideology – they have carried. Neo-liberalism, remember, is not a 'thing' in itself; it is, rather, a view of the world, of supposedly what 'reality' is actually about. What we need to be aware of in the following sections, then, are the ways in which the key institutions of neo-liberalism, in particular the World Bank and the International Monetary Fund, have succeeded in making their collective view of the world appear not just the dominant one, but the only correct one. This has implications for the way in which we view developing countries and processes of urbanization within them. For example, Figure 6.3 shows a World Bank sub-division of developing countries into 'low-income economies', 'middle-income economies' and so on, but when looking at it we need to exercise caution for at least two reasons.

First, there is a need to resist the temptation to follow this smoothing of real differences in histories and trajectories amongst countries within any category. Second, we need to recognize something of the purposes and implications of such categorization. In this way a map like Figure 6.3 can be understood as part of a wider ideological process of defining and grouping countries and thereby interpreting their 'problems', with economic growth being at the centre. As we shall see in section 5 this sort of ideological footwork served to legitimate, particularly in the 1980s and 1990s, the imposition of 'market' solutions to the problems of developing countries – notably that of economic growth – and the role of private finance in making that solution possible.

We need to be alert, too, to some of the central claims of neo-liberalism, not least because they have formed the ideas out of which approaches to urbanization have come to be tailored. Neo-liberalism aims to give *universal* legitimacy to an idea of freedom based on the individual and the *private sector*, an idea which dates back to eighteenth-century liberal ideology. The second claim is that the basis for successful economic growth and thus individual human happiness lies in the *free operation of markets*, an idea first aired by Adam Smith in his *An Enquiry into the Causes of the Wealth of Nations* published in 1776. You might remember that traces of this free-market zeal ran through the description of the World Bank's Juan Carlos Aguilar in Extract 6.1,

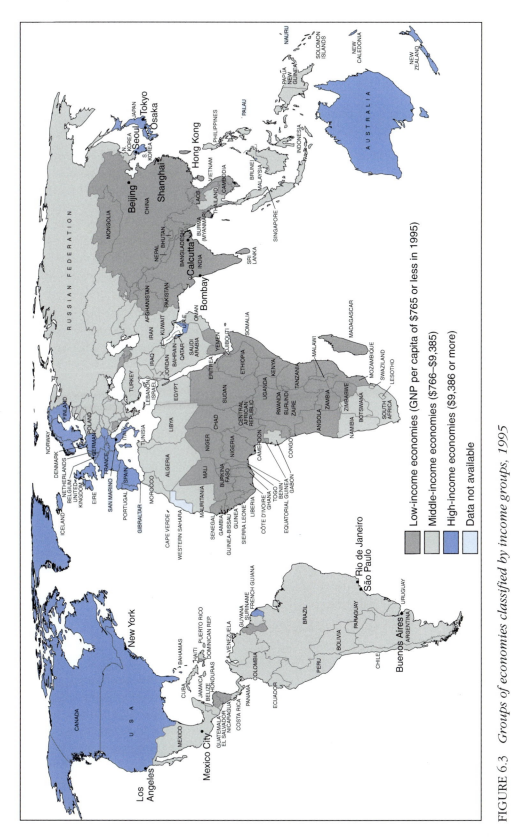

FIGURE 6.3   *Groups of economies classified by income groups, 1995*

with his 'awesome self-belief in free markets' and how these could provide economic growth for Bolivia and the city of La Paz. Lastly – again with its basis in Smith's thoughts – *government intervention*, it is believed, *hinders* the operation of markets and thus impedes economic growth. Simple though all this may seem, it is a set of ideas that has had far-reaching implications and *influences* when implemented through policy and circulated by certain international institutions.

An emphasis has been placed on influences because in many ways we are continuing elements of a theme introduced in the previous chapter, namely, the theme of 'cities of power and influence'. Recall that in that chapter certain cities – New York, Tokyo and London – are described as centres of active promotion and control of flows and private finance. So, in what follows you might like to think about how the 'power' of international institutions, such as the World Bank and the International Monetary Fund, actually *works* through networks of interconnections, passing through these three cities, and why to talk of influence is perhaps a more appropriate way to describe this power.

● For while it is helpful to think about what makes certain cities powerful, here it is useful to turn this question and to think about how certain cities become subject to powerful forms of influence emanating from the big institutional centres.

3.1 THE ROLE OF THE WORLD BANK AND THE IMF

There are two key institutions of neo-liberalism that we need to be concerned with at the outset. Both were established as part of the Bretton Woods Agreement at the end of the last World War and both are headquartered in Washington, DC, in the USA. The *first* is the World Bank (formally known as the International Bank for Reconstruction and Development). The World Bank has primarily been concerned with providing developing countries with loans predominantly for development projects. Generally these are long-term loans not gifts; they have to be repaid. Such loans (as well as those made to some of the poorest developing countries through the International Development Association, part of the World Bank group) provide scope for these institutions to exert particular influences.

Further scope to influence developing countries has come through the *second* key institution, the International Monetary Fund (IMF). The IMF – the 'sister of the World Bank' – traditionally focuses on the short-term economic stabilization of member countries, with the aim of ensuring that a country's economic management of its balance of payments – in effect its account with the outside world – is in order. When a country is in balance-of-payments difficulties, in other words (and at the risk of oversimplification) when it finds that it owes more than it can pay for, the IMF can also provide loans; this again allows influence to be exerted through the setting of loan conditions. In recent times this influence has touched upon pricing policy, financial sector reform, trade liberalization, privatization, tax structures and levels of government expenditure (Bird, 1996).

Yet how was it possible in the first place for such institutions to be in a position to exert such influence – an influence that has passed through into ways of thinking about cities, as we shall see in the next section? To answer this we need to be alert to the effects of the 'debt crisis' for developing countries in the 1980s and the policy implemented as a consequence. The debt crisis, which impacted predominantly on – and still affects – most non-oil-producing Latin American countries, as well as countries in Africa, has its roots in the liberalizing international finance system of the 1970s and the dramatic hike in oil prices in the early 1970s. Box 6.1 provides a brief background to this period. The significance of this period is that it initiated a process of readjustment to the economic policy prevailing in developing countries and thus altered the context for urbanization.

Box 6.1 Oil, debt and moves towards a neo-liberal agenda

Our briefest of histories allows us only to note the following key moments. In the 1970s soaring oil prices meant that oil-producing – OPEC – countries found themselves awash with money, much of which was quickly recycled by western banks through the newly emerging euromarkets (part of the new international financial system we noted in section 1.1) and loaned on what were then favourable terms to developing countries. At the time, these countries were struggling to maintain their economic policy of import substitution and were facing escalating costs for oil and other basic imports. Many followed the path of borrowing cheaply from international financial markets in order to meet the rising import bill. For the most part these loans were entered into by willing borrowers and lenders. With the second oil price rise in 1979, however, it became clear that many of these developing countries would soon be facing severe economic difficulties as lending terms and conditions altered for the worse for borrowers.

For most of this period western economies were experiencing economic stagnation. This in turn affected their demand for the exports of developing countries. The latter were thus caught between a rock and a hard place: the price of their loans was rising dramatically, while their ability to earn income from exports into international, predominantly, developed country, markets was decreasing almost as rapidly. The inevitable happened: developing countries, starting with Mexico in 1982, began to default on their commercial loan repayments. To make matters worse, economic stagnation amongst developed countries signalled a crisis of faith in Keynesianism. Thus was neo-liberal thinking re-born, signalled by the rise of monetarism and the ascendancy of the new right in western Europe and the USA, championed by Margaret Thatcher in the UK and Ronald Reagan in the USA. Neo-liberal thinking quickly gained ground and began to feed into and shape the nature of the links between developed and developing countries.

3.2 ... AND THEIR POLICY INFLUENCE

What is most significant for our purposes is the way in which the World Bank and the IMF attempted to implement policy to correct the debt crisis. The dual aim of this policy, summarized in Box 6.2, was to prevent further default on developing country debt (*stabilization*) and to implement this through macro-economic reform (*structural adjustment*) in such a way that developing countries could be turned towards a more open market orientation in their macroeconomic policy thinking. Sometimes executed with the assistance of local élites in developing countries, who encouraged and eased the achievement of policy goals (Green,

---

Box 6.2 Three steps to building capacity for economic growth

The influence of international institutions works through three interrelated policy stages:

- stabilization
- structural adjustment, and
- export-led growth.

Stabilization seeks to correct imbalances in a country's economy, in particular its balance-of-payments deficit and, under IMF tutelage, to ensure that loan repayments to northern commercial banks, the IMF and the World Bank are maintained. The correction of domestic budget deficits in an effort to stem inflation is a supplementary IMF aim. This includes particular and restrictive policies for public expenditure (including welfare spending), interest rates, foreign exchange, credit and wages.

The primary aim of *structural adjustment* is 'to build capacity for sustained economic development' (World Bank, 1997a, p.7). This is to be achieved through the deregulation of markets. Administered and co-ordinated more by the World Bank than the IMF, an important element is decreasing the role of the state and 'unleashing' the private sector. Privatization and a 'flexible' workforce are thus key policies aimed at reducing structural impediments. To help facilitate this process, structural adjustment loans are made available to developing countries. In effect this is policy-based lending serving to strengthen the Bank's influence (George and Sabelli, 1994, p.55).

Further stimulation to global integration comes in the form of *export-led growth* (rather than the past policy of import substitution). Export-led growth promotes the price mechanism as the efficient allocator of resources within a country. While opening developing countries to flows of global capital markets – seen as a good thing within the neo-liberal agenda – the same measures have resulted in higher prices of essential commodities such as food and have lowered incomes and employment rates. Urban settlements have been particularly badly hit (HABITAT, 1996).

Sources: Green, 1995, pp.4-5; George and Sabelli, 1994, pp.55–6, 147; Herbold Green and Faber, 1994, pp.1–8; HABITAT, 1996, pp.7–8

1995; Harris, 1996; Nafziger, 1993) – remember the *tensions* between the 'different worlds' of the unemployed tin-miners and the smartly dressed men and women with their US postgraduate degrees in La Paz reflected in Extract 6.1? – the implementation of IMF and World Bank policy has had, and continues to have, profound results. Its main impact has been to liberalize markets.

● Through debt collection and the liberalization of global financial markets, 'free-market' ideas began to flow through these newer forms of interconnection between developing countries and international institutions like the World Bank, together with private finance organizations, such as commercial banks.

But how did these ideas come to *settle* and take hold in developing countries?

The successful settling of these ideas in developing countries has *first* been achieved through (and has been assured by) the system of loan conditionalities that have attached to the debt rescheduling of the 1980s and beyond, whereby debtor countries are rewarded with debt relief if substantial market-oriented economic reform is implemented. Various IMF/World Bank initiatives such as the Baker Plan of 1985, the Brady Plan launched in 1989 and the programme for debt relief for 'heavily indebted poor countries' (HIPC), initiated in 1997, all involve World Bank/IMF surveillance and commercial bank approval of a country's macroeconomic record (Corbridge, 1993, p.43; World Bank, 1997c, p.57).

The *second* way the new context has been settled comes through the influence of liberalized private-sector finance (International Monetary Fund, 1997; Kincaid, 1988; Bouton and Sumlinski, 1996).

ACTIVITY 6.2   Before reading any further you might like to take a look at Tables 6.1 and 6.2 and think about how processes of urbanization might be affected by the different flows of 'offical development finance' and 'private finance' (noted in Table 6.1) and the highly unequal way in which these flows have moved into developing countries (Table 6.2). For example, what does the decline in official finance and the rise in different forms of private finance signal in terms of the growing influence of a neo-liberal agenda on developing countries? You might like to refer to Figure 6.3 in order to trace these flows through to actual places. ◆

As Table 6.1 shows, not only has private-sector finance grown in importance for developing countries as official aid has declined in the 1990s, but with the policy emphasis on privatization and the encouragement to employ foreign investment, macroeconomic policy needs to be established so as to send – in the words of a World Bank report – a message of 'commitment and certainty' to the private sector (Holden and Rajapatirana, 1995, p.71). Movement to the dominant rhythm established by the key institutions of neo-liberalism is thus almost unavoidable. For one thing, the rise of private flows, to around 244 billion dollars in 1996, signals a shift to a particular financial environment in which national economic policy decisions are made. Cities, too, are unable to escape the influence of this environment, as we shall soon see.

TABLE 6.1 *Financial flows to developing countries[1], 1990–96 (billion dollars)*

	1990	**1992**	**1995**	**1996[2]**
Total flows	100.6	146.0	237.2	284.6
Official development finance	56.3	55.4	53.0	40.8
Grants	29.2	31.6	32.6	31.3
Loans	27.1	23.9	20.4	9.5
Total private flows	44.4	90.6	184.2	243.8
Debt	16.6	35.9	56.6	88.6
Commercial banks	3.0	12.5	26.5	34.2
Bonds	2.3	9.9	28.5	46.1
Others	11.3	13.5	1.7	8.3
Foreign direct investment	24.5	43.6	95.5	109.5
Portfolio equity flows	3.2	11.0	32.1	45.7

Notes: Low- and middle-income developing countries are defined as those having 1995 per capita incomes of less than $765 (low) and $9,385 (middle), respectively.
[1] Aggregate net long-term resource flows.
[2] Preliminary.

Source: World Bank, 1997b

TABLE 6.2 *Net private capital flows to developing countries (billion dollars)*

	1990	**1992**	**1994**	**1996[1]**
All developing countries	**44.4**	**90.6**	**161.3**	**243.8**
Sub-Saharan Africa	0.3	-0.3	5.2	11.8
East Asia and Pacific	19.3	36.9	71.0	108.7
South Asia	2.2	2.9	8.5	10.7
Europe and Central Asia	9.5	21.8	17.2	31.2
Latin America and Caribbean	12.5	28.7	53.6	74.3
Middle East and North Africa	0.6	0.5	5.8	6.9
Income group				
Low-income countries	11.4	25.4	57.1	67.1
Middle-income countries	32.0	64.8	104.2	176.7
Top five country destinations[2]				
China	8.8	21.3	44.4	52.0
Mexico	8.2	9.2	20.7	28.1
Brazil	0.5	9.8	12.2	14.7
Malaysia	1.8	6.0	8.9	16.0
Indonesia	3.2	4.6	7.7	17.9

Notes: Low- and middle-income developing countries are defined as those having 1995 per capita incomes of less than $765 (low) and $9,385 (middle), respectively.
[1] Preliminary.
[2] Country ranking is based on cumulative 1990–95 private capital flows received. Private flows include commercial bank loans guaranteed by export credit agencies.

Source: World Bank, 1997b

For example, where private-sector finance plays a growing role in *processes of urbanization,* such as in infrastructure provision – the provision of urban transport, waste disposal, power, airports, water supply and so on (World Bank, 1994) – perceptions of efficient macroeconomic management are of growing significance. As you will see in the following chapter, today's interconnections enable private-sector financial organizations, such as the US-based credit rating agency Standard and Poor's, to scrutinize a country's economic performance and prospects, and to issue a 'credit rating' for that sovereign state.

With governments now encouraged to make use of private-sector finance, the conventions of the international private finance market must be adhered to. Of major concern, then, is to ensure that credit rating agencies are satisfied with a country's current and projected economic environment. The degree to which they are satisfied is signalled to potential lenders through the rating given to a proposed financial package, such as a government raising funds through, for instance, the international bond markets. There are private finance markets where loans can be raised by corporations and agreements for periods of 5–25 years; the pricing, packaging and monitoring of these loans will almost inevitably take place through global financial centres such as New York and London. It is in the interests of governments, therefore, to establish and *maintain* an economic environment perceived as being favourable to the private sector. The better the perception, the greater the chance that a high rating will be given to a bond issue and thus to a lower cost of borrowing: a high rating signals a reduced likelihood of, say, a government defaulting on a loan repayment or of currency devaluation wiping out the returns to investors. Over the life of the bond, governments will, of course, want to retain or improve their credit rating. A country's policy-makers must therefore ensure a favourable rating through stable and efficient macroeconomic policies to ensure that infrastructure projects are funded through international capital markets (Ferreira and Khatami, 1996, p.26).

● Today's interconnections expose cities to the scrutiny and criteria of private-sector finance and the dominance of New York, Tokyo and London. Through these practices, cities have little choice but to operate along the lines laid down by the dominant rhythm of neo-liberalism.

Emphasis is placed on private-sector finance and its mechanisms of surveillance, as their characteristics and ways of working hold further and related implications for neo-liberal cities. For, as Table 6.2, illustrates, whilst international institutions such as the World Bank and the IMF seek to merge many different histories into a story of one urban future and to impose on developing countries the dominant rhythm of neo-liberal economic growth in order to achieve that goal, a significant stumbling-block remains.

● Unequal histories made through the unevenness of past interconnections, together with the processes of scrutiny adopted by private-sector organizations and what they prioritize as 'sound economic mangement', necessarily implies that unequal flows of private-sector finance result. Note the significant differences in flows to 'Sub-Saharan Africa' and 'East Asia and the Pacific'.

This makes the big story difficult to follow for some countries and their cities, such as those in Sub-Saharan Africa. Recall the points we made around Figure 6.3 and note the substantial differences in flows of private finance to low- and middle-income countries. Both groups of countries are supposed to follow the same neo-liberal story, yet with markedly different private-sector resources. Even for those seemingly more successful countries of East Asia, such as Indonesia, and cities like Jakarta, the future can never be read as certain. As the remaining sections serve to remind us, today's interconnections, shaped increasingly by a liberalized financial system, are always potentially unsettling.

4 *Dominant rhythms*

From the standpoint of this chapter, which has at its centre the connections of money and finance, the emphasis on viewing these connections as dynamic is intentional, for it moves cities into the 'space of flows' introduced in the previous chapter. For a moment, then, let us briefly reconsider the space of flows and the role of global cities – notably the financial centres of New York, London and Tokyo. Here the focus is on these centres as the *producers of dominant rhythms*, as the producers of a time rhythmed, so to speak, by the social relations of international finance housed in these global cities.

Global cities are the centres of webs of financialized information; they are the control centres for the growing financial flows which operate in simultaneous time, empowered by the new information communication technologies which link markets in different time-zones and serve to integrate national and 'global' financial markets. These centres are capable of reacting to any set of economic events in a deregulated global economy, acting to direct and redirect flows of finance from one market or investment opportunity to another. As this suggests, this is a very *particular time, one that is rhythmed through an ongoing calculation of financial profitability.*

And as the conditions of this profitability are heavily influenced by economic factors such as rates of inflation, wage-levels, growth rates, exchange rates and the relative stability of these factors, then some degree of control must be exercised over their movements if these flows are to be attracted and settled in ways that can be turned into beneficial ends, rather than unsettling countries and cities.

Whether these flows are disruptive or beneficial will depend on how a country manages its economy, its so-called macroeconomic management. The establishment of the 'correct' macroeconomic management, as we will see in sections 5 and 6, is something that has grown in importance among developing countries since the early 1980s. Via macroeconomic management – through changing government expenditure and rates of taxation (fiscal policy) and altering interest rates (monetary policy), for example – a national government can attempt to alter, for instance, average price levels, money wages, employment rates and general economic growth. All these factors work together to influence the perceptions held by private-sector investors working through financial centres as to the degree of risk and thus the relative attractiveness of directing flows of money and finance into any one country, to a particular use within that country and the most suitable time-period over which to invest. Through recognizing the effects of interaction amongst the various components within 'economic flows' – interest rates, exchange rates, stable price levels and wages and so on – we can begin to see how dynamic are present economic connections and how dominant rhythms can work through them in simultaneous time, moving between and through countries and cities with potentially destabilizing effects, *recomposing city rhythms* – through the dramatic rise in the cost of living in short periods of time – sometimes in quite unexpected ways.

It is through an appreciation of this rhythmic interaction channelled through today's economic interconnections that we may be able to appreciate how the time–spaces of cities are now being filled and thus the consequences for urbanization. The point here is that 'time' itself is not the organizer of urbanization. A temporal line stretching between 'now' and some point in the future does not determine the form that urbanization will take. That form is heavily influenced by how a city and its nation-state interconnect with the wider global economy and by the nature of those relationships. By recognizing *that time can be rhythmed* in certain ways by dominant social relations (for our purposes we have focused on the influence of the social relations of international money and finance), then how a city is exposed to a dominant rhythm becomes an important influence on its urbanization process.

To take this a little further, projecting the growth and size of the cities says little, if anything, about what life within those cities will be like between 'now' and the year 2015. How that time is filled, how the processes of urbanization in Shanghai, Mexico, São Paulo, Dhaka, Lagos and so on, evolve, will be shaped through webs of interaction between each city, its nation-state and the wider global economy, much of which will be monitored and orchestrated by the centres of dominant rhythm, those highlighted in the work of Castells and Sassen in the previous chapter, such as New York, London, Tokyo. Each city will evolve in a sense through the linear 'big time' of now and 2015, yet each city will contain a range of rhythmic interactions which are dependent on a host of economic/financial interconnections continuously shifting back and forth between the city, the city's host country and the wider global economy, much of which will be mediated and monitored from within the dominant financial centres of New York, Tokyo and London.

One way of appreciating the types of impact which have their roots in changing economic interconnections is to understand them as the outcome of a change in rhythm; a change which provokes new movements, a new pace and intensity in everyday city life, which is most noticeable in the altered 'daily routines' flowing through the home. (**Allen (1999)** sub-section 2.1 and **Massey (1999a)** section 2 both talk about the rhythms and space of the city.)

● Thinking about connections in this way helps to recognize that the introduction of a particular economic policy, such as structural adjustment, is not a simple 'one-off' event. The implementation of such a policy serves to reconfigure times and spaces as a result of its impact on social relations – the day-to-day of city life.

To speak of rhythms within economic connections helps us to think about why the mix of times and spaces within a city is organized in a certain way, why they move to one rhythm and not another, and how the nature of today's interconnections can lead to a change in a city's rhythms. In this way the disruption caused to daily city lives can be understood as emerging out of that city's exposure to and interconnection with the particular social relations of today's global economy and associated policy of structural adjustment. The imposition of these presently dominant ideas on a city, working their way through particular flows of money, finance and economic policy, disrupts the times, the daily routines, of its various public and private spaces, just as a change in rhythm would alter the movements of a dancer.

In this sense the rhythm metaphor helps us to appreciate the real spatial and temporal impact on city spaces of otherwise dry economic policy, such as flexibility in work practices, privatization and so on. From Mexico (Esobar Latapi and Gorizates de la Rocha, 1995) through Latin America (Janvry and Sedonlet, 1993) to Africa (Jamal and Weeks, 1988; Weeks, 1992), the imprint of the dominant market-oriented ideology of today's global economy can be traced through to the day-to-day use of the city as the impact of the debt crises and structural adjustment alters the daily routines of the city's households. In this way, economic interconnections can 'come alive' and be understood to have real impacts through the way they re-rhythm cities.

To talk of economic rhythms of a city is thus

● to move from a static notion of connection between a country and the 'global economy' or between a city and the wider macroeconomy, to a more *dynamic notion of connections.*

Any city may be understood not simply as a place, a fixed location connected almost rigidly to another fixed location, but as a bank of times and spaces – the many and different households within Mexico, for example – all of which are set in play through the influences that are contained in webs of relations which connect it to the wider global economy.

A change in economic regime – the move to neo-liberalism – is not, then, a simple shift in direction, neatly executed, but a compelling rhythmic motion moving cities and their spaces to an inescapable beat.

The purpose of thinking in terms of economic rhythms, in a sense, is to listen geographically to the agenda and consequences of neo-liberalism and the growing power of private finance as they merge into the cities of the South. By listening geographically, we can begin to think of these interconnections and what they carry as rhythms: we can begin to see that neo-liberalism has its own motives which can rhythm city spaces and times, creating new intensities and experiences. In this way we shall see how two entwined forces, one ideological – neo-liberalism – the other abstract – money and finance – can arrive in cities and be transformed into immediate realities. Such immediate temporal and spatial realities, as we shall consider shortly, include sharp rises in the cost of housing, the decline in availability of social services and rocketing food prices and the sudden freezing of construction work. These outcomes are strongly influenced by particular connections between a country and the wider global economy. Our aim is to explore the nature of these connections, of these specific relational webs, and how they work. It is to this that we must attune our ear.

5 *Cities of neo-liberalism*

Our interest is therefore to examine how the dominant economic rhythm of neo-liberalism establishes the context for change in the daily life of cities of developing countries. This section begins by outlining the key influences in urban policy that have stemmed from and been encouraged by neo-liberalism. In many ways this is about recognizing the rhythm change that has taken place in cities of developing countries introduced earlier.

What is central to the thinking which informs policy connections between urbanization and macroeconomic performance is the statistically supported suggestion that there is a close relationship between the growth of large cities and national economic development. For example, according to a World Bank sponsored report published in the mid 1980s, 'one third of Thailand's gross domestic product (GDP) originates in Bangkok; ... more than 60 per cent of the manufacturing firms in the Philippines are located in Manila; and nearly 50 per cent of Sri Lanka's employment in commercial, financial and transport services is accounted for by Colombo' (Sivaramakrishnan and Green, 1986, p.34, cited in Stern, 1993, p.127). Similar figures for Latin America and Africa show the urban generation of economic growth. For instance, in Africa just under 30 per cent of formal sector employment in Kenya was accounted for by Nairobi in 1989 which at that time had only 5 per cent of the country's population. Similarly, Harare, with roughly 8.6 per cent of the country's population, accounted for 36.6 per cent of the total recorded income and over 26 per cent of the country's formal employment in 1984 (Kenyan and Zimbabwean government data, cited by Stern, 1993, p.127). Indeed, by the end of the 1980s the World Bank claimed that over half of all GDP in the greater majority of developing countries could be traced to urban areas. Informed by this supposedly statistically impressive background, international institutional thinking turned its attention during the 1980s and 1990s to reformulating urban policy.

As was suggested in the previous section, the links between neo-liberalism and cities, particularly those in developing countries, needs to be set in the context of the growing influence since the 1970s of international institutions and multilateral agencies such as the World Bank and the IMF, as well as bilateral aid agencies that include the USAID (Pugh, 1995). The ideological persuasion exerted increasingly by these and other institutions mirrors the types of influence at work at the level of the global economy discussed earlier. The conditions attached to loan agreements, that were noted in section 3.2 for example, can affect approaches to urban policy within debtor countries, as will become clear a little later. For the moment, let us run through a brief history of urban policy inspired by multilateral agencies that will help to make sense of current urban policy and thinking.

The World Bank began its urban policy formulation in the early 1970s and has moved from a piecemeal project focus to much broader thinking that incorporates the private sector in dealing with urban issues. During the 1980s policies and programmes began to focus on *integrating urban policy and national macroeconomic policy*, and on using private-sector finance to fund urban investment, such as infrastructure projects. Such integration involved fitting urban policy into the context of a reworked macroeconomic policy.

In 1986 the World Bank, the United Nations Centre for Human Settlements (UNCHS) (HABITAT), and the United Nations Development Programme (UNDP) joined forces and founded the (New) Urban Management Programme (NUMP). This was a two-phase programme, the first phase running from 1986 to 1990 and the second from 1991 to 1997 (Jones and Ward, 1994, p.37). Through establishing the appropriate institutional structures, the purpose of the NUMP *Phase one* supposedly was to help to 'improve' approaches to land management, municipal finance, infrastructure and, since the early 1990s, health and poverty (Stern, 1993, p.10). What should be remembered, however, is that the main objective of the NUMP institutional structure is to foster a 'new consensus' about certain urban issues. In a sense this involves coaxing governments into viewing their cities in particular ways; to identify the main 'problems' within cities; and to come up with the best means and techniques of solving those problems (Jones and Ward, 1995, p.63). The promotion of this approach to urban policy in developing countries and the language of the market – 'efficiency, competition and involvement of the private sector' – was just as apparent in *Phase two* of NUMP. What is of interest in this is that it indicates how consensus thinking can take hold so that a city, its problems and their solution are approached only from one angle. Soon it becomes impossible to think outside of an orthodoxy. To question the orthodoxy is to fail to realize that the gains from the policy in question can only be achieved in the *long term*. This is a view which sees time – the linear, long term – as the organizer of urbanization, a view criticized in section 4. Such an approach misses the influences flowing through today's economic connections, influences that may just as easily *unsettle* cities as deliver the gains claimed by policy.

ACTIVITY 6.3   While reading through the remainder of this section, make a mental note of the *similarities* between the way in which macroeconomic policy reform was talked about in section 4 and how the need for reform of urban policy is presented by the World Bank. ◆

Much of the thinking evolved over the 1980s culminated in an influential policy document published in 1991 by the World Bank entitled *Urban Policy and Economic Development: An Agenda for the 1990s* (World Bank, 1991). One of the main aims of the policy document was to develop urban policy within the context of the wider macroeconomic policy already favoured and being implemented in one form or another in many developing (as well as developed) countries. The ideal role of government at both national and municipal level is viewed as being that of enabler, facilitating the development of market-based solutions to a range of issues from infrastructure to housing finance, through to the promotion of entrepreneurial activity within communities. As the document states, within an analysis of 'fiscal, financial and real sector linkages between urban economic activities and the macroeconomy', the aim is to '… propose a policy framework and strategy that will redefine the urban challenge in developing countries'. Two central parts to such a redefinition are the need for the developing countries, the international community, and the World Bank '… to move towards a broader view of urban issues, a view that … emphasizes productivity of the urban economy and the need to alleviate the constraints on productivity'. And with '… urban poverty increasing, the productivity of the urban poor should be enhanced by increasing the demand for labour and improving access to basic infrastructure and social services' (p.3). Urban issues are thus located within the practical confines and 'broader objectives of economic development and macroeconomic performance' (World Bank, 1991, p.4). Indeed, as two World Bank employees (one of whom was the principal author of the 1991 World Bank urban policy document) made explicit,

> Structural adjustment policies at the macro level are intended over the longer term to create an enabling policy environment for more productive urban economies. Such an environment would increase the efficiency of firms and households, and would thus support the economy-wide adjustment and the resumption of growth. For many countries, however, these policy changes require a corresponding urban adjustment to support national economic adjustment goals. Such a process should result in more flexible institutional regulatory regimes at the city level to adjust to new macroeconomic realities. It would affect the production of goods and services and the broad context for investment, savings, resource mobilization and capital formation in urban areas.
>
> (Cohen and Leitmann, 1994, p.120)

What is striking about this comment is, *first,* the implicit reduction of 'the city' – almost regardless of its history and current insertion into the global economy, and what goes on within it – to general laws of supply and demand, in other words to market forces. *Second,* that urban productivity is achievable seemingly only through markets and market reform and that all of these measures are compatible with other aspects of city life. *Third,* that what is good for macroeconomic policy may not be the best for urban policy seems to be overlooked. For example, while a chapter in the 1991 World Bank policy paper recognizes the social costs of structural adjustment on the urban poor, the solution is to 'improve productivity' – 'increasing demand for the labour of the poor', 'alleviating structural constraints', and so on. The causes of urban poverty and policies for its alleviation are set within an economic framework that at once connects cities in a conceptually smooth way into broader change in the global economy, yet at the same time seems to pass over the tensions that emerge between policy directed at alleviating poverty and policy aimed at establishing a macroeconomic environment suited to a neo-liberal agenda. The solutions to urban policy seem to be the outcome of looking at a city as a collection of economic agents: increasing urban productivity will, almost without question, reduce poverty. Such a policy is happily advocated for all cities with little regard for the different ways each has been and is presently connected into the wider global economy. *Fourth*, talk of cities having to 'adjust to new macroeconomic realities' is quite striking and seems to me to suggest an implied (if not explicit) attempt to construct a global story of inescapable economic change into which cities are drawn.

● Regardless of differences amongst countries and cities, policy is presented in a way that presents a unilinear, predetermined path of urbanization. This path is shaped and constrained ultimately by macroeconomic policy which itself is finely tuned to fit with the imperatives of the global economy – particularly private finance and its surveillance.

It is to the effects of exposing cities to these new global economic rhythms that we now turn.

6 Life underneath a dominant economic rhythm

What we have seen so far is the growing influence of private finance working within an environment of neo-liberalism and the manner in which urban policy within a growing number of cities of developing countries has been influenced by the effects of macroeconomic adjustment formed in relation both to different recent economic histories, such as the debt crisis of the 1980s, and the globalization of private finance. With these points in mind, this section is concerned with tracing the effects of these latest interconnections. The section focuses on three readings in order to explore life under dominant rhythms in three different cities – Harare (Zimbabwe), Dar es Salaam (Tanzania) and Bogotá (Colombia). Recalling the emphases of section 4, the readings will be used to suggest how life in these cities has been altered by processes of economic adjustment – how, in other words, the time of adjustment has been filled and altered by the nature of recent economic interconnections and the impact of their rhythms on the immediate realities of these very different cities. The readings also offer a glimpse at the possibilities for alternative networks to be put in play by those affected by the shock of the impact of the latest interconnections.

6.1 THE IMPACTS OF A DOMINANT ECONOMIC RHYTHM

The three readings which follow describe the varied effects of structural adjustment policies within three cities, two in Africa and one in Latin America. In both pieces on cities in Africa, what is noteworthy are the ways in which economic rhythms, working their way through structural adjustment programmes, have a variety of immediate effects. What is remarkable, too, is how swiftly these rhythms reorganize life within cities and in so doing can lead unexpectedly to the emergence of new indigenous rhythms which may form the basis for alternative networks that can be made to work under dominant, imposed rhythms – despite their relative power.

ACTIVITY 6.4   You should now turn to Readings 6A and 6B which are at the end of this chapter. When working your way through the readings, make a note of what strikes you most about the influence of structural adjustment on city life and how the various histories of each city constrain their respective abilities to reap the supposed benefits of such policy. ◆

I think the readings are interesting in a number of ways, notably in the suddenness and breadth with which structural adjustment affected life in both cities. Structural adjustment programmes and 'correct' macroeconomic policy, including import liberalization, devaluation, and reductions in public expenditure, are shown clearly to have disrupted the indigenous rhythms of both cities. For example, what took place in Dar es Salaam is a helpful reminder that unlike the official accounts of urbanization (some of which we touched upon in section 5), the country's past, and in many ways unequal, encounters with the global economy are seen to have significant influence on the way both the country and its cities cope with structural adjustment. To take one example from Reading 6B, look at the paragraph beginning with 'Participation in the …': this describes the informalization of much of the economy that resulted from the adoption of the structural adjustment programme, and suggests how the programme set in motion a new economic dynamic, filling the private spaces and times of the city with a more intense rhythm, signalling new imperatives which transformed relations and obligations within households and altered the rhythms between the city and country. And while Tanzanian government policy was clearly part of the problem, it seems reasonably clear to me that the country's economic context and consequently that of its cities has been opened up to the unsettling rhythms of global finance, international commodity markets and the preferences of the IMF.

What is also clear, particularly in the case of Harare, is that the effects on cities of increased competition and the growth of private-sector financial flows, for example, expose cities – and the lives of those within them – to sharp and profound upheaval. The outcome of Zimbabwe's Economic Structural Adjustment Programme (ESAP) is not just surprising for the 'huge increases in the cost of living', the hike in school fees, the jump in electricity charges and bus fares that occurred, but also because of the new geography of rhythms, of altered daily city life, that emerged. The alarming rises in household expenditure

FIGURE 6.4
Harare, Zimbabwe

FIGURE 6.5
*Shell stall, Dar es
Salaam, Tanzania*

shown in Table 1 of Reading 6A can in a sense be read in this way: as an
'economic' representation of the impact of the rhythms of structural adjustment.
Yet unlike a pure economic reading of the table, we can view these price rises not
as a single event but as setting in play new rhythms which re-choreograph the
spaces of Harare. In a similar way, the sequence of events leading to the closure
of small clothing firms, sparked by the decline in purchases of new clothing and
school uniforms, show the openness of cities to relational webs that can
superimpose a dominant rhythm on cities, unsettling day-to-day life. Moreover,
by listening geographically to this rhythm we can imagine how the 'retrenchment
of 3000 workers and closure of several small companies' created new intensities,
conflicts and disturbances within households, as people tried to accommodate
the neo-liberal motives of economic interconnections which in turn produces an
altered geography of city spaces and times. As the Readings show, these
interconnections are not static; their dynamism produces effects which flow
through cities – a process which the rhythm metaphor helps us to capture.

The economic flows are not 'all determinate', however. As the reading on Dar es
Salaam makes clear, despite everything, new opportunities have emerged from
beneath the dominant rhythm imposed through structural adjustment. Some of
the women from the city have resourcefully managed to innovate and redirect
some of the motives of neo-liberalism – the calls for labour productivity increases
through informalization and flexible work practices – and the upheavals such
policies cause. This has been achieved through the active creation of new
networks. From amidst the conflicting demands caused by the effects of jumps in
the cost of living and rising male unemployment, and the tensions that followed,
these women, sharing the same house or apartment block, began to 'establish new
information and resource networks', and in doing so turned the drive for
informalization to their advantage.

In contrast to these two readings, the next offers an account of what many in the
World Bank and the IMF would call a success story of neo-liberalism.

ACTIVITY 6.5   Now study Reading 6C and, while reading it, consider whether Bogotá is a suitable exemplar of neo-liberalism. Note, too, the way in which interpretations of economic 'success' or 'failure' are presented. ◆

This account of the effects of liberalization is instructive for two reasons – reasons that you may have picked out for yourself. *First,* it is a reminder of the different starting-points that cities have as they interconnect with a world of neo-liberal flows and influences, and thus the different trajectories they are able to follow. This is to highlight the simple but important point made in the introduction: that the mix of influences within today's economic interconnections can unsettle processes of urbanization in marked and sometimes unexpected ways, to their supposed advantage – as in the case of Bogotá – or to their disadvantage – as in the cases of Harare and Dar es Salaam. Similarly, past interconnections can alter substantially a city's trajectory. Colombia's avoidance of the debt crisis, for example, meant that it was in a very different economic and social position to either Tanzania or Zimbabwe, with consequent effects on city life under the influence of neo-liberalism. In other words, how the country was bound into wider relational webs produced different outcomes within the spaces of Bogotá.

The *second* point that this reading makes, relates to how important data and interpretations of data are to the promotion and telling of global stories. We saw in section 5 how the World Bank had utilized a reading of certain data to demonstrate the importance of cities to the production of economic growth. In this reading, the writer makes us aware of the possibilities for more than one story to be read from data on income levels and trends in poverty. 'Impressive data sets' therefore may not be all that are needed to begin an assessment of the 'success' or 'failure' of neo-liberal orthodoxy, particularly if we want to understand the effects of such orthodoxy on the multiple times and spaces of a city. For while it is telling to read that '... many more workers were working in 1992 to produce the same share of GDP [Gross Domestic Product, a key indicator of economic growth] as in 1970' , what does this tell us about the altered economic connections and influences between these two dates, and their impact on the city? It is helpful to imagine into this 'economic fact' the changed work practices – the practices generating these data – and the daily rhythms they produce. *Employment,* for instance, seems more precarious – 'jobs are not held for very long'; no doubt provoked by *devaluation,* agricultural change has set in play new migration flows from the countryside to the city; and while incomes may have risen, this has its roots in flexible and *longer working hours* – 'every adult is a great deal busier'.

Even if read as a success story, Bogotá has not escaped some of the shifts in rhythm associated with neo-liberalism. Once again, we see how the influences – to reorganize labour processes, to devalue goods to make them more competitive on world markets, and so on – flowing through today's neo-liberalized economic connections, work their way into city life, to produce a new geography of urban rhythms (but, in this case, still delivering more or less the same economic output as in the 1970s, or so it would seem).

FIGURE 6.6   (top) *An aerial view of the central area of Bogotá, Colombia*
(above) *The southern barrios of Bogotá*

7 *Conclusion*

This chapter has located the cities of developing countries in the dynamic interconnections, the wider relational webs, of today's global economy. In doing this, we have developed an understanding of 'globalization' and 'liberalization' of the world economy and used this to appreciate better the full implications of these twin processes for the context of urbanization and urban life in 'cities of the south'. The aim of the chapter was to develop an understanding of cities and their mix of times and spaces as the outcome of relational webs. By questioning the characteristics of those webs, of what exactly are the *influences* flowing through them,

● it is clear that these interconnections carry *dynamic processes* which move ideology and economic criteria between cities and in doing so shape the context of urbanization.

We have noted that increasingly both these criteria and the context that they help to shape are established by a current orthodoxy of neo-liberalism. Key international institutions, such as the World Bank and the IMF, have helped to carry the neo-liberal message and have exerted their influence on these countries and their cities, so as to establish a favourable framework for increasing flows of private-sector money and finance – the 'essential' ingredients for 'successful' urbanization.

Our account has prioritized a relational way of thinking about the imperatives of neo-liberalism and private finance, how these interconnect with and affect cities in developing countries. Moreover, we have recognized how dynamic these interconnections are. It is a dynamic that results from the practices of private finance, a dynamic produced by ongoing calculations of profitability. We have noted, too, how these flows of private finance are co-ordinated through the financial centres of global cities, most notably Tokyo, London and New York.

● These global cities can be understood as producers of dominant rhythms, rhythms that work through relational webs to bind an increasing number of cities in developing countries to the influences of neo-liberalism.

The reach of this *influence* has extended greatly following the implementation of 'solutions' to the debt crisis, which began to unfold within these countries from the beginning of the 1980s. As we saw in section 3, the implementation of structural adjustment programmes has been encouraged by the World Bank and IMF and has accelerated the adoption of neo-liberal policy. The adoption of this policy has connected and exposed these countries to the practices of the private sector and to the discipline of the rhythm of private finance. Together they have helped to establish the context for urbanization through the very particular ways they work their way through and according to the framework for economic management put in place under the influence of the World Bank and the IMF. The chapter also suggested how similar influences, fostered by the World Bank, established urban policies that further impose and refine a particular context for urbanization, one supposedly suited to the imperatives of a neo-liberal global economy. It is in this sense that we talked of 'neo-liberal cities'.

Through recognizing the particular dynamic of today's economic interconnections, it then becomes clearer how these flows and their influences settle within cities, *un*settling the daily routines, the day-to-day of the spaces and times making up these cities. We have thus moved from static notions of connections, to appreciating such connections and webs as carriers of dominant rhythms. This became clearer through the examples of the impact of structural adjustment on the cities of Bogotá, Harare and Dar es Salaam. Here we could see how the immediate effects of the imposition of new economic criteria, a new economic context, quickly established a new geography, a new choreography, of urban rhythms, as public and private spaces and times had to accommodate the dominant rhythms of neo-liberalism. In this way 'dry' and 'remote' economic policy pushing for economic growth through informalization of work practices, the call for devaluation and fiscal tightening and so on, could be traced through to the dramatically altered daily city routines.

The chapter began by questioning how 'accelerated globalization' could affect processes of urbanization, particularly in developing countries. What the chapter has attempted to show is that

● connecting cities into the very particular flows of global finance and neo-liberal ideas exposes them to unsettling rhythms which not only have immediate effects within a city, but carries the potential for further disruption.

While the key institutions of neo-liberalism such as the World Bank and the IMF suggest the inevitability of a unilinear, unified neo-liberal urban future, it seems clear that this is unlikely. Not only does such drive to build a consensus fail to recognize fully the effects of past interconnections and the unequal trajectories that this produces, the claim seems to overlook how the criteria of private finance necessarily involve the continual switching of flows from one investment opportunity to another. This can perhaps be read into figures showing flows of different forms of private finance into (and out of) different categories of developing country, as we saw in section 3. And as the neo-liberal framework seeks to transform cities and their various spaces into investment opportunities, city life and daily routines all seem set to experience an unsettling impact as flows of investment move through them.

References

Allen, J. (1999) 'Worlds within cities' in Massey, D. *et al.* (eds).

Bird, G. (1996) 'The International Monetary Fund and developing countries', *International Organization*, vol.50, no.3, pp.477–511.

Bouton, L. and Sumlinski, M.A. (1996) *Trends in Private Investment in Developing Countries 1970–79,* IFC Discussion Paper 31, Washington, DC, International Finance Corporation/World Bank.

Cohen, M.A. and Leitmann, J.L. (1994) 'Will the World Bank's real "new urban policy" please stand up?', *Habitat International,* vol.18, no.4, pp.117–26.

Corbridge, S. (1993) *Debt and Development*, Oxford, Blackwell.

Escobar Latapi, A. and Gorizates de la Rocha, M. (1995) 'Crisis, restructuring and urban poverty in Mexico', *Environment and Urbanization*, vol.7, no.1, pp.57–75.

Ferreira, D. and Khatami, K. (1996) *Financing Private Infrastructure in Developing Countries,* World Bank Discussion Paper 343, Washington, DC, World Bank.

George, S. and Sabelli, F. (1994) *Faith and Credit: The World Bank's Secular Empire,* Harmondsworth, Penguin.

Gilbert, A. (1994) *The Latin American City*, London, Latin American Bureau.

Gilbert, A. (1997) 'Work and poverty during economic restructuring: the experience of Bogotá, Colombia', *Institute of Development Studies Bulletin*, vol.28, no.2.

Green, D. (1995) *Silent Revolution: The Rise of Market Economics in Latin America*, London, Cassell.

HABITAT (1996) *An Urbanizing World: Global Report on Human Settlements 1996*, Oxford, OUP/United Nations Centre for Human Settlements (UNCHS).

Harris, N. (1996) 'Introduction' in Harris, N. (ed.) *Cities and Structural Adjustment*, London, UCL Press, pp.1–2.

Herbold Green, R. and Faber, M. (1994) 'Editorial: the structural adjustment of structural adjustment: Sub-Saharan Africa 1980–1993', *IDS Bulletin,* vol.25, no.3, pp.1–8.

Holden, P. and Rajapatirana, S. (1995) *Unshackling the Private Sector: A Latin American Survey*, IBRD/World Bank.

International Monetary Fund (1997) *World Economic Outlook: A Survey, May 1997*, Washington, DC, IMF.

Jamal, V. and Weeks, J. (1988) 'The vanishing rural–urban gap in sub-Saharan Africa', *International Labour Review*, vol.127, pp.271–90.

Janvry, A. and Sedonlet, E. (1993) 'Market, state, and civil organizations in Latin America beyond the debt crisis', *World Development*, vol.21, no.4, pp.659–74.

Jones, G.A. and Ward, P.M. (1994) 'The World Bank's "new" urban management programme', *Habitat International*, vol.18, no.3, pp.33–51.

Jones, G.A. and Ward, P.M. (1995) 'The blind men and the elephant: a critics reply', *Habitat International*, vol.19, no.1, pp.61–72.

Kanji, N. (1995) 'Gender, poverty and economic adjustment in Harare, Zimbabwe, *Environment and Urbanization*, vol.7, no.1, pp.37–55.

Kincaid, G.R. (1988) 'Policy implications of structural changes in financial markets', *Finance and Development,* March, pp.2–5.

Massey, D. (1999a) 'Cities in the world' in Massey, D. *et al.* (eds).

Massey, D. (1999b) 'On space and the city' in Massey, D. *et al.* (eds).

Massey, D., Allen, J. and Pile, S. (eds) *City Worlds*, London, Routledge/ The Open University (Book 1 in this series).

Nafziger, E.W. (1993) 'Debt adjustment and economic liberalisation in Africa', *Journal of Economic Issues*, vol.xxvii, no.2, pp.429–39.

Pugh, C. (1995) 'Urbanization in developing countries: an overview of the economic and policy issues in the 1990s', *Cities*, vol.12, no.6, pp.381–98.

Sivaramakrishnan, K.C. and Green, L. (1986) *Metropolitan Management: the Asian Experience*, New York, Oxford University Press for The Economic Development Institute of the WorldBank.

Stern, R. (1993) '"Urban management" in development assistance', *Cities,* vol.10, pp.125–38.

Tripp, A.M. (1994) 'Deindustrialization and the growth of women's economic associations and networks in urban Tanzania' in Rowbotham, S. and Mitter, S. (eds) *Bread and Dignity*, London, Routledge, pp.139–57.

Weeks, J. (1992) *Development Strategy and the Economy of Sierra Leone*, Vermount, St. Martin's Press.

World Bank (1991) *Urban Policy and Economic Development: An Agenda for the 1990s,* Washington, DC, World Bank Policy Paper.

World Bank (1994) *World Development Report*, Washington, DC, World Bank.

World Bank (1997a) *Global Economic Prospects and the Deveoping Countries*, Washington, DC, World Bank.

World Bank (1997b) *Global Development Finance*, Washington, DC, World Bank.

World Bank (1997c) *World Development Report*, Washington, DC, World Bank.

READING 6A
Nazneen Kanji: 'Gender, poverty and economic adjustment in Harare, Zimbabwe'

Introduction

In early 1991, Zimbabwe embarked on a major Economic Structural Adjustment Programme (ESAP), scheduled to run until 1995, with the objectives of sustaining higher medium and long-term economic growth and reducing poverty (Government of Zimbabwe, 1991). Some adjustment measures had already been introduced but the adoption by the Zimbabwean government of this programme marked the beginning of a new, more intensified phase in the implementation of structural adjustment. Although the government claims that ESAP is home-grown, it shares the main characteristics of programmes implemented in more than 40 countries in Africa, Latin America and parts of Asia in the 1980s. The programme includes further cuts in consumer subsidies, severe cutbacks in government spending (including the social sectors), extensive liberalization of price and import controls, and promotion of exports, particularly the expansion of non-traditional exports, that is, manufactured goods …

… The research was carried out in Harare, the capital of Zimbabwe, which has a population of 882,000. Kambuzuma, a high-density suburb located ten kilometres from the city centre near the industrial area, was selected for the study. The term 'high-density suburb' refers to townships built in the pre-independence period to house the black population. After independence, better-off black families moved to lower-density suburbs, previously exclusively white. Some better-off black families did remain in high-density suburbs due to the housing shortage in Harare so that there is a considerable income range within these settlements. High-density suburbs, however, remain exclusively black.

Kambuzuma is a well established and stable settlement with varying income levels and tenure status, and is fairly typical of most high-density suburbs. It is not the poorest of locations since the capital has at least two poorer settlements with squatters and recent migrants. In Kambuzuma, a high proportion of men are employed in industry. The areas of investigation included employment and income-generating activities; household expenditure, debt and savings; domestic work; use of social services; leisure activities and involvement in social organizations …

…

ESAP has not been successful to date, even using the narrow economic criteria of the international finance institutions and the government. The droughts of 1990–91 and 1991–92 undoubtedly had a negative impact on macro-economic growth rates but many economists argue that the policies of extensive import liberalization and currency devaluations have played their part in undermining economic growth in Zimbabwe … In 1992, debt-servicing rose to 27.2 per cent of export earnings (World Bank, 1993), rather than declining to 20 per cent as predicted by the government (Government of Zimbabwe, 1991).

ESAP and the cost of living

Since the introduction of ESAP, prices of food and basic commodities have escalated. The removal of subsidies and the de-control of prices have resulted in a huge increase in the cost of living. This was exacerbated by escalating costs of imported inputs for manufactured products due to the devaluation of the Zimbabwean dollar. The cost of living for lower-income urban families rose by 45 per cent between mid-1991 and mid-1992 … whilst that for higher income groups rose by 36 per cent (CSO, 1992) …

ESAP and social services

In January 1992, primary school fees were introduced and secondary school fees in urban areas were raised. This, in addition to levies already set by individual schools and approved by government …

The costs of other basic services also rose in this period. In November 1991, electricity charges rose by 15.5 per cent having already been increased by 20 per cent in July 1991. This was justified in terms of the higher costs of imported inputs due to devaluation … Urban bus fares were increased in November 1991, a basic 55 cents bus fare rose to 80 cents. Payments for various services, therefore, make competing demands on low incomes.

Changes in household expenditure

Table 1 shows average monthly household expenditure on regular items for 1991 and 1992. Whilst food and total regular expenditure applies to all 100 households [in the research sample], the figures for the other items relate to the number of households which actually incurred those expenses.

TABLE 1 *Changes in regular household expenditure (Z$)*

Item	Average monthly sum (Z$)		Percentage rise
	1991 US$1=Z$2.8	1992 US$1-Z$4.9	
Food	284 (N=100)	382 (N=100)	35
Food at work	42 (N=65)	58 (N=56)	38
Transport	57 (N-66)	69 (N=54)	21
Electricity	39 (N=62)	49 (N=64)	26
Other fuel	12 (N=34)	14 (N=47)	17
Water	6 (N=86)	6 (N=92)	0
Rent	97 (N=40)	128 (N=40)	32
Rates	40 (N=60)	48 (N=60)	20
Education	54 (N=76)	60 (N=81)	11
Money to family	62 (N=61)	88 (N=56)	42
Other expenses	69 (N=32)	87 (N=39)	26
Total regular expenditure	542 (N=100)	727 (N=100)	34

Table 1 gives a picture of the magnitude of rises in absolute terms for various items included in the household budget. However, expenditure *fell* in the real terms, when compared to the rise in the cost of living. The real change in total regular expenditure is minus 7.6 per cent, since the cost of living rose by 45 per cent. Whilst Table 1 shows the decline in expenditure for the sample as a whole, disaggregating the sample provides a clearer view of which groups faced the most severe problems. Per capita regular expenditure rose by 47 per cent in the highest 25 per cent income group, representing a real increase of 1.4 per cent, whilst it only rose by 27 per cent in the lowest 25 per cent income group, representing a real decline of 12.4 per cent. The more severe effect of price rises on the poorest households are clearly illustrated by the way they were forced to make the greatest cutbacks on basic items.

Changes in food expenditure and consumption

… Nine per cent of very poor households actually spent less money on food in 1992 than they did in 1991, despite the enormous price rises.

Household structure also influenced changes in per capita food expenditure. Spending in real terms declined by 1.3 per cent in male headed households whereas it declined by 13.4 per cent in female headed households. The latter group, which are over-represented in lower-income percentiles and under-represented in the higher percentiles, therefore made much greater cuts in consumption. Another factor which may account for this difference is that female household heads were more reluctant to borrow

money and fewer (as compared to male heads) had the option of taking loans or salary advances from employers. They were, therefore, more likely to opt for cutting back on consumption to make ends meet …

Changes in spending on services

Lodger households faced enormous rent rises during 1991 and 1992, mainly as a result of attempts by main householders to buffer themselves from the increased cost of living, including local council rate rises for basic services. Many lodgers felt that they no longer had any hope of owning their own 'stand' in current sites-and-service projects because they could not afford the down-payment and/or the monthly repayments. In any case, the huge gap in supply and demand meant that rents were rising and overcrowding was rife (Potts and Mutambirwa, 1991). For households which did not maintain strong links with their rural areas of origin, this provoked enormous anxiety for the future.

Fewer households were able to pay for transport to work and a number of men and women had begun walking to work or had arranged lifts which cost less than the bus service …

…

Changes in spending priorities

Table 2 shows the changes in monthly expenditure on clothes and school uniforms for households (number shown in brackets) incurring these expenses. The percentage changes in absolute and real terms (compared to the overall rise in the cost of living of 45 per cent) are also shown.

TABLE 2 *Changes in expenditure on clothes and school uniforms (Z$)*

Item	Average monthly sum		% absolute change	% real change
	1991 US$1=Z$2.8	1992 US$1-Z$4.9		
Clothes	$64 (94)	$64 (72)	0	-31
School uniforms	$18 (73)	$16 (65)	-11	-39

As the figures show, most households had cut down drastically on new clothing and school uniforms, logically seeing this as the first area in which cuts could be made with the least repercussions. The number of households which could not afford to buy any clothes went up from six in 1991 to 28 in 1992. The decline in the demand for clothing and textile products was manifest at the national level and resulted in the retrenchment of about 3,000 workers and the closure of several small companies in Harare and Bulawayo, the second largest city. The Zimbabwe Clothing Council estimated that the local market was 20 per cent of normal as a result of drought in rural areas and the rise in the cost of living in urban areas. The loss could not be made up by exporting more because of the worldwide recession in the clothing industry.

References

Central Statistical Office (1992) 'Consumer price index for lower-income urban families', CSO, Harare, 30 September.

Government of Zimbabwe (1991) *Zimbabwe : A Framework for Economic Reform (1991–1995)*, Harare, Zimbabwe Government Printers.

Potts, D. and Mutambirwa, C. (1991) 'High-desnity housing in Harare: commodicifcation and overcrowding, *Third World Planning Review*, vol.13, no.1, pp.1–25.

World Bank (1993) *World Development Report* , Oxford, Oxford University Press.

Source: Kanji, 1995, pp.37–44

Taken from *Environment and Urbanization*, a twice-yearly journal about urban development and the urban environment in Africa and Asia and Latin America published by the Institute for Enviornment and Development (IIED). The journal seeks to inform thinking and advance learning about critical themes and to ensure recognition for Southern authors (practitioners and academics).

For more information, contact the Human Settlements Programme, IIED, 3 Endsleigh Street, London WC1H 0DD, UK, telephone: (0171) 388 2117; fax (0171) 388 2826; e-mail: humans@iied.org. See also the web page http://ww.iied.org/human/envurb.html (as at November 1998).

READING 6B
Aili Mari Tripp:
'Deindustrialization and the growth of women's economic associations and networks in urban Tanzania'

Causes of deindustrialisation

Like many African countries, Tanzania suffered in the 1970s from drops in export commodity prices; increases in import prices, especially of oil; and rising interest rates. Tanzania also suffered in this period from drought, from its war with Uganda and from the breakup of the East African Community, Tanzania's economic consortium with Uganda and Kenya.

These external causes of crisis were compounded by internal policies, including the heavy investment in industrial programmes at the expense of agriculture, in a country where 85 per cent of the population is engaged in agricultural production. Unfavourable government prices for producers of export crops resulted in shifts away from the production of cash crops to the production of food crops and to sales of both export and domestic crops on unofficial markets rather than through state-run crop authorities, thus undermining the state's main resource base.

The crisis that ensued took many different forms. The trade balance, which had been in surplus in the 1960s, had fallen into deficit by the 1970s. In the late 1970s, Tanzania experienced a severe and chronic foreign exchange crisis, which subsequently affected agricultural and industrial production, both of which are heavily dependent on imported inputs ...

The crisis limited the government's options significantly and forced it to initiate a series of economic reform programmes. The first two programmes which were initiated (the 1981 National Economic Stabilisation Programme and the 1982 Structural Adjustment Programme) were largely unsuccessful. Two Economic Recovery Programmes (ERP) followed in 1986 and 1989 ...

In spite of the economic reform programmes, Tanzania's foreign exchange crisis persisted, while pressures from foreign donors mounted, forcing Tanzania to negotiate an agreement with the International Monetary Fund (IMF) in 1986 after a six-year stalemate in talks. The IMF agreed to a standby arrangement subject to various criteria, which included substantial devaluations; restrictions on the amount of credit that could be transferred from the banking system to government institutions; limits on the accumulation of new debt; and controls on new external borrowing and on the overall budget deficit.

...

After 1985 both exports and imports grew, but the total import bill was still three times greater than export receipts in 1990 ... By 1988 Tanzania's debt had risen to almost twice its GNP and ten times its export earnings ...

One of the most alarming consequences of the Economic Reform Programme, however, was the manner in which government expenditure was redistributed. Social, public and welfare services as a percentage of government expenditure had decreased substantially in the 1980s and by 1986 they had dropped to their lowest point in twenty years. The amount spent on health, education, housing and public, community and social services was cut in half between 1981 and 1986, from 21 to 11 per cent of total government expenditure. Attempts were made to reverse these cuts with the 1989 Economic Recovery Programme–Priority Social Action Programme.

Consequences of deindustrialisation

...

In sum, the ERP adversely affected industrial performance in a number of ways. Tight monetary policies hindered industries from obtaining loans for financing working capital; trade liberalisation eroded the domestic market for various industries; devaluation increased the price of directly imported inputs; and raising producer prices stimulated the production of crops but resulted in sharp increases in the cost of inputs (Mbelle, 1989, pp.9–10).

Urban Tanzanian women's responses to deindustrialisation

For workers, this process of deindustrialisation meant sharply declining real wages and layoffs. There had been an 83 per cent decline in the real income of wage earners between 1974 and 1988 (Bureau of Statistics, 1989). Prices increased 5.5 times from 1982 to 1988, while wages only doubled in this same period. For a low-income household food expenditure alone exceeded the minimum monthly salary eight times. Because wage earnings could no longer support the household, the burden fell largely on women to sustain the household through informal small businesses. Because of low wages and discriminatory hiring practices, women were excluded from formal employment. As real incomes dropped for men, women were therefore more likely to initiate income-generating activities, frequently making them the main breadwinner in the family. In 1976, wages constituted 77 per cent of the total household income, while other private incomes accounted for only 8 per

cent of the total household income (Bureau of Statistics, 1977). In contrast, by 1988, according to my survey in Dar es Salaam, informal incomes constituted approximately 90 per cent of the household income, with wage earnings making up the remainder.

Thus, women, who in the past had contributed relatively little to the urban household income, were now critical to the very survival of the household through their involvement in projects. For low-income women these projects ranged from making and selling pastries, beer, paper bags, and crafts, to tailoring, hair braiding and even urban farming.

By the late 1980s it was women themselves who were often rejecting wage employment and seeking more beneficial incomes from self-employment. Women as a proportion of the employed workforce in Dar es Salaam had risen from 4 per cent in 1961 to 11 per cent in 1967 and 20 per cent in 1980. By 1984 the proportion had dropped to 19 per cent and if my survey is any indication it remained at 19 per cent up to 1987 (International Labour Organisation, 1982, p.143). Between 1953 and 1980 few women left their jobs; according to my survey, however, the numbers more than doubled between 1980 and 1987. The most common reason both men and women (45 per cent) gave for leaving was low wages. The second most frequent response (17 per cent) was layoffs due to factory cutbacks or closures. Women were hit especially hard by these layoffs. For example, in 1985 when 12,760 workers were laid off in a retrenchment exercise, most of them were women (Tibaijuka, 1988, pp.31–2).

As one woman hairdresser who also sold charcoal and fried buns put it:

'In the past women took care of the children and didn't know how to make money for themselves. Now women have to find work [self-employment] because this is the way things are progressing in the country. You can't wait for your husband to give you money. You have to go out and find work for yourself. That is why you find women today doing all kinds of things like making *maandazi* [buns], selling *mitumba*, sewing, everything.'

(interview with author, 15 August 1987)

Women's changing role is also borne out by comparing various surveys. A 1970–1 study found 66 per cent of women in Dar es Salaam with no source of income; 13 per cent of the women were wage earners; and 7 per cent self-employed (Sabot, 1979, p.92). In contrast, my 1987–8 survey found 66 per cent of women self-employed while 9 per cent were wage-employed. Similarly, another survey of 134 women in Dar es Salaam in 1987 showed that 70 per cent of all women had projects while 5 per cent were employed

(Tibaijuka, 1988, p.37). Sixty-five per cent of those who started small businesses between 1982 and 1987 were women, while in the previous five years women made up only 28 per cent of those starting businesses. Of the self-employed women 67 per cent began their enterprises between 1980 and 1987, while only fifteen per cent started in the 1970s. Women workers, like men, frequently had sideline businesses because wages were so low relative to the income they could obtain from a small enterprise.

Participation in the informal economy had resulted in massive reversals in dependencies and reversals in the direction of resource flows at many different levels: away from the state towards private solutions to problems of income, security and social welfare; away from reliance on wage labour to reliance on informal incomes and farming; and a gradual shifting of immigration patterns with increasing new movements out of the city into the rural areas. Similarly, relations and patterns of obligation in the household were being transformed with greater resource dependencies on women, children and the elderly, where only a decade earlier urban women had mostly relied on men for income, children on their parents and the elderly on their adult children.

The pervasiveness of urban women's involvement in business was unprecedented in a country like Tanzania. This major change went relatively unnoticed by the government, scholarly and donor circles since much of it was informal and therefore unrecorded. Moreover, women's informal work was often considered unimportant and petty because of the bias towards official employment. But gradually men were, at least privately, beginning to take note of these changes. As the head of the National Bank of Commerce told me in 1987: 'Women have become the largest private sector in Tanzania, but no one knows what they do' (interview, 11 October 1987).

One indication of men's acceptance of women's increased involvement in micro-enterprises was the fact that 44 per cent of women with projects had received starting capital from their husbands. Several men indicated their apprehension, especially about the fact that women were making so much more than they. One way for women to avoid open conflict with their husbands over their new access to cash was not to disclose their full earnings to their spouse. This was also a way of ensuring control over their earnings, which had to be stretched to cover the household expenses.

The new economic importance of women to the household was also changing women's perceptions of themselves. As one woman, who had left employment as a secretary at a state-owned company to start a hairdressing salon in 1984, explained:

'I started a business because I did not want to use my body to get money. I had four children to support and my salary was not enough. I wanted to use my own hands, brains and wanted to aim high. Whatever a man can do, I can do better. If you start aiming high, the sky is the limit... . Most women are in business and are no longer dependent on men.'

(interview with author, 18 August 1987)

...

Creating new networks

...

Urban life presented new opportunities for women to meet new people and establish information and resource networks. For example, living together in one house, as many families did, encouraged women neighbours to start projects together. It was not uncommon to find women living in the same house sharing a project like making local beer, while maintaining separate accounts. In one apartment complex, I visited four separate apartments where women were involved in sewing as a business and I was told that there were more women sewers in the building. They were all friends and had inspired each other to begin their projects to supplement their husbands' meagre wages. They helped each other out with their businesses and insisted that they did not compete with one another. Other women found their formal jobs a place to exchange ideas and support for their projects. The hairdressing salon was also another frequently mentioned place for meeting other women to discuss business.

Working arrangements with friends and neighbours were of various kinds and served many different purposes. Friends would accompany each other to a restaurant or bar, one perhaps selling fried fish and chips and the other selling *makongoro* soup ...

...

Conclusions

The decline of large-scale industry in Tanzania throughout the second half of the 1970s and the 1980s resulted in declining real wages and layoffs, putting greater pressures on household members to pursue income-generating projects. Similar patterns were found in many African countries to one degree or another. Urban women thus moved from a position of relatively little involvement in income generating activities, to being, in many cases, the main economic support for the household ...

...

In order to understand women's response to the impact of deindustrialisation upon relationships in

Tanzanian society it is important to look not only at such visible formal organisations, but also at those organisations embedded in the daily lives of women. Through such associations, women are transforming their lives in their day-to-day struggle to survive and at the same time changing the political landscape of the country.

References

Bureau of Statistics (1977) *Household Budget Survey: Income and Consumption 1976–77*, Ministry of Finance, Planning and Economic Affairs, Dar es Salaam.

Bureau of Statistics (1989) *1988 Population Census: Preliminary Report*, Ministry of Finance, Planning and Economic Affairs, Dar es Salaam.

International Labour Organisation (1982) *Basic Needs in Danger: A Basic Needs Oriented Development Strategy for Tanzania*, Addis Ababa, International Labour Organisation .

Mbelle, A. (1989) 'Industrial (manufacturing) performance in the ERP context in Tanzania', *Tanzania Economic Trends*, vol.2, no.1, pp.7–9.

Sabot, R.H. (1979) *Economic Development and Urban Migration: Tanzania 1900–1971*, Oxford, Clarendon Press.

Tibaijuka, A.K. (1988) *The Impact of Structural Adjustment Programmes on Women: The Case of Tanzania's Economic Recovery Programme*, Economic Research Bureau, University of Dar es Salaam.

Source: Tripp, 1994, pp.140–4, 154–6

READING 6C
Alan Gilbert: 'Work and poverty during economic restructuring: the experience of Bogotá, Colombia'

Introduction

Over the last 20 years 'globalisation' has continued apace and has consolidated an already highly integrated world economy (Dicken, 1986; Mittelman, 1994; Ould-Mey, 1994; World Bank, 1995; ILO, 1995). Major changes have occurred in the operation of the world's capital markets, the location of productive activity, the role of transnational companies, the ways in which business enterprises operate and the functioning of labour markets. Globalisation has demanded that national and local economies be restructured in order to compete internationally; a process often known as liberalisation. Liberalisation has a major impact on employment and welfare. Those who approve of globalisation and liberalisation argue that these processes have brought major benefits in terms of employment creation and reduced poverty in Third World countries (World Bank, 1995; Dornbusch and Edwards (eds), 1991). In contrast, the critics of globalisation argue that greater international competition and internal liberalisation have led to higher unemployment, excessive casualisation of labour markets, economic instability and deteriorating living conditions for large numbers of people (Sassen, 1991; Friedmann and Wolff, 1982; Green, 1995; Iglesias, 1992).

…

The choice of Bogotá, as an exemplar of globalisation is both appropriate and inappropriate. It is apt insofar as it is the capital of a country that is widely regarded as an exemplary model of a well-managed Third World economy. In addition, recent national governments have been reforming the Colombian economy following neo-classical economic guidelines. At its crudest, recent national government policy has attempted to turn the Colombian economy into a Latin American version of an Asian 'tiger'. An attempt has been made to make the economic system more efficient and to reduce the role of the state. Colombia's economy has managed to grow successfully for many years and has been broadly following neo-liberal management principles since 1990. I would suggest that what happens in Colombia in terms of employment, poverty and urban living conditions represents a best-case test of the New Economic Model (Scott, 1995).

At the same time, Colombia is an inappropriate exemplar of globalisation and the New Economic Model insofar as it never plumbed the depths of the debt crisis in the way characteristic of most other Latin American countries. Both luck and good management meant that the national economy fared rather well during the 1980s. Recent attempts at liberalisation and modernisation are being superimposed on an economy that experienced little in the way of IMF-imposed austerity. Colombia is also an inappropriate exemplar in the sense that it is an extremely complicated country to understand. It is highly regionalised, there is a vast black economy, political and social violence are rife, and considerable areas of the country lie beyond the control of government.

Partly because of this complexity I am choosing to concentrate on Bogotá. What is happening in the national capital is a great deal easier to analyse than what is occurring in the country as a whole. It also has a further advantage for my purpose.

Since Bogotá has prospered even more than the Colombian economy as a whole, it represents a best-case scenario within a best-case scenario. If the poor have fared well in Bogotá, then the current optimism about the benefits of appropriate macro-economic management can be accepted less reservedly, at least in those countries where economic growth is associated with the liberalisation process. On the other hand, if the evidence shows that Bogotá's poor have gained little, the neo-liberal approach needs to be questioned forcefully. If liberalisation does not improve living standards in an economy that has been growing rapidly, it is likely to have an extremely negative impact in less successful Latin American countries. In short, within Latin America, the experience of Bogotá offers a partial test of the effectiveness of trickle-down economics in an expanding economy.

Trends in poverty

Colombia and its capital largely escaped the 'lost decade' [of the 1980s] (Morley, 1992). As a result, poverty seems to have become less serious. Data on both income and the satisfaction of basic needs clearly suggest that poverty fell dramatically in Colombia during the 1970s and 1980s. Such a trend occurred in the countryside, in the urban areas and in the seven major cities. It was also true for Bogotá.

The quality of life in Bogotá undoubtedly improved in many respects between 1970 and the early 1990s. Life expectancy rose by five years between the early 1970s and the early 1990s and the infant mortality rate fell from 50 per thousand live births in 1971 to 22 in 1993 (Rinaudo *et al.*, 1994, p.28). Literacy rates also rose impressively, from 85 per cent in 1973 to 96 per cent in 1993 (ibid.). The proportion of homes built out of flimsy materials fell from 7 per cent in 1973 to 3 per cent in 1993. Per capita incomes also rose and, between 1971 and 1993, the city's gross domestic product rose at an annual rate of 2.2 per cent. Poverty also fell; the proportion of *bogotanos* living in poverty declining from 57 per cent in 1973 to 17 per cent in 1991, those living in misery from 26 to 4 per cent (Londoño de la Cuesta, 1992, p.15). Nevertheless, far too many *bogotanos* continued to live in poverty. In 1991, some 800,000 people still lacked certain basic needs and 200,000 lived in misery (Londoño de la Cuesta, 1992, p.15).

Income levels

Between 1970 and 1986, the value of the minimum salary in Colombia rose fairly consistently and increased by between one-third and one-half over that period depending on the month of calculation. Since then its real value has gradually declined and its value in June 1995 was 12 per cent lower than it had been in June 1986. Similarly, industrial wages are higher today than they were in the 1970s, although the difference is much less marked than with the minimum salary (Urrutia, ed., 1991). Since the middle 1980s, real industrial wages have increased slightly.

This generally positive picture needs to be qualified, however, in three important respects. First, the proportion of workers earning less than the minimum wage has increased rapidly during the 1990s; from 18 per cent in 1990 to 23 per cent in 1995 … While domestic service workers did much better, industrial workers did much worse. Since the real value of the minimum wage declined by 7 per cent between March 1990 and March 1995 the increasing proportion of all workers earning less than that level is very serious. Second, between 1976 and 1995, the proportion of workers earning less than two minimum salaries in Bogotá rose from 50 to 58 per cent. Thus, the proportion of workers earning low incomes has actually increased, particularly since 1990. Finally, labour's share of GDP increased from 1970 until the early 1980s and then declined. If we compare the situation in 1970 with that in 1992, labour's share of GDP remained basically stable. Under the forces operating under globalisation such an outcome might be viewed positively. The problem with such an interpretation is that the proportion of workers in Colombia's population increased greatly. As such, many more workers were working in 1992 to produce the same labour share of GDP as in 1970.

The overall picture, therefore, is that most workers in Bogotá are earning more today than they were 20 years ago, but are earning lower wages than they were ten years ago.

Conclusions

The evidence on employment trends in Bogotá suggests that many of the fears about economic restructuring are unjustified. Despite liberalisation and a vast increase in the number of people seeking work, unemployment rates have fallen. Jobs have been created in a variety of sectors although during the 1990s, principally in the informal sector. However, the trend towards informality has not been consistent across sectors; manufacturing and financial services have been casualised but many more formal jobs have been created in commerce and construction. The quality of work has probably not been greatly affected by these changes except in terms of work stability. Workers in Bogotá do not hold their jobs for very long.

Feminisation has been operating very strongly in Bogotá and would seem to have had little to do with either liberalisation or increasing poverty. The labour market has changed and jobs are much less stratified by gender than they were. Today, some women work because there are good jobs open to them; others work because they have no choice. The household survival strategy hypothesis seems to explain only a limited part of Bogotá's experience. There seems little real doubt that since 1970 poverty in Bogotá has become both less common and less serious. Incomes have increased, although since 1990 a rising proportion of the workforce has earned less than the minimum wage. Shelter conditions have generally improved, although there is a major question mark about the quality of service delivery after 1985. Despite these overall improvements, the highly unequal distribution of income has remained unchanged. By general Latin American standards Bogotá's is not a bad record, especially given the continued arrival in the city of numbers of migrants displaced by rural violence and agricultural change. Economic growth has managed to create jobs and household incomes have risen.

However, one important caveat must be made about the social record. If there is less poverty, it is principally because children form a smaller proportion of the population. There are many more adults in the 1990s than there were in the 1970s. Because of this change in the age structure, more *bogotanos* are working. While the ability of the economy to generate jobs for this vast expansion in the number of workers is impressive, it is clear that increasing numbers of workers are toiling in insecure forms of employment. If more homes have televisions and fewer children go hungry, it is not because individual workers' incomes have risen; the 1990s have seen a decline in the real value of the minimum wage and an increase in the number of workers earning that wage. If household incomes have risen, it is principally because more

adults are working. Reduced poverty has been brought about by the huge rise in labour participation. [Fewer] people are hungry, but every adult is also a great deal busier.

Given the polarisation of views about the desirability of economic restructuring and the New Economic Model, we are all likely to respond to the Bogotá experience in different ways. Some readers will no doubt conclude that to be poor in Bogotá is far preferable to being poor in Lima or Mexico City where real incomes have plummeted and poverty has increased dramatically since 1980. In this respect, Bogotá is a positive example of what restructuring and sensible macro-economic policy can bring to Latin America. Its record offers the hope that economic growth does genuinely 'trickle down'. It creates jobs and can reduce poverty. No doubt other readers will be concerned about Bogotá's flawed social record. Basic needs provision is currently in decline despite continued economic growth. Similarly, while welcoming the decline in poverty, many will lament that the poor have mainly improved their living standards because more of them are working than ever before. If this is the best that can be achieved during thirty years of almost sustained economic growth, it is arguably a dubious achievement.

Clearly, no simple conclusion can be drawn because the case either way is inconclusive. While many of the trends I have demonstrated are clearcut, their causes are much less obvious. What Bogotá's rather impressive data set does demonstrate is that only some of the more general conclusions suggested by the restructuring literature are valid. Both supporters of neoliberal orthodoxy and its critics have been provided with ammunition. Perhaps both should reflect that the data provide strong prima facie evidence that some of each side's basic arguments may be flawed.

References

Dicken, P. (1986) *Global Shift: Industrial Change in a Turbulent World*, London, Harper and Row.

Dornbusch, R. and Edwards, S. (eds) (1991) *The Macro-economics of Populism in Latin America*, Chicago, IL, Univesity of Chicago Press.

Friedmann, J. and Wolff, G. (1982) 'World city formation: an agenda for research and action', *International Journal of Urban and Regional Research*, vol.6, pp.309–43.

Green, D. (1995) *Silent Revolution: The Rise of Market Economics in Latin America*, London, Latin American Bureau.

Iglesias, E.V. (1992) *Reflections on Economic Development: Toward a New Latin American Consensus*, Wsashnington, DC, Inter-American Development Bank.

ILO (1995) *World Employment: an ILO Report: 1995*, Geneva, International Labour Office.

Londoño de la Cuesta, J.L. (1992) 'Problemas, instituciones y finanzas para el desarrollo de Bogotá: algunos interrogantes' in Departamento Nacional de Planeación (ed.), *Bogotá: Problemas y Soluciones*, Bogotá, pp.13–38.

Mittelman, J.H. (1994) 'The globalisation challenge: surviving at the margins', *Third World Quarterly*, vol.15, pp.427–43.

Morley, S. (1992) 'Policy, structure and the reduction of poverty in Colombia: 1980–1989', Washington, DC, Inter-American Development Bank Working Paper, Series 126.

Ould-Mey, M. (1994) 'Global adjustment: implications for peripheral states', *Third World Quarterly*, vol.15, pp.319–36.

Rinaudo, U. Molina, C.G., Alviar, M. and Salázar, D. (1994) *El Futuro de la Capital: Estudio Propsectivo del Desarrollo Social y Humano*, Misión, Bogotá Siglo XXI.

Sassen, S. (1991) *The Global City: New York, London, Tokyo*, Princeton, NJ, Princeton University Press.

Scott, C.D. (1995) 'The distributive impact of the new economic model in Chile', paper presented to conference on The new economic model in Latin America and its impact on income distribution and poverty, Canning House, 11–12 May.

Urrutia, M. (ed.) (1991) *40 Años de Desarollo: Su Impacto Social,* Bogotá, Biblioteca Banco Popular.

World Bank (1995) *World Development Report 1995*, Oxford, Oxford University Press.

Source: Gilbert, 1997, pp.24–5, 30–1, 32–4

CHAPTER 7
Cities and economic change: global governance?

by Nigel Thrift

1 *Introduction*

The World Economic Forum, founded in 1971, is a not-for-profit forum, perhaps best known for the World Economic Summit it runs each year in Davos, Switzerland. It is dedicated to 'entrepreneurship in the global public interest'. In the advertisement shown in Figure 7.1 opposite, it is looking for 'global community builders', who can act as catalysts and facilitators in creating 'new powerful global communities'.

The advertisement makes three points. First , and most straightforward, the global community builders form part of a bid by the World Economic Forum to *govern* the global economy by bringing together a coalition of nine powerful constituent groups, including leaders of industry, heads of state, the media, cultural leaders and regional and city leaders (see Figure 7.2).

FIGURE 7.2 *The World Economic Forum*

Second, these global community builders are labouring to produce an account of what the global economy consists of, and in producing this description it is possible to argue that they are also, in part, *producing what is there to be governed.* After all, although we may think of the global economy as a coherent, bounded set of activities which can easily be apprehended and ordered, the reality, as Chapters 1, 5 and 6 have already shown, is rather different. The global economy is a blizzard of transactions – continuous, never-ending transactions – which require continuous human labour and technological expertise to keep going. Take the case of the Ford Motor Company in Europe. At any one time, Ford has about 12,000 tonnes of components in transit around Europe – on

**COMMITTED TO
IMPROVING THE STATE
OF THE WORLD**

The World Economic Forum is expanding its activities into the digital age. As a result and to maintain high standards in serving our members, we are now looking for several truly exceptional and highly qualified professionals to join our organisation as

Global Community Builders

The World Economic Forum is based on the principle that the great economic and social challenges of humankind cannot be met by governments or business alone. A strong business–government alliance is needed, empowered by the integration of the world's best experts, and made transparent to the public by the media.

The World Economic Forum has become the foremost global partnership of leaders from business, government, academia and the media committed to improving the state of the world. Our role is to be a catalyst and facilitator for continuous, highest-level, personalized, value-added interaction for global integration among our constituency. We are incorporated in Geneva as an independent and impartial not-for-profit foundation.

The World Economic Forum particularly serves the following communities on a global scale:

- 1000 *foundation members*
  the foundation members of The World Economic Forum are the 1000 leading global companies in over 70 countries, which are represented at the chief executive or chairman level. The combined revenues of member companies has been calculated to be well over US$ 4000 billion.

- *GGC members*
  the World Economic Forum has created a special membership category for Global Growth Companies (GGCs). The world's industrial activities are increasingly determined not only by the traditional multinational companies but also by smaller entrepreneurial enterprises that very often act as niche players on a global scale.

- *Industry Governors*
  exclusive groups composed of the chief executives of the most dynamic companies in the world's key industrial sectors.

- *Global Leaders for Tomorrow (GLTs)*
  some 600 men and women under the age of 43 who will shape the future. They come from business, politics, public interest groups, the media, the arts and sciences.

- *Informal Group of World Executive Leaders (IGWEL)*
  heads of state and government, ministers and officials from international organisations who are regular discussion partners.

- *Media Leaders Club*
  editors-in-chief and commentators from the most influential media groups.

- *Forum Fellows*
  over 400 academics and experts from political, economic, scientific, social and cultural fields.

- *Cultural Leaders*
  some 100 distinguished figures from the world of arts.

- *Regional and City Leaders*
  heads of some of the world's most dynamic and successful regions and mega-cities.

Our different activities, particularly the Annual Meeting of our members in Davos, Switzerland, have gained a worldwide reputation as the pre-eminent gatherings of leaders from the private and public sector.

The time has now come to combine face-to-face with electronic and virtual dimensions of interaction into new powerful global communities.

The World Economic Forum has developed, and successfully presented at the 1997 Annual Meeting in Davos, WELCOM (the World Electronic Community), a secure, private, high-level, reliable network for active, multipoint, face-to-face global videocomunication.

The World Economic Forum is also actively involved in a pioneering project (Think Tools) using technology to support genine thinking processes as required in understanding and shaping complex structures and dynamics. The project uses the most advanced results of modern science for entirely new forms of planning and decision-making, both in politics and industry.

Candidates must have the following background and personal profiles

a) *academic background and professional experience/skills*
   - solid academic background and education from a recognized university
   - several years of practical work experience in industry/consulting (private sector) and/or with academic or (non-) governmental institutions
   - fully computer literate, fascination and experience in the use of latest technology supporting virtual communication

b) *cognitive and conceptual skills*
   - ability to think analytically, in a structured and systematic way and in strategic terms
   - innovativeness and creativity; the ability to absorb complex constellations quickly and to find creative solutions that take the discussion to a new level

c) *communication and moderation skills*
   - networking power at top level worldwide, ability to create trust and to trigger continuous, personalized and value-added interaction
   - ability to motivate and steer a group selecting the relevant issues and maintaining pace and coherence of the interaction process
   - natural authority; ability to fascinate people and to make them listen and contribute to the interaction process
   - strong communication skills including fluency in English and at least one additional language

d) *personality*
   - self-motivated, hard-working, organized and flexible; not afraid of operational/administrative details and of working under pressure
   - able and willing to work as part of a team in a multicultural environment; at ease within a lean and flat organization

Senior Community Builders
will take over full strategic, operational and financial responsibility for one or more communities. They have substantial expertise to provide an immediate added-value and have already served in capacities which establish them as full peers inside their community. Today, they hold a senior position in industry, the public sector or the academic world (e.g. top managers, highest-level former government members, well-known professors, etc.).

Junior Community Builders
will assist one of our Executive Board members in his task and gradually take over full operational responsibility for their community. They have several years of practical experience and can immediately add substance and value in the community-building process.

Your workplace will be Geneva, Switzerland or Cambridge, Massachusetts. We will offer you, in the framework of an international organization, attractive working conditions. Only if your background and experience fully correspond with the requirements, please send a cover letter with a one-page summary of your curriculum vitae, marked 'confidential', to the address below. Holders of valid working permits for either Switzerland or the USA will be treated with preference. In case of interest on our side we will react immediately.

Thomas Scherer Human Resources 53 chemin des Hauts-Crêts,
1223 COLOGNY/GENEVA Switzerland

FIGURE 7.1   *An advertisement by the World Economic Forum for 'global community builders',*
The Economist, *19 April 1997*

lorries, trains and ships – moving between its plants. Now think about interactions like this – with all the accompanying phone calls, faxes, e-mails, memos, filing and general paperwork – powered up by many thousands of times, and channelled through all kinds of government, financial and other institutions – from customs agents and tax authorities to banks and lawyers. What we can see, straightaway, is a situation of almost unimaginable complexity.

It follows that we have to turn to certain conventions to describe this situation, practical 'shorthand' stories that both describe what the global economy is and how it should be governed. Of these conventions perhaps the most persuasive currently are those that go under the general title of *neo-liberalism*. As you have seen from Chapter 6, these neo-liberal conventions are a set of stories about the appropriate ways and means of governing the world economy, stories about what governments, industries, financial systems and the like must be like and must do to become ever more economically competitive in an ever more mobile world.

The third point made by the advertisement is that one of the constituencies with which the global community builders must make contact is regional and city leaders. Why? Because, increasingly, *cities and regions* are regarded as *actors in their own right* in the global economy. According to the conventions of neo-liberalism, they must strive to make themselves more economically competitive, thereby making themselves into 'better' places.

This chapter is therefore about

●   how cities and regions have increasingly become both the objects of, and players in, the governance of the global economy.

Notice that I use the term 'governance'. This is because I want you to think about how the global economy is governed in a much broader sense than is often implied by the term 'government', which is usually associated only with the activities of nation states. The term governance suggests that the global economy is increasingly governed by all manner of institutions – like the World Economic Forum – and not just nation-state institutions.

ACTIVITY 7.1   Using the previous chapters in this book, make a list of the international organizations that have an influence – direct or indirect – on the governance of cities and regions. (When I made up such a list I was hard put to find an organization which did not have some influence.) Then choose two major cities, one in the developed world and one in the developing world, and try to assess the degree of influence that each international organization you have listed has in the two cities you have chosen. In most cases I think you will find that this influence is much greater in the developing country city. What does this tell you about the governance of cities in the global economy? ◆

The issue of governance is explored in each of the chapter's six main sections. In section 2, I will briefly describe the growth of the conventions of neo-liberalism since the Second World War as a new form of what the French philosopher, Michel Foucault, called 'governmentality'. In the rest of the chapter we will look at how these conventions have played out in practice in and through cities. My argument is that these conventions are rooted in a specific urban geography which itself acts as one of the means of governance of the global economy as well as being one of the main objects upon which neo-liberal conventions are meant to operate.

In section 3, therefore, we begin by looking at one of the most obvious forms of governance of the global economy, one that stems from the rules and regulations produced and policed by national economic institutions. This is the case of international commercial law and the international frameworks of rules of economic exchange. This law is administered from a quite specific set of cities – the so-called global cities that we have already met in previous chapters. Even in this case, however, we will see how international law firms have informally extended this framework through institutions such as international commercial arbitration centres, which are answers to a deficit in international state regulation.

In section 4 we move on to forms of governance of the global economy which are clearly not the preserve of nation states but which are clearly powerful – what Rosenau (1992) calls 'governance without government'. The example I have chosen here is credit-rating agencies, which provide assessments of the risk of lending to sovereign governments, companies and, increasingly, cities. These privately owned companies have immense influence, and they also have a distinct urban geography which is crucial to their functioning – an urban geography which is also centred on global cities because of the need to be able to gather and process information rapidly.

In section 5 we look at how cities are expected to be run in a neo-liberal global economic order in which governance is diffused across many governmental and non-governmental institutions. We shall focus on the means by which cities can be made more competitive and, at the same time, at the chief institutions that are retailing the conventions of competition: government itself, city leaders, management consultancies, and city networks. Then, in section 6, we shall consider some of the newest forms of governance of the global economy. These are networks of people and firms which rely upon *mobility* between cities – what Ong and Nonini (1997) call 'ungrounded empires' and what Andrew Leyshon and I (1994) have called 'phantom states'. The urban geography of these forms of governance – which I will typify by the case of the overseas Chinese – is distinctive but also transitory; this is both a strength and a weakness.

All the forms of governance we will have considered to this point in the chapter are means by which power can be exerted by the powerful, mainly on behalf of the powerful. However, what of other forms of power? In section 7, I suggest that institutions that represent constituencies which may be economically weak can also have a global reach and can themselves become forms of governance of the global economy. I underline this point by considering the rise of 'alternative' financial institutions in places as diverse as Bangladesh and Vancouver, and their effects on cities and regions around the world.

2 *Neo-liberalism*

Michael Pryke has already introduced you to the term 'neo-liberalism' and some of its meanings in Chapter 6. In this chapter I want to extend your understanding of this term by arguing that neo-liberalism is a set of conventions or stories about the right ways to do things in order to succeed economically – as a firm, a country, a person or, as we shall see, a city. Over time it has become the chief means through which the global economy is understood and – through the gradual build-up of institutions like the International Monetary Fund and the World Bank – governed. Before going on to explain the chief tenets of neo-liberalism, I want to expand on the notion of neo-liberalism as a set of conventions by stressing that it is not just a passive script which institutions simply act out; rather it is an activity which spreads across all sorts of human practices. The French philosopher, Michel Foucault, coined a new term to describe this notion of governance-as-activity. He called it 'governmentality' (or more accurately govern-mentality) in order to signify two things: first, the ongoing activity of government as carried out through all manner of 'forces (legal, architectural, professional, administrative, financial, judgmental), techniques (notation, computation, calculation, estimation, evaluation), and devices (surveys and charts, systems of training, building forms) that promise to regulate decisions and actions of individuals, groups, organizations' (Rose, 1996, p.43). Second, Foucault wanted to signify the way in which all this governmental activity is grounded in particular styles of political reasoning (the mentality in governmentality) as 'conceptions of the objects to be governed – nation, population, economy, society, community – and the subjects to be governed – citizens, subjects, individuals' (Rose, 1996, p.43). Together, these different sides of the same coin can be thought of as intellectual techniques for 'rendering reality thinkable and practicable, and constituting discourses that are amenable – or not amenable – to reformulating life' (Rose, 1996, p.43). Neo-liberalism, then, is one particular way in which government is made possible.

As Michael Pryke has made clear in Chapter 6, neo-liberalism has had a long ideological bloodline which has its beginnings in the market liberalism of the eighteenth and nineteenth centuries. These earlier forms of liberalism believed that the state should not interfere in the economy either because the effects were likely to be different from those intended, or because the interference might prove actively harmful.

This *laissez-faire* doctrine argued for the limitation of the exercise of political sovereignty over the market. More than this, it acted as a positive justification for market freedom on the grounds that the state can benefit – become richer and therefore more powerful – by governing less.

Modern forms of liberalism, which developed after the Second World War and which are usually known as neo-liberalism, take this doctrine forward but they *add* significantly to it by telling a slightly different story. In this story, it is:

the responsibility of political government to *actively* create the conditions within which entrepreneurial and competitive conduct is possible. Paradoxically, neo-liberalism, alongside its critique of the deadening consequences of the 'intrusion of the state' into the life of the individual, has none the less provoked the invention and/or deployment of a whole army of organizational forms and technical methods in order to extend the field within which a certain kind of economic freedom might be practised in the form of personal autonomy, enterprise and choice.

(Barry *et al.,* 1996, p.10)

Neo-liberal governmentalities became powerful means of telling the story of the global economy after the global economic crises of the 1970s. Propelled by economic thinkers like Friedrich von Hayek and Milton Friedman, and then by visionary politicians like Ronald Reagan and Margaret Thatcher, these practices and ideas gradually became dominant means of both constituting what the global economy is, and how it should be governed, around the world. In effect, they constituted a whole new set of *beliefs*. Indeed, what writers like Hayek and Friedman, and politicians like Reagan and Thatcher, set out to do was precisely to change people's beliefs about how the world is – or should be. As Margaret Thatcher recently put it, 'it started with beliefs'. She paused. 'That's it. You must start with beliefs. Yes, always with beliefs' (quoted in Yergin and Stanislaw, 1998, p.124).

Neo-liberal practices and ideas have not come into the world naked, however. They have had to be developed and then nurtured through various forms of governance. Four of these forms of governance have been particularly important.

First, national governments have become strongly involved in promoting their national economies, and in making them more competitive. Development agencies, investment bureaux and the like have sprung up everywhere. Meanwhile, there have been wholesale restructurings of workforce training and educational systems with the goal of producing workforces that are continually

FIGURE 7.3
Margaret Thatcher and Ronald Reagan: putting neo-liberal beliefs into practice

able to compete in the global economy through a career of 'lifelong learning'. Second, national governments have ceded certain economic functions to the private sector. The most visible of these have been the waves of privatization that have washed across the world. But other economic functions have also gone private, especially in the financial sphere. Financial markets such as foreign exchange, though they are markets for national currencies, function outside national control (though we should not make too much of this; national governments still regulate the markets, and decisions by national governments are crucial to their functioning). Many national governments have also made their central banks independent or quasi-independent, in order to produce economic conditions that will please the financial markets. Third, a new layer of global governance has come into existence which is based on functions that might previously have been the functions of national governments. For example, credit-rating, which I will address later in the chapter, is in part a private sector means of judging how well the finances of national governments are doing. Fourth, new kinds of mobile economic community have come into being which use national territories to their advantage without having any fixed territory of their own. These 'ungrounded empires' (Ong and Nonini, 1997) or 'phantom states' (Leyshon and Thrift, 1994) are growing in importance – we shall return to them later.

What is very striking about neo-liberal governmentality is its *geographic* nature. Thus, much of almost any national government's economic agenda is now based upon getting particular pieces of national territory to shape up, to become fit to survive in the global economy, working out through new infrastructure, workforce training programmes and the like. Many of the new private or privatized economic control functions are based on making judgements about particular pieces of territory, while the new kinds of mobile economic activity are trying to rejig space as a series of partial sites which they can flit into and out of for their own purposes and as a series of spaces in between these sites – on aeroplanes, along telecommunications links, and so on – through which and within which concentrated interaction can take place.

Even more striking is the role that *cities* play in these spaces. Cities are

- both the spaces *upon* which neo-liberal practices of governmentality and industry operate, and the spaces *through* which they operate. They are both hunter and hunted, in other words.

In the following sections of the chapter I want to consider some of the ways in which cities figure in the neo-liberal global economic order. What we will see is cities caught up in an enormous 'global shift' (Dicken, 1998) which both slots them into a neo-liberal economic order and, at the same time, makes them into some of the chief motors of this order.

3 *International regimes*

Finance and trade do not operate in a vacuum. The exchange of money and goods requires all sorts of preconditions to be met, of which two of the most important are rights over property and enforceable contracts. Commercial law is in large part about the practice of these two preconditions. Since the end of the Second World War, as the volume of cross-border finance and trade has increased dramatically, so the need for international commercial law has increased, as has the range of possible disputes. For example, a recent crucial issue has been what property rights can be held over knowledge – through copyrights, patents and so on. Yet national laws on these issues are still often remarkably different in both scope and scale.

In other words, the international legal system is a patchwork of practices, which often do not fit well together. There is a series of international conventions and agreements: the UN Commercial Law, which came into effect on 12 January 1988, the Uniform Customs of the International Chamber of Commerce, INCOTERMS, ECE standard conditions, the Hague Convention on Uniform Law for the International Sale of Goods and on contracts for such transactions, the Vienna Agreement on Unified Commercial Law, various European Union directives, and so on. Together, these connections, conditions and agreements still only add up to 'tiny islands of unified law in a sea of national legislation' (Gesser and Schade, 1990, p.262). They leave enormous gaps and, even when international law can be applied, wide variations in judicial practice still exist. In this fluid situation, international lawyers and other experts have thrived:

> one can spot the serried ranks of grey-suited experts, international lawyers, corporate tax accountants, financial advisors, and management consultants. These new professionals present themselves as appointed to the noble task of modernizing and rationalizing the management of organizations … Profiting from the lack of involvement of the politicians, and from the absence of wider social forces, the international and multi-professional networks – which may be informal, uncontrollable, even chaotic to a degree, and without any legitimacy other than that based on technical expertise, and without any motive other than profit, even if it does go by the rather modest name of the professional fee – constituted by these new functionaries actually serve to provide some regulation of economic activity.

(Dezalay, 1990, pp.283–4)

International law is administered through a set of cities which act as places of legal dispute, negotiation and settlement and which are a fundamental part of the nascent international legal system. Chief among these cities are global cities like New York and London, where large specialist international law firms (now often multinational corporations in their own right) tend to gather. Indeed, because of the imperfect state of international commercial law, it is possible to argue that it is in cities like these that much international law is produced and dispersed. Nowhere is this made clearer than in the growth of international commercial arbitration centres, which have been set up to arbitrate in commercial disputes:

> Over the past twenty years, international commercial arbitration has been transformed and institutionalized as the leading contractual method for the resolution of transnational commercial disputes. Again, a few figures tell a quick and dirty story. There has been an enormous growth in arbitration centres. Excluding those concerned with maritime and commodity disputes – an older tradition – there were 120 centres by 1991, with another 7 established by 1993; among the more recent are those of Bahrain, Singapore, Sydney, and Vietnam. There were about a thousand arbitrators by 1990, a number that had doubled by 1992. In a major study on international commercial arbitration, Yves Dezalay and Bryand Garth [1995] find that it is a delocalized and decentralized market for the administration of international commercial disputes, connected by more or less powerful institutions and individuals who are both competitive and complementary. It is in this regard a far from unitary system of justice, perhaps organized, as Dezalay and Garth put it, around one great *lex mercatoria*, which might have been envisioned by some of the pioneering idealists of law.
>
> (Sassen, 1996, p.15)

It is important to point out that, so far as the world of international commercial law is concerned, there is still a strong linkage to the nation state: commercial law is simply trying to do the job of governance that states should be able to do.

4 *Private governance*

In contrast to international commercial law, which is half in and half out of the nation state system, some elements of the governance of the global economy now lie almost entirely outside the hands of national governments. Nowhere is this point made clearer than in the example of the international financial system. Since the 1960s, parts of the international financial system have increasingly become privatized, for three reasons. First, new financial activities have come into existence which have never been subjected to government control, for example the derivatives markets referred to by Michael Pryke in Chapter 6. Governments have simply not been able to keep up with the regulation of these new activities (some of which, it might be added, have been specifically designed to avoid state regulation). Second, extant government activities have been arranged in order to comply with financial market wishes. For example, many central banks have become either independent (like the German Bundesbank) or semi-independent (like the Bank of England) in ways which mean they can single-handedly (and single-mindedly) police important financial market criteria like inflation. Third, neo-liberal conventions have spread into government economic policy, providing a sympathetic backcloth against which the international financial system can do business (Leyshon and Thrift, 1997).

Of course, the international financial system still has to have rules and protocols if it is to run in an orderly fashion, and, to an extent, the international financial system has invented these itself. Perhaps the most prominent and powerful example of this invention of private sector institutions of governance by the international financial system is the credit-rating agency.

Credit-rating agencies are non-state forms of global authority based on judging debt security. They had their beginnings in the early part of the twentieth century in the far west of the USA, after a series of disastrous speculations – failed railroads, land schemes and other property deals – which suggested the need for some reliable means of judging investments. Nowadays, there are really only five major rating agencies. The two largest are both US-based: Moody's Investors Service (Moody's) and Standard and Poor's Rating Groups (S&P). There are two other large US-based agencies: Fitch Investors Service and Duff and Philips. Then there is IBCA, a London-based agency, which has now merged with a French agency in order to create a European rating agency. Finally, there is a host of small, domestically based agencies working out of countries like Japan, France, Canada, Israel, Brazil, Mexico and South Africa.

What do rating agencies do? Their main task is to make judgements on the likelihood that borrowers such as corporations, financial institutions, governments and municipalities will repay the principal and interest on loans, commercial paper and other forms of financial instrument. The agencies produce a letter symbol representing a ranking on a scale from most to least creditworthy. S&P provide four categories of 'investment grade' (the best risks) from AAA to BBB, and seven of 'speculative grade' (less good risks) from BB to D (for default).

Moody's rank from Aaaa to Baa3 and Ba1 to C respectively (both agencies have other scales for certain kinds of financial instrument). Rating agencies maintain surveillance over the ratings they give, and send out warnings to investors when they consider that developments may lead to a revision of an existing rating, whether upwards or downwards. A change of rank can lead to dire consequences. However,

> Even where the actual magnitude of a downgrade is minor, as was the case when Standard and Poor's reduced the ratings on the Government of Canada's C$9 billion debt denominated in foreign currency, the impact on the reputation of the issuer can be immense. As *The [Toronto] Globe and Mail* commented at the time,
>
> > While the downgrade is expected to increase the Liberal government's borrowing costs only marginally in the near future, [private] investment officials said the action sent alarms throughout the international market that Canada's financial health is eroding. 'It's a warning bell that the country's finances are deteriorating. This will only add to investor worries about Canada', a New York investment executive said.
>
> For the developing countries, ratings provide perhaps the supreme seal of approval in their struggle to obtain development funds at a less than exorbitant cost.
>
> (Sinclair, 1994, p.150)

This privatized system of judgement has very definite links to cities. To begin with, in the United States and around the world, cities are increasingly issuing their own bonds in order to raise money, and these are given a ranking by the rating agencies. Cities are therefore becoming a part of the rating system. Furthermore, the rating agencies are located in global cities, particularly New York, where Moody's, Standard and Poor, and Fitch all have their headquarters. Duff and Philips is based in Chicago. It is clear that credit-rating is a US affair. Though Moody's and Standard and Poor have small branch offices in other global cities – London, Paris, Frankfurt, Tokyo and Sydney to be precise – Fitch does not. Only IBCA has its headquarters outside the USA – in London – with branch offices in Paris, Madrid, Tokyo and New York. Why are the rating agencies so concentrated in a very small number of global cities? Partly because, as we have already seen in Chapter 5, these cities are at the intersection of the flows of information that credit-rating agencies most need (Leyshon and Thrift, 1997), but also because credit-rating is chiefly a US activity.

This fact suggests an important caveat concerning arrangements about global governance. Though it might appear as if a new private system of governance has come or is coming into being in various parts of the global economy, we need to qualify this statement. This does not mean that no national interests are involved. Quite the reverse. As Moran (1991) has noted in the course of research on the international financial system, and as the example of international law shows, what is being produced more often than not are private systems that are most in tune with US corporate interests, often helped along by the US government. So private regulation is always intertwined with national ambition: 'governance without government' is too pat a phrase; rather it is governance *and* government.

5 *Cities in the global economy*

What should cities do in a global economy that is increasingly run by institutions like credit-rating agencies, which can exert strong economic disciplines to converge on particular economic solutions? You might think that this is a difficult question to answer. After all, as **Massey, Allen and Pile (1999)** make clear, cities do not start from the same point; they have different historical trajectories, different mixes of people, different economic, social and cultural assets, different politics, and different means of imagining their futures. However,

● neo-liberal governmentality *has* made sense of the urban future – and of how to get to it – in ways which are sufficiently adaptable that all cities can be included, and positioned. That sense of what cities should be like is based on the idea of *competition*.

In neo-liberal thinking, cities must compete, and compete hard, to secure a better position in the global urban hierarchy because 'cosmopolitan companies have the world as their playing field and their choice of the world's best places. Communities must determine how to attract support and hold them in order to ensure a viable economy and quality of life that links local to global success' (Kanter, 1995, p.199). Cities become rather like people fostering their workplace careers: they must constantly educate themselves; they must make themselves attractive; they must work on the assets they have and try to develop others; and they must maximize their potential. Then, if they are really lucky, they will be able to make their way to the top of the heap, becoming one of the world's 'best cities'.

The Harvard Business School academic and management guru, Rosabeth Moss Kanter (1995, p.200) considers, for example, that the most successful cities are places where 'people learn better and develop faster than they otherwise would' by being good at one of the 'three Cs': concepts (new ideas and technologies), competence (high quality products) or connections (international trade). Cities, then, can become either good thinkers, good makers or good traders. But how can cities get to become what Kanter calls 'globally excellent' or 'world class' or, simply, 'best-in-world'?

The answer is that cities must be *managed* and not, as in previous eras, planned. In using the word 'managed', what is clearly being suggested is that cities should be managed *like a business*.

ACTIVITY 7.2   Now turn to Reading 7A, 'Best-in-world cities', in which Rosabeth Moss Kanter describes the five actions that she thinks are necessary for cities to become 'best-in-world'. Notice the prevalence of management ideas and terminology. The city is clearly to be treated as a coalition of business partners which can tackle urban problems by being managed as a business. When you have studied the reading, answer the following questions:

- What are the key similarities and differences between a city and, say, a large business?
- Do you think a city *can* be run like a business?
- Do you think a city *should* be run like a large business? ◆

The kinds of ideas that Kanter puts forward in Reading 7A are powerful because, increasingly, there is an *institutional infrastructure* in cities and the world which both relies upon them for its existence and is responsible for setting them into particular cities. There are four main elements in this accumulating infrastructure.

5.1 GOVERNMENT AND BUSINESS

The first is the result of the increasing mixture of the business of government with business. We can see the general trend all around us, in the increasing reliance of government on business – from the people who are drafted into government with business education and business backgrounds to the government task forces led by senior business people and the increase in public–private partnerships.

In Britain, for example, Clarke and Newman (1997) have written of the shift from a welfare state based on bureaucratic administration and professionalism to a 'managerial state' based on a managerialism which has, at least in part, arisen from the application of business ideas to the government of the nation state: 'The legitimacy of managerialism draws on a strongly articulated juxtaposition of the failings of other means of coordinating and controlling social welfare. The qualities of management [are] contrasted with the problems of the old regime and its dominant forms of organization power: bureaucracy, professionalism and political representation' (Clarke and Newman, 1997, p.65). Table 7.1 shows how Clarke and Newman sum up the supposed distinctions between the old and new approaches.

TABLE 7.1

Bureaucracy is:	*Management is:*
rule-bound	innovative
inward-looking	externally oriented
compliance-centred	performance-centred
ossified	dynamic
Professionalism is:	*Management is:*
paternalist	customer-centred
mystique-ridden	transparent
standard-oriented	results-oriented
self-regulating	market-tested
Politicians are:	*Managers are:*
dogmatic	pragmatic
interfering	enabling
unstable	strategic

Source: Clarke and Newman, 1997, p.65

ACTIVITY 7.3   Consider Table 7.1 as an *argument* that cities are better run by managers than bureaucrats, rather than as a straightforward presentation of the facts of the matter. Then ask yourself the following questions:

- Who are the main beneficiaries of this argument: urban citizens in general, politicians, or business people?
- Does the question really matter if cities are better run? ◆

Similar changes have occurred in numerous other countries around the world, both developed and developing. This shift can also be found taking place in the government of cities. Thus, cities are increasingly run as business enterprises providing services to customers, with, on the whole, much greater involvement by business people, either as representatives on various committees and trusts, as participants in public–private projects, or as members of civic partnerships competing for state funding. The agenda is clear: the aim is not only to deliver services more effectively but also to create 'urban growth machines' – economic powerhouses that can make it to the top of the urban heap.

5.2   CITY LEADERS

The second element of the new institutional infrastructure is a change in how city leaders are imagined. In times past, in many (though not all) countries, city leaders were often largely ceremonial figures, expected to dress up and host the latest local awards ceremony or town twinning dinner. But the new vision of cities as important cogs in a neo-liberal global economic order sees them as more like chief executive officers, managing their city. Thus, around the world, the position of mayor, or equivalent, has become more important as a focus for the notion of cities as a managed entity. It is no coincidence, then, that since 1994 the World Economic Forum has had a special category of regional and city leaders. At the 1998 annual meeting, for example, there was a 'Meet the Mayor' session, moderated by Raymond Barre (Mayor of Lyon, and a former French Prime Minister) and with contributions by Richard M. Daley (the Mayor of Chicago), Mikhail Passock (Governor of the Novgorod Region of Russia), Jaime Raviret de la Fuente (Mayor of Santiago, Chile) and Fernando de la Rua (Governor of Buenos Aires, Argentina). As if to demonstrate the new-found importance of such city leaders, London will have an elected, city-wide mayor for the first time in 2000. Other UK cities are expected to follow suit.

5.3   MANAGEMENT CONSULTANTS

The third element of the new institutional infrastructure is management consultants. Management consultants are now a major economic force in the business world (see Table 7.2) but they have also become a major force in many states, riding on the back of neo-liberal policies like privatization. For example, in mid 1995, in a series of parliamentary questions, the Conservative government admitted that it had spent at least £320 million on management consultants (and this was almost certainly a fraction of the real figure). More generally, management consultants from companies like Price Waterhouse and McKinsey & Co have become forces in the land, supplying expertise and, increasingly, people to government. For example, in Britain, the roll call would include Margaret Hodge, a former consultant to Price Waterhouse and now a Labour government employment minister, and William Hague, a former 'McKinseyite' and currently leader of the Conservative opposition (O'Shea and Madigan, 1997).

TABLE 7.2   *The top twenty management consultants in the world (by revenue)*

	Revenue ($m, 1996)	Staff numbers	
		Consultants	Partners
Anderson Consulting	5,300.0	37,389	1,036
Ernst & Young	2,010.4	10,657	n/a
McKinsey & Co	2,000.0	3,994	587
KPMG	1,836.0	10,673	888
Deloitte Touche Tomatsu International	1,550.0	n/a	n/a
Coopers & Lybrand	1,422.0	8,511	564
Arthur Andersen	1,379.6	n/a	n/a
Price Waterhouse	1,200.0	8,900	470
Mercer Consulting Group	1,159.2	n/a	n/a
Towers Perrin	1,001.3	6,500	635
Booz-Allen & Hamilton	980.0	5,300	228
AT Kearney	870.0	2,300	200
IBM Consulting Group	730.0	3,970	n/a
American Management Systems	812.2	6,042	n/a
Watson Wyatt Worldwide	656.0	3,730	n/a
The Boston Consulting Group	600.0	1,550	n/a
Gemini Consulting	600.0	1,550	n/a
Hewitt Associates	568.0	n/a	n/a
Aon Consulting	490.0	2,700	n/a
Bain & Co	459.4	1,350	145

In turn, management consultancies have disseminated neo-liberal conventions which have had important effects on the government of cities. A good example is the well-known management guru, Michael Porter. Like Rosabeth Moss Kanter, Porter was originally a Harvard Business School academic. He became interested in governance in the 1980s, when the then President, Ronald Reagan, asked him to serve on the President's Commission on Industrial Competitiveness. At this time, the primary wisdom was that the globalization of trade was making nations less and less important:

> Indeed Robert Reich, then Porter's colleague at Harvard, later put the case in its most sophisticated form in *The Work of Nations* in 1991, arguing that education and training were the only real source of national competitive advantage, because labour is the least mobile of all the factors of production.
>
> (Micklethwait and Wooldridge, 1996, p.321)

Michael Porter took a somewhat different tack, however. He was impressed by the large role that nation states and national differences seemed to play in successful global competition, and his best-selling, 800-page book, *The Competitive Advantage of Nations* (1989), was in effect a manual about how to turn nations into successful business operations. In particular, he argued that *competitive advantage* in a globalizing system

> is created and sustained through a *highly localized process*. Differences in national economic structures, values, cultures, institutions, and histories contribute profoundly to competitive success. The role of the home nation seems to be as strong as, or stronger than ever. While globalization of competition might appear to make the nation less important, instead it seems to make it more so. With fewer impediments to trade to shelter uncompetitive domestic firms and industries, the home nation takes on growing significance because it is the source of the skills and technology that underpin competitive advantage.
>
> (Porter, 1989, p.19, emphasis added)

Given Porter's stress on the *localized* nature of competitive advantage, it is no surprise that he should place a heavy emphasis on geographical clusters of industries as crucial motors of national economic successes: 'competitors in many internationally successful industries, and often entire clusters of industries, are often located in a single town or region within a nation' (Porter, 1989, p.154). In turn, it is no surprise that Porter should have increasingly turned his attention to the competitive role of cities and regions, in effect providing a set of 'recipes' to help cities and regions to make their way up the global hierarchy.

Through the vehicle provided by a consultancy firm, Monitor, which Porter co-founded in Cambridge, Massachusetts in 1983 and which now employs more than 600 consultants, as well as through a series of other like-minded consultancy firms, Porter's ideas about competitive advantage have made their way out into the world. For example, Monitor has written reports on the competitive position of many countries (including Canada, New Zealand, Portugal and Scotland) and of cities like Boston, all of which were intended to help its clients win out in a vast global game of reaping competitive advantage. As the Monitor publicity material puts it:

> Today, winning is not done – cannot be done – in relaxed, familiar settings at a comfortable distance behind the lines. It is the work of pioneers. It happens where uncertainty still reigns and on terrain that has not yet been thoroughly mapped. In other words, it happens 'on the frontier' – of new ideas, new technologies, new management approaches, new modes of analysis, new forms of organization, new ways to compete, new metrics of success, and of course geographies new to top-level strategy consulting. This, after all, is where the future is being shaped. This is where the new solutions that drive new learning first get designed. And this is where we are uniquely positioned to harness and apply new knowledge about how to win.
>
> (Monitor, Internet website)

In Reading 7B, which is from an article written by Porter for the *Harvard Business Review*, he applies his ideas on how to make territories globally competitive to the inner city. Note, in particular, how Porter contrasts the old model of inner-city economic development, based on public-sector subsidy and inward-looking business, with a new get-up-and-go inner-city economy based on the private sector, wealth creation and expert orientation. In the old model, government leads regeneration; in the new model the private sector takes on the leading role.

ACTIVITY 7.4   Now turn to Reading 7B, 'The competitive advantage of the inner city' by Michael Porter. According to Porter, 'businesspeople, entrepreneurs, and investors must assume a lead role' in the regeneration of the inner city, and 'community activists, social service providers, and government bureaucrats must support them' (p.71). This vision of the future of the inner city is heavily economic and sees no role for community-based organizations except as a support to a business-led agenda.

● Think back to Chapters 1, 3 or 5: what elements of urban life does this account miss?

● Can these elements be superimposed on Porter's approach, or does the fact that they are lacking suggest that Porter's approach has some fundamental flaws? ◆

I personally find Porter's approach rather depressing; it is as if cities were just an opportunity to create more business opportunities. You might equally argue, though, that this is the view of someone who has a job, is relatively well-off and is therefore able to have the luxury of making such a judgement.

5.4   CITY NETWORKING

The fourth element of the new institutional infrastructure, city networking, has already been noted in Chapter 5. One of the key conventions of the new neo-liberal practices is that cities, though they must compete individually, can also compete better if they gather together in *networks*, in strategic alliances which act both as ways for cities to learn about each other's successes and failures and as ways to mount new initiatives.

These networks are particularly prevalent in Europe, since 'to participate in Europe, at whatever level, is increasingly a matter of being "on the network" ' (Barry, 1996, p.27). All manner of inter-city co-operation and association programmes now exist, from town twinning at the lowest level to various EU initiatives aimed at specially created networks of European cities.

In turn, what we see is a new form of governance coming into existence in Europe which both embraces neo-liberal conventions and tries to reach beyond them:

> In the new model, Europe is to be organized through the use of alliances which cross over and disrupt existing territorial divisions. Moreover, individuals and corporations operating outside of the formal territorial boundaries of the Union will be allowed more or less access to the European networks on the basis of a growing number of co-operation and association agreements. The European sphere of influence is to be extended by networking, externally as well as internally.
>
> (Barry, 1996, p.35)

What each of these four new institutional arrangements represents is more than just a change of ideas. They are part of a general shift in *who* runs cities. What we can see taking place is that

● a group of business people – especially the managers of large transnational corporations and successful entrepreneurs – and those allied with business people, are being given the means to have a much greater influence than formerly on the way cities are run. This is becoming the age of the 'managerial city'.

We should not make too much of this shift, which is much more pronounced in some cities (for example in the USA and the UK) than others (for example in many developing countries). But for many cities it represents an important change in the economic, social, cultural and political balance of power.

6 *Ungrounded empires*

So far we have seen how neo-liberal conventions have spread to most nation states through enforcers like credit-rating agencies, and how they have also made their way into the government of cities through institutions which help these beliefs to settle in. You might think, given the account so far, that neo-liberalism is becoming completely dominant in the global economy. However, that would be a false assumption, since

● other forms of governmentality are making competing claims on the global economy, and other networks of governance are making connections alongside the dominant rhythm of neo-liberalism.

In this and the following section we shall look in turn at two of these other claims. One is clearly still the preserve of the relatively powerful, but the other claim is quite different, being mainly the result of efforts by those without much power to make their way in the global economy.

A number of writers now believe that new 'mobile' economic forms are appearing which presage new, more 'mobile' forms of government that are economically powerful but outside the territorial control of states. These new forms still depend on space, but on a space of circulation between locations, rather than the locations themselves. Furthermore, the locations between which the new more mobile forms of government circulate are cities.

These 'ungrounded empires' (Ong and Nonini, 1997) or 'phantom states' (Leyshon and Thrift, 1994, 1997) are networks that span several cities and depend on the development of various practices of crossing and interaction within them. In other words, they gain their power precisely from their mobility and their ability to outrun normal national forms of regulation (Appadurai, 1996). These are 'empires that constantly change shape, being constructed … in the ether of airports, international time zones, management labour contracts, mass media images, virtual companies and electronic transactions, and operating across all recognizable boundaries' (Nonini and Ong, 1997, p.20).

There are two prime examples of such empires at present. One is certain parts of the international financial system, which Michael Pryke has already addressed in Chapter 6. The other, which has already been mentioned in Chapter 5 and which I will concentrate on here, is the 56 million overseas Chinese who live chiefly in the cities of the Pacific basin (see Figure 7.4). On one estimate (Seagrave, 1995), their combined GNP is about $450 billion – about the same size as China itself, now that Hong Kong is included in China – and they control liquid assets (excluding securities) of approximately US$2 trillion (Long, 1998). Put another way, although the overseas Chinese account for, at most, 4 per cent of China's population, their national income is about two thirds as big. Therefore

● the overseas Chinese represent an enormous conglomeration of economic and cultural power which uses *a system of cities* as its base, rather than just one or two, chiefly in order to avoid the government of nation states (Thrift and Olds, 1996). In other words, it is this system that is constitutive of power, rather than just the cities themselves.

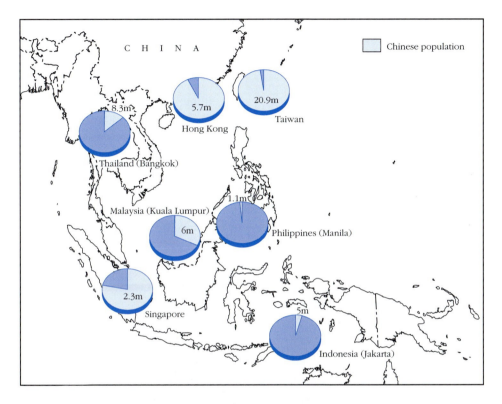

FIGURE 7.4 *Overseas Chinese populations in Asian countries, 1995*

One common feature of the overseas Chinese, both historically and in the present, has been the development and strategic use of flexible *social networks* (based on bonds of kinship, clan or language group, and friendship) at a variety of geographical scales. These networks, which often transcend national boundaries, have been used to create what Seagrave (1995) has called 'an empire without borders'. Nation states, in contrast, are involved in a constant effort to make people into the citizens of a nation by linking their allegiances to a particular territory. They do this

through various disciplinary practices as diverse as saluting a flag, marching in an army parade, being entered as an item in a government dossier, dancing in a national folklore festival, or learning to speak a national language. These practices require for their effects that subjects be defined probabilistically – for example, as citizen, or alien, immigrant or indigene – in ways with knowable spatial referents. For instance, citizens go to police stations and voting registrars while undocumented workers do not;

indigenes enter government schools not open to most students of immigrant descent. In short, the modern nation state requires for its effects that the subjects its discourses and disciplines construct be generally *localizable*.

But, overseas Chinese 'transnationalists' will try to elude the localizations placed on them by nation states by, above all, moving between national spaces, playing off one nation-state regime against another, seeking tactical advantage – knowing that it is easier to become a citizen here than there, that there are more legal and political rights in country X than in country Y, and so on.

(Nonini and Ong, 1997, p.23)

Thus the aim of the overseas Chinese is to avoid the localizations required by nation states by circulating in a set of 'partial sites' which resist or avoid the disciplinary practices of the nation state, replacing them with networks of family, clan or *guanxi* (the Mandarin for connections, or personal relationships with friends and business acquaintances) spread out across the cities of the Pacific basin. As Nonini (1997, p.208) says, Chinese transnationalists 'seek to resist the localizations … required of them as disciplinable subjects, while appropriating for their own causes the effects of the localizations these regimes have upon other disciplinable subjects' by using their social networks as a way of 'walking on two roads, not one'.

FIGURE 7.5   *The wreckage of overseas Chinese housing in Jakarta, 1998, following anti-government riots: overseas Chinese are often envied by others and can make easy targets*

For example, a Malay-Chinese businessperson may have a business in Kuala Lumpur, but might also set up a business in New Zealand. He has permanent residence in New Zealand but has not given up his Malaysian citizenship. His oldest son is studying computer science in New Zealand and will then go to the United States for graduate training on computing. His second child, a daughter, is studying commercial studies, also in New Zealand. His third child, a son, is still to embark on a university education (Nonini, 1997). Each child is the first step in a potential transnational movement and relocation, and the establishment of a strategic family network able to span a number of countries and cities.

FIGURE 7.6
An overseas Chinese 'astronaut' floating in space, with lifelines to Hong Kong, Canada and Australia

No wonder that, in the Pacific basin, the most affluent overseas Chinese are often likened to astronauts, as in the cartoon shown in Figure 7.6. Here an overseas Chinese astronaut floats in space but with lifelines to Hong Kong,

Vancouver and Sydney. Even some of the less well-off overseas Chinese are attempting to produce practices of movement that give them some measure of freedom from natural constraints:

> Thus pliable strategizing by Chinese transnationalists, focused on sending grown children for education and eventual employment overseas, provides an apt and friendly response to … constantly shifting economic opportunities … It serves as a means to secure their adult children's entry into global and national professional elites and operates to gather information about emerging markets, market trends, new techniques in production, new technologies of advertising, communication and transport, and much more. Above all else, it expands the universe of potential *guanxi* relationships accessible to small-scale Chinese capitalists from/in/of Malaysia across national borders. This is the basis of what I call 'middling' transnationalism.
>
> (Nonini, 1997, p.213)

Indeed, even the poorest overseas Chinese have now become involved in this constant circulation:

> Transnational reversals by Malaysian Chinese working-class men are called 'airplane jumping'. Reunited by Malaysian Chinese contractors working on behalf of Japanese and Taiwanese firms and flown to Japan and Taiwan, where they enter on tourist visas and then overstay, these men set out to spend from one to three years labouring as illegal workers to earn the relatively high wages, by Malaysian standards, paid by these firms. They live in dormitories or houses with the other men from Malaysia, at times cheek by jowl with Chinese and non-Chinese from other countries in Asia.
>
> (Nonini, 1997, p.216)

As overseas Chinese social networks have extended back and forth across the Pacific through large numbers of interactions, like the shuttle on a loom gradually building up a pattern, so they have had a marked effect on a number of Pacific basin cities. A good example is provided by Vancouver, Canada's third largest city and principal port (Hutton, 1995; Blomley, 1997). With a city region population of over 1.8 million, Vancouver has been rapidly making its way up the global urban hierarchy, in part because it is a very popular place of residence for transnational Chinese, offering all the right conditions for a transnational existence: a welcoming, Canadian government, high economic growth, good communications, a high-quality housing market and an increasingly sizeable Chinese community. But the arrival, since the 1970s, of large numbers of overseas Chinese has not been without its stresses and strains. In Extract 7.1 Katharyne Mitchell portrays the process of cultural adjustment as a sometimes fraught affair, especially in the city's housing market.

FIGURE 7.7 *Vancouver's waterfront: much of the high-quality housing on or near the waterfront is owned by overseas Chinese*

EXTRACT 7.1
Katharyne Mitchell: 'Overseas Chinese: global citizens, local commitment?'

The friction caused by the massive changes in Vancouver's urban environment was expressed in a number of different ways. In 1989 a series of articles in the major local newspaper, the *Vancouver Sun*, focused on the connection between Hong Kong investments in real estate and the rapid rise of house prices. In an article entitled 'The Hong Kong Connection: How Asian Money Fuels Housing Market', reporter Gillian Shaw discussed the large percentage (85 per cent) of Asian buyers in offshore purchases, and noted that 'the majority of those were from Hong Kong'. She also stressed that many of the new buyers were paying for the houses *in cash* (Shaw, 1989). Other article titles in the series included 'Computer Shopping for B.C. Property'; 'Investment Anger Confuses Hong Kong'; and 'Money Is King in Hong Kong: Entrepreneurs Find Paradise in the Streets of Hong Kong'. In many of these articles there was a strong emphasis on identifying 'essential' Hong Kong Chinese characteristics; these were categorized as a tendency to speculate in the housing market for profit, an aggressive drive toward material wealth and its subsequent ostentatious display, and a general disregard for the natural environment.

These essentialized representations had an incendiary effect on Vancouver's residents, who were becoming increasingly anxious about housing affordability, neighborhood character and the loss of local control over urban development. Hong Kong Chinese investors and home buyers were increasingly targeted as the agents responsible for unwanted neighborhood change, and many urban social movements became imbricated [*sic*] in a racial discourse. Much of the growing antagonism against Chinese buyers was exacerbated by the wealth of many of the new immigrants and by some of these immigrants' purchases of 'monster' houses.

Several of the social movements that were established during this period were expressly concerned with these kinds of neighborhood change. The Kerrisdale Concerned Citizens for Affordable Housing protested the demolition of walk-up apartments and houses, the Kerrisdale-Granville Homeowners Association battled the destruction of neighborhood trees, and the Shaughnessy Heights Property Owners Association attempted to amend zoning laws to limit the size of new houses that could be built in the neighborhood. Although these movements were publicly concerned with local control over the urban environment, the strong feelings against large-scale Hong Kong investment in commercial and residential real estate was made clear in association meetings and in a number of interviews with me.

In addition to state-sponsored efforts to redouble the emphasis on multiculturalism and cultural harmony in Canada in the 1980s, several private institutions were also quickly established to counter these growing urban frictions. The Laurier Institute, a non-profit agency concerned with 'promot[ing] cultural harmony in Canada' and 'encourag[ing] understanding among and between people of various cultures', began a well-publicized campaign to inform Vancouver residents that recent inflation in the housing market was *not* the result of speculation by nonlocal (that is, Hong Kong Chinese) buyers. The series of reports emphasizing this theme were picked up and reproduced with great fanfare by both a local newspaper, the *Vancouver Sun*, and a national paper, the *Globe and Mail*. By this point in time, the *Vancouver Sun* had been heavily criticized for its recent series of articles on Hong Kong and was eager to appear both objective and conciliatory concerning the issue (see Mitchell, 1993). Vancouver's major newspapers, the Laurier Institute, and a number of other private organizations thus joined forces in the attempt to neutralize negative impressions of the impact of Hong Kong real-estate investments on Vancouver's rapid urban transformation.

The barrage of information that suddenly appeared from these sources and others in 1989 and the early 1990s represented efforts within the public sphere to discipline British-Canadians to the new global forces impacting their lives. Transnational flows and their local effects were both neutralized and normalized in a language of cultural harmony and general blamelessness. In one Laurier Institute report of 1989, the author wrote of rising house prices in westside neighbourhoods: 'If we seek someone to blame for this increase in demand, we will find only that the responsible group is everyone, not some unusual or exotic group of residents or migrants. In fact, there is no one to blame: the future growth in housing demand is a logical and normal extension of trends in the nation's population' (Baxter, 1989).

With this kind of rhetoric, the construction of the 'enlightened' Canadian urbanite of the Pacific Coast was advanced by both state leaders and private (business) concerns.

References

Baxter, D. (1989) 'Population and housing in metropolitan Vancouver: changing patterns of demographies and demand', Laurier Institute report.

Mitchell, K. (1993) 'Multiculturalism, or the united colors of capitalism?', *Antipode*, vol.25, no.4, pp.263–94.

Shaw, G. (1989) 'The Hong Kong connection: how Asian money fuels housing market', *Vancouver Sun*, 18 February.

Source: Mitchell, 1997, pp.241–3

ACTIVITY 7.5   The immigration of overseas Chinese into certain Canadian cities is relatively unusual because it is the movement of a comparatively well-off group of people of colour. Having read Extract 7.1, set down the arguments for allowing an overseas Chinese person to build a 'monster house' in a traditional (and well-off) English-style Vancouver suburb, and the arguments against.

● Do you think the arguments against are racist and elitist?

● Or is the matter different when immigrants may have as much economic power as those they have come to live alongside? ◆

7 *But, but, but ...*

Let me now return to where I started this chapter: the World Economic Forum. So far you have been given a vision of the global economy as the stamping ground of rich and powerful corporations and individuals, and of cities as participants in a global competition for their favours. Bodies like the World Economic Forum appear to set the seal on this vision, intent as they seem to be on turning the world into business and business into the world.

However, there are three 'buts' that qualify this vision of a neo-liberal ascendancy over cities. The first of these 'buts' is one that has been repeated many times in this series: cities are not all the same. They have different economic, social, cultural and political histories. These histories and geographies both make it more or less easy for neo-liberal conventions to settle and thrive, and mean that when these conventions do settle they often take on distinctively local forms. The second 'but' relates to the first one: rarely is a city the stamping ground of just one set of beliefs and practices. As we have already seen with the case of the overseas Chinese, other forms of governance are both possible and successful. Finally – the third 'but' – attempts are being made by those who feel they have been shut out by neo-liberal and other forms of governance to construct different forms of power based on relations of co-operation and association.

Vancouver is a good place in which to note these three 'buts'. Vancouver has been the site of a long and involved political history, in which certain neo-liberal-inspired conventions (often emanating from the state government) have made their way into the city's governance. Then again, Vancouver has also become a key node in the overseas Chinese 'ungrounded empire'. These two forms of governance have in turn been challenged by various members of the local population. The result is a rich smorgasbord of different political organizations, attitudes and policies, as the city council elections of 1990 showed:

> Jim Green, community organizer in the Downtown Eastside for a decade, was the mayoralty candidate for the COPE [Committee of Progressive Electors]/NDP [New Democratic Party] Unity Slate. An American draft-dodger, union shop steward, and representative of the poor (a group increasingly threatened by the polarizing tendencies of a world city), Green focused the sympathies of those critical of unregulated growth, gaining much support despite COPE's reputation as a party to the left of the NDP. His election slogan, the 'Neighborhood Green', clearly alluded to the twin issues which had led to electoral success for reform groups in the early 1970s: protection for the neighborhoods and the environment. His opponent was the NPA [Non-Partisan Association] incumbent, Mayor Gordon Campbell, a 1970s liberal who had become a 1980s conservative,

and who had approved unprecedented development in his 1988–90 term of office. A developer and businessman, young, personable, well-educated, and well-travelled, he epitomized the free market internationalization of the world city. The election was fought essentially around the issue of development and housing affordability, and some novel alliances among groups newly marginalized and politized were created. Elderly tenants in middle class Kerrisdale sought advice from Green's Downtown Eastside Residents Association. In Strathcona, poor Chinese residents were attentive to the pro-growth entreaties of the new Chinese business class, which supported the NPA. Aided by an advertising budget its opposition could not match, the NPA barely won an election that the party had expected to lose. Reflecting the deeply ambivalent attitudes of Vancouverites, both the NPA and COPE elected five candidates to council – the single new NPA council number was a Hong Kong-born banker. The controlling vote in the 1990–3 council [was] held by Mayor Campbell who was returned with a small majority – 54 per cent of popular support.

(Ley, Hiebert and Pratt, 1992, p.265)

7.1   ALTERNATIVE ECONOMIC NETWORKS

In the rest of this chapter I want to dwell in a little more detail on the last 'but', for attempts are being made around the world by those shut out from wealth and power

- to construct economic networks that are based on co-operation and association, rather than unfettered competition, and which can mount a challenge not just to the economic power of the rich and powerful but also to how we think about 'the economic'. In other words, these networks are trying to change prevailing neo-liberal conventions.

Appropriately enough, I will start my tale back in the vicinity of Vancouver. I want to focus on the town of Courtenay on Vancouver Island, British Columbia in 1983, and especially on a self-employed Canadian, Michael Linton, who had been set an economic puzzle. Courtenay had been hit by a shortage of money following the closure of the key, externally owned industry in the town. The resulting economic slump meant that few residents had sufficient money to buy Linton's goods and services. How else might they pay?

Linton decided to try an idea which had last been attempted in North America and Europe in the 1930s depression – a local currency system which he called LETS (local exchange trading system):

LETS are local systems of production, multilateral exchange, and consumption, articulated through a local currency – a single-purpose money – independent of, but often related to, the prevailing national currency. This currency is frequently named after some local association and so a geography of the LETS economy is central at the outset and is expressed in terms of its most profound expression – the means of exchange.

Membership is open to all, usually on payment of a membership fee. Details
of the offers and needs of members within the system, normally expressed
in terms of products or labour power, are published in a regularly issued
directory distributed to all members. Transactions are effected by agreement
between transacting parties. They may occur either at a prepublished price
or by negotiation and they are expressed in terms of the local currency or in
terms of a mix between the local currency and the national currency unit.
However, no money changes hands: transactions are recorded on cheques
forwarded to and logged by the accountant or treasurer to the system.
Details of all transactions are then published on a regular basis so that all
members of the system have knowledge of the full transactional activity and
trading balance (debit or credit) of all other members

(Lee, 1996, p.1378)

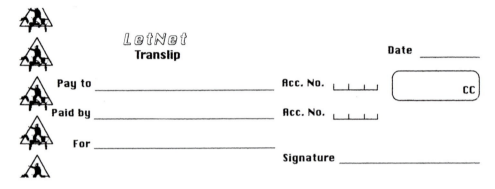

FIGURE 7.8   *Example of a LETS 'cheque'*

Linton's idea has spread from Canada to the USA, Australia, New Zealand and
many European countries, including Denmark, Finland, France, Germany, the
Netherlands, Norway, Spain, Sweden and Switzerland (see Figure 7.9 opposite).
In the UK, the first LETS was established in Norwich in 1985, but it was not until
the early 1990s that the idea really began to take off. There are now more than
400 LETS schemes in operation in the UK.

LETS provide a practical demonstration, however low-key, that there are
alternatives to competitive neo-liberal governmentalities based on co-operation
and association:

> prospective members have to get involved in the active search for local
> information and direct contact with the co-ordinators of local groups, and
> membership is, by definition, active and participatory. Members cannot
> remain anonymous; their names, their telephone numbers, and the nature of
> their involvement – defined in terms of what services or products they can
> offer to the system and what needs they feel may be met through it – are
> published in the regularly published *Directory*. Even more significantly, the
> transactions ... through which LETS operate are made known on a regular
> basis to all members as details of their trading accounts, including the nature
> of the transactions and the overall deficit or credit, are published throughout
> the system.

(Lee, 1996, pp.1379–80)

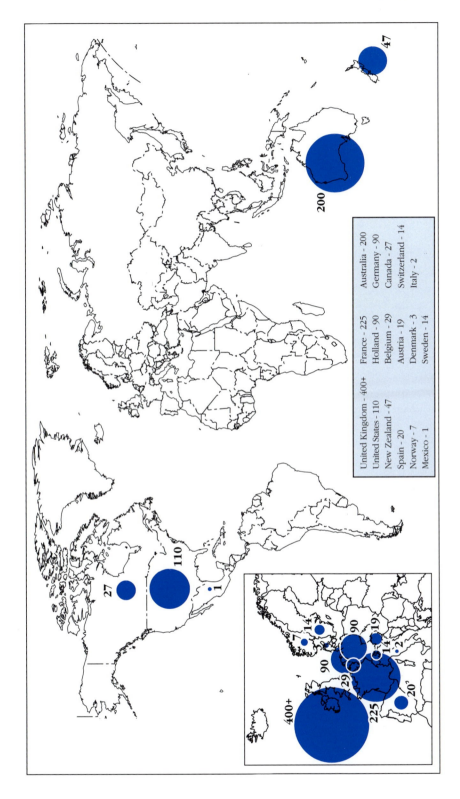

United Kingdom - 400+	France - 225	Australia - 200
United States - 110	Holland - 90	Germany - 90
New Zealand - 47	Belgium - 29	Canada - 27
Spain - 20	Austria - 19	Switzerland - 14
Norway - 7	Denmark - 3	Italy - 2
Mexico - 1	Sweden - 14	

FIGURE 7.9   *LETS schemes around the world*

LETS, then, hold out the promise of a different kind of economy, one in which social, ethical and environmental considerations are able to have a much greater say. But LETS are small-scale. Since they are autonomous local organizations, there are real practical constraints on how big they can grow. This is not the case for some other 'alternative financial institutions'. Of these institutions, perhaps the most successful have been those based on micro-credit – the granting of small-value loans to individuals and small businesses – which are one of the main financial phenomena of the 1990s.

There are still arguments about exactly where and how the micro-credit movement started, but it is generally agreed to be the result of two separate initiatives, one in Bangladesh and one in the USA.

The idea for the Grameen Bank in Bangladesh came from Muhammad Yunus, an economics Professor at Chittagong University. Yunus saw that the rural poor were bedevilled by the absence of cheap credit. Their only recourse was to family or loan sharks because they had no collateral. Starting in 1976, Yunus acted as a guarantor for loans to impoverished village dwellers. The idea that wealthy intermediaries could lend to the less well-off, in effect assuming the risk that the banks would not, was so successful that in 1983 Yunus decided to start up his own bank. The Government of Bangladesh initially owned 60 per cent of Grameen and the borrowers the rest. Today the 1.4 million borrowers, each with a mandatory share, own 88 per cent of the bank. Most of the loans to the bank have come from aid agencies through the Bangladesh government and are used to make loans to borrowers which average £100 in size and can go up to a maximum loan, always unsecured, of £300.

> [Grameen] charges simple interest at a rate of 20 per cent a year, compared with compound interest of thirteen to sixteen per cent at Bangladesh's commercial banks. Principal is repaid first, so that a borrower of Taka 2500 would typically repay Taka 50 a week over 50 weeks. Interest, calculated weekly on the diminishing principal, is repaid only after the principal is paid off, making for an effective interest rate of ten to twelve per cent, Grameen sources say.
>
> Home loans are repayable over ten years at the same weekly rates. The maximum home loan is Taka 12,000, which will build a tin-roofed house.
>
> Borrowers are formed into groups of five – the basic unit – and are indoctrinated in Grameen social values, known as the 'sixteen decisions' …
>
> Grameen is effective partly because it is self-policing. Rather than bank officials it is the members themselves, meeting weekly, who approve the loans. Group members are residents of the same village, perhaps next door neighbours. It is this familiarity that provides transparency, guaranteeing that the recipients are truly those who need it most. A bank official attends the meeting, but group members decide who receives a loan, and they assume responsibility for ensuring its repayment.
>
> (Kamaluddin, 1993, p.38)

Not surprisingly, perhaps, Grameen Bank's costs are low and its default rate is minimal – 2 per cent. But the real key to Grameen's success is its target borrowers: 92 per cent of them are women. The bank targets them because, quite simply, they are more reliable than men; their priorities are more likely to be ordered by their children's needs and they are therefore unlikely to squander loans in the way that men might do.

The loans are used for modest ventures to buy chicks for eggs, to establish a village shop, to buy a car or, nowadays, even to establish a mobile phone business under a project by Grameen to provide a mobile phone for each of Bangladesh's 68,000 villages (100 million out of Bangladesh's 120 million people have no access to telephones).

Of course, for all the rhetoric, there are problems. About one-third of borrowers never become any better off, and Grameen Bank loans are sometimes used – in defiance of the rules – either as collateral for further loans from money lenders, producing a vicious circle of debt, or as contributions to dowries. Nevertheless, studies of the Grameen Bank show that it has clearly caused some overall improvements in nutrition in Bangladesh, has brought about structural changes in local economic relationships that have benefited women, and, in general, has 'effected change like termites, hollowing out structures quietly from within' (Todd, 1996, p.221).

The Grameen model has become the iconic foundation of the micro-credit movement, which has been copied in 58 countries to date. In particular, it has spread to many of the cities of the developed as well as the developing world. For example, Chicago's successful Full Circle Fund is modelled on Grameen (Counts, 1996).

The USA has also seen, quite independently, the rise of a series of other kinds of micro-credit networks, and in particular community development banks and community development loan funds, which accept money from socially minded investors and make loans to fund housing rehabilitation, non-profit housing development, and small businesses. Community development banks have grown rapidly in US cities, though their performance is mixed. For example, Estey (1995) found that the Chicago-based South Shore Bank, the first community development bank founded in 1973, had a suspect financial performance and that its record of delivery on its social goals had been indifferent. Community development loan funds are various non-bank forms of local lending which date from the mid 1980s. The industry is still quite small, but growing fast. At the end of 1995, the industry had $108 million in outstanding loans, and $204 million in capital; a third of the capital was permanent, the rest borrowed. In the nine years from 1986 to 1995, members of the National Association of Community Development Loan Funds financed 56,243 housing units, 73 per cent of them permanently affordable for low-income residents, and it created or preserved 11,313 jobs, 59 per cent of them for 'low-income people', 51 per cent for women and 37 per cent for minorities. Their loan loss was only 0.92 per cent of loans made.

The first British community development loan fund, the Aston Reinvestment Trust, was launched in Birmingham in 1997. It was modelled on the Philadelphia Community Development Loan Fund, with a small but robust capital base. It has chosen to operate in four niche areas in the community: micro-credit for self-employment and small businesses, equitable utility payment and insurance schemes, housing repairs and energy self-sufficiency improvements, and mortgage resources for the over-indebted.

8 *Conclusion*

I have argued in this chapter that global economic change is increasingly the preserve of cities involved in often cut-throat competition with one another, in part as a result of the rise of neo-liberal governmentalities whose conventions stress that cities are the chief agents of economic change in the global economy. This argument can be furthered in two ways.

First, the apparent increase in the economic importance of cities can be taken as a sign of a far wider shift. Indeed, the geographer Peter Taylor (1996, 1997) has argued that some global cities are now becoming so powerful that they may well become independent city-states, rather like Singapore. For example, London might break free from the rest of the UK because 'London's location within the UK polity is a disadvantage in [the] world city competition. This is the reason that the idea of an autonomous [London] may be moving from the realm of political impossibility to merely being improbable' (Taylor, 1997, p.769).

ACTIVITY 7.6   Make lists of the main points that might convince you to vote for or against the secession of London from the United Kingdom. Write from the point of view of someone living in London and then someone living elsewhere in the UK. The chapters in this book should give you some sense of the kinds of questions you need to consider in making these lists. For example:

- Is London more connected to other world cities than cities in the rest of the UK?

- Does London obtain most of its wealth from international activities?

- In what ways is London a *British* (or English) city? ◆

Second, global cities can be seen as the strategic sites within which hierarchies of power and influence are both most striking *and* most open to challenge. For example, Sassen (1996) and many of the contributors to Eade (1997) argue that it is within these cities that new forms of citizenship are now being forged which are both more reliant on mobility and more open to the juxtaposition of different cultures. The invention of LETS and micro-credit can both be seen, in their different ways, as possible economic contributions to this redefinition of citizenship as a more global-local concern.

It may or may not be the case that global and other kinds of cities really can be, simultaneously, both sites of economic power *and* the focus of social movements which are the most able to dispute and disrupt this power. However, in the final part of this chapter I have tried to show that cities do not have to be judged just in terms of their ability to take part in and win an urban economic arms race. In particular, cities can act as stimulants to the construction of new relations of co-operation and association of the kind that Ash Amin and Stephen Graham also wrote about in Chapter 1. In order for some cities to win the urban economic arms race, others must lose. But it may be that the construction of new relations of co-operation and association can benefit the inhabitants of *all* cities. That certainly is the hope.

References

Appadurai, A. (1996) *Modernity at Large: Cultural Dimensions of Globalization*, Minneapolis, University of Minnesota Press.

Barry, A. (1996) 'The European network', *New Formations*, no.29, pp.26–38.

Barry, A. *et al.* (1996) 'Introduction', in Barry, A., Osborne, T. and Rose, N. (eds) *Foucault and Political Reason: Liberalism, Neo-liberalism and Rationalities of Government*, London, UCL Press.

Blomley, N. (1997) 'Property, pluralism and the gentrification frontier', *Canadian Journal of Law and Society*, vol.12, pp.187–218.

Clarke, J. and Newman, J. (1997) *The Managerial State: Power, Politics and Ideology in the Remaking of Social Welfare*, London, Sage.

Counts, A. (1996) *Give Us Credit: How Muhammad Yunus's Micro-Lending Revolution is Empowering Women from Bangladesh to Chicago*, New York, Random House.

Dezalay, Y. (1990) 'The big bang and the law: the internationalization and restructuration of the legal field', *Theory, Culture and Society*, vol.7, no.2–3, pp.279–94.

Dezalay, Y. and Garth, B. (1995) 'Merchants of law as moral entrepreneurs: controlling international justice from the competition for transnational business disputes', *Law and Society Review*, vol.29, no.1, pp.27–64.

Dicken, P. (1998) *Global Shift: Transforming the World Economy*, 3rd edition, London, Paul Chapman.

Eade, J. (ed.) (1997) *Living in the Global City: Globalization as Local Process*, London, Routledge.

Estey, B.C. (1995) 'South Shore Bank: is it the model of success for community development banks?', working paper 95–072, Harvard Business School.

Gesser, V. and Schade, A. (1990) 'Conflicts of culture in cross-border legal relations', *Theory, Culture and Society*, vol.7, no.2–3, pp.253–78.

Hutton, T. (1995) 'Vancouver', *Cities*, vol.11, pp.219–39.

Kamaluddin, A. (1993) 'Grameen Bank', *Far Eastern Economic Review*, October.

Kanter, R.M. (1995) *World Class: Thriving Locally in the Global Economy*, New York, Simon and Schuster.

Lee, R. (1996) 'Moral money, LETS and the social construction of local economic geographies in Southeast England', *Environment and Planning A*, vol.28, pp.1377–94.

Ley, D., Hiebert, D. and Pratt, G. (1992) 'Time to grow up? From urban village to world city, 1966–91', in Wynn, G. and Oke, T. (eds) *Vancouver and its Region*, Vancouver, University of British Columbia Press.

Leyshon, A. and Thrift, N.J. (1994) 'A phantom state? The detraditionalization of money, the international financial system, and international financial centres', *Political Geography*, vol.13, no.3, pp.299–327.

Leyshon, A. and Thrift, N.J. (1997) *Money/Space, Geographies of Monetary Transformation*, London, Routledge.

Long, S. (1998) 'Special report: the overseas Chinese', *Prospect*, April, pp.60–5.

Massey, D., Allen, J. and Pile, S. (eds) (1999) *City Worlds*, London, Routledge/The Open University (Book 1 in this series).

Micklethwait, J. and Wooldridge, A. (1996) *The Witch Doctors: What the Management Gurus Are Saying, Why It Matters and How to Make Sense of It*, London, Heinemann.

Mitchell, K. (1997) 'Transnational subjects: constituting the cultural citizen in the era of the Pacific rim capital', in Ong, A. and Nonini, D. (eds).

Moran, M. (1991) *The Politics of International Finance*, Basingstoke, Macmillan.

Nonini, D.M. (1997) 'Shifting identities, positioned imaginaries: transnational traversals and reversals', in Ong, A. and Nonini, D. (eds).

Nonini, D.M. and Ong, A. (1997) 'Chinese transnationalism as an alternative modernity', in Ong, A. and Nonini, D. (eds).

O'Shea, T. and Madigan, F. (1997) *Dangerous Company: The Consulting Powerhouses and the Businesses They Save and Ruin*, London, Nicholas Brealey.

Ong, A. and Nonini, D. (eds) (1997) *Ungrounded Empires: The Cultural Politics of Modern Chinese Transnationalism*, New York, Routledge.

Porter, M.E. (1989) *The Competitive Advantage of Nations*, London, Macmillan.

Porter, M.E. (1995) 'The competitive advantage of the inner city', *Harvard Business Review*, May–June, pp.53–71.

Rose, N. (1996) 'Governing "advanced" liberal democracies', in Barry, A. *et al.* (eds) *Foucault and Political Reason: Liberalism, Neo-liberalism and Rationalities of Government*, London, UCL Press.

Rosenau, J.N. (1992) 'The relocation of authority in a shrinking world', *Comparative Politics*, vol.24, no.2, pp.250–63.

Sassen, S. (1996) *Losing Control? Sovereignty in an Age of Globalization*, New York, Columbia University Press.

Seagrave, S. (1995) *Lords of the Rim*, London, Bantam.

Sinclair, T. (1994) 'Passing judgement: credit rating processes as regulatory mechanisms of governance in the everyday world order', *Review of International Political Economy*, vol.1, pp.133–59.

Taylor, P.J. (1996) 'World cities and territorial states: the rise and fall of their mutuality', in Knox, P. and Taylor, P.J. (eds) *World Cities in a World-System*, Cambridge, Cambridge University Press.

Taylor, P.J. (1997) 'Is the United Kingdom big enough for both London and England?', *Environment and Planning A*, vol.29, pp.766–70.

Thrift, N.J. and Olds, K. (1996) 'Refiguring the economic in economic geography', *Progress in Human Geography*, vol.20, no.2, pp.311–37.

Todd, J. (1996) *Women at the Center: Grameen Bank Borrowers After One Decade,* Boulder, Colorado, Westview Press.

Yergin, D. and Stanislaw, J. (1998) *The Commanding Heights*, New York, Simon and Schuster.

READING 7A
Rosabeth Moss Kanter: 'Best-in-world cities'

Actions for quality and excellence: becoming the best-in-world (B-I-W)

• *Develop regular areawide civic forums for communication and mutual understanding between business and government.* Use these forums to develop visionary goals, as well as action plans to remove obstacles to a healthy business climate and community life. Form regional councils linking public officials and business leaders across the area to create a shared agenda, working together in their own jurisdictions to implement it.

• *Identify the area's core skills and the investments required to increase the area's stock of world class concepts, competence, and connections.* Create a vision of excellence that will show how those key global assets will increase in the future.

• *Define B-I-W performance standards for every aspect of the community that supports its core skills –* from airports to schools, from recreational opportunities for foreign business travellers to incubators for new technology companies. Compare the area's performance against the world's best places in both private industry and the public sector. Benchmark against comparable places to illuminate strengths and weaknesses; create action plans to enhance the strengths and eliminate the weaknesses.

• *Celebrate excellence.* Create achievement awards, and use the criteria for receiving them to help educate organizations about best practices. Publicize role models. Produce tools for performance improvements that can be disseminated widely across sectors, creating a common vocabulary for the community as well as spreading high standards.

• *Promote a customer focus in government, business, education, and service industries.* Use it as an attraction for tourists, conventions, business travellers, and new outside investors.

• *Extend education in quality problem-solving skills to the schools.* Offer this to the adult community as well, in evening programs at the schools. This would prepare students for jobs of the future, provide a common educational experience for both parents and children, and improve the at-home dialogue critical to school performance.

• *Launch or reinforce 'best partner' quality campaigns to educate companies throughout the area about world standards (such as Baldrige quality criteria, ISO 9000, value-added through customer-supplier partnerships) and criteria for success in global markets, complete with awards and recognition.* Create international road shows to market area companies exemplifying 'best partner'.

Actions to enhance business-to-business networking and partnering

• *Develop mechanisms to facilitate small-business collaboration.* Possibilities include purchasing, marketing, importing, exporting, ideas, exchange of personnel at slack times, sharing of facilities. Use directories, computer networks (bulletin boards), resource kits, local trade shows, seminars, and forums.

• *Create one-stop business service centres dispersed throughout the region that combine information from both the public and private sectors.* Provide information on government resources and support services, assistance for meeting regulatory requirements, and information from industry associations, civic groups, free trade zones, World Trade Centres, and other resources. These centers can act as places for small businesses to initiate a search for collaborators.

• *Develop databases for local suppliers and their skills.* Introduce new companies from outside the region to local companies that could support them. Ensure that local companies get information about potential procurements from large customers.

• *Hold regular minitrade shows at technology parks and research labs to connect developers of new concepts with potential entrepreneurs and suppliers who will commercialize them.*

• *Create welcoming kits for newly incorporated businesses that give them trial memberships in civic associations and offer them the chance to announce their existence to the membership.* Create networks to help smaller businesses make contacts that are inclusive of newcomers. Partnering requires active initiative; businesses cannot wait for it to happen spontaneously. Small businesses must take the initiative to find partners and sell their skills. While it is not the role of government to form business collaborations, government resources could help jumpstart collaborations in industries where this is difficult – or for populations where this is difficult.

• *Help minority companies join networks as equal players.* Small minority businesses emanating from the inner city need a link that connects them to other places. Create financing pools or government-backed loan guarantees for small businesses that invest in the inner city. Identify investment opportunities in the inner cities that can attract non-minority businesses as partners for inner-city firms. Dollars tend to be available for the inner city for philanthropy but not for business activities; for this reason many of the strongest leaders in the black community of some cities head social sector enterprises rather than businesses. Encourage these leaders to form job-creating businesses.

• *Ensure that mentoring programs for minority businesses make them 'world-ready' from the start.* Include connections with companies that operate internationally as well as domestically; educate them about world class standards; and identify strengths that would make them appropriate local partners for new companies coming to the area.

Actions to become world-ready, foreign-friendly, and globally linked

• *Identify the criteria used to select new locations that are important to international companies in industries related to the city's core skills.* Use those criteria to identify improvements that community must make – not only to attract outside investment, but also to encourage current companies to expand activities in the area rather than taking them elsewhere.

• *Actively market the area to potential international investors, but look beyond multinational giants.* Find midsize outside companies that are already links in a supply chain into the area and could relocate to serve large customers who would otherwise purchase outside the area – an 'import substitution' strategy. Seek investors who will make long-term commitments to the area, bring innovation and new knowledge, and become involved in civic affairs – in the way that European middle-market companies contributed to Spartanburg. Find companies that are performing one function in the area, such as sales, and encourage them to perform others.

• *Follow the trade routes.* Identify the key trade connections of major companies in the area and develop strategies for transportation, telecommunications, and language skills that build on existing ties. For example, if Germany and Korea receive the highest number of exported products from the area, then work on bringing German and Korean airlines to the airport, expanding routes for cargo and passengers.

• *Examine the points of entry to the city to see how they look to outsiders.* This was what the Atlanta Chamber of Commerce did in asking about the experiences of international visitors to Atlanta. Develop amenities geared to international business travelers and not just to tourists – activities that make them feel comfortable and welcome or places where they can find items from their home, culturally appropriate entertainment, and resources to facilitate their visit, such as those that Seattle's international press centre offers. Tourists bring onetime dollars, but business travelers bring continual investment.

• *Build ties in other countries with civic associations and economic development bodies as the basis for exchange of information about joint venture, plant location, and export market possibilities.* Send local executives as temporary ambassadors to other countries to take advantage of their business trips and help them sell the region.

• *Create databases of products developed by local companies that are candidates for export markets.* Bring them to trade fairs in other countries. Exchange information with economic development agencies in other countries to help link producers to international distributors or joint venture partners.

• *Multiply the payoff from trade missions.* Offer slots to groups of companies; ask those who participate to teach others and share connections; ask large companies to sponsor a small-business executive who could not otherwise afford to go.

• *Offer trade finance help to smaller companies before they enter foreign markets.* Tiny grants to seed the business are often more helpful for international trade than large loans after the company already has the business.

• *Identify international companies with extensive international knowledge and connections.* Determine how to tap their expertise and relationships for synergies with other area companies – for example, a manufacturing executive bringing knowledge of the German apprenticeship system; a bank with long-standing Latin American connections opening doors in that region for other local companies.

• *Bring newcomers and expatriates into local business councils to add an outside perspective or international presence.* Make them feel welcome. Determine how to use foreign companies as teachers for community institutions and in public school classrooms – a role Honda played in Marysville.

• *Set up mentoring relationships linking large international companies in the area with small companies seeking international opportunities.* Encourage the local press to report about the international activities of area companies.

• *Develop registers, directories, and information clearinghouses to help connect emerging companies with potential international partners.* Collect and provide data on companies capable of and interested in serving globally oriented, world class customers.

• *Publicize local educational resources for learning about international opportunities, companies, and business practices.* The resource pool is extensive: universities, foreign students, language institutes, World Trade Centers, foreign Chambers of Commerce, translators, expatriates, and retired executives are all possibilities. Use foreign students in the area as a source of knowledge about their countries; they are eager to get work experience, and they often come from prominent families with excellent connections. Invite foreign dignitaries or speakers to visit the schools, meeting with schoolchildren.

Actions to create employability security and spread workplace education

•   *Support 'learning alliances' to offer training, compare best practices, share success stories, and provide internships across companies.* Nashville's Peer Learning project, for example, is a management development consortium of local companies, organized into groups of a dozen companies – with only one company in each group – which share educational programs and exchange best practice ideas, supported by a local university.

•   *Develop human resource and job banks linking networks of related companies.* Help companies to see their network of suppliers, customers, and venture partners as a resource for skills development, personnel exchanges, internships, or job placements. Encourage companies to invite speakers together, create training programs together, and help their people's skills become known to one another.

•   *Offer customized job training for new companies bringing large investment to the area.* Use this training to increase the skills of the potential pool of workers, before they are chosen, to upgrade more people than the number to be hired immediately.

•   *Bring ideas to and from the workplace.* Offer 'open visit' days in which managers and workers can visit excellent companies in the area and ask questions about their activities and approaches. Create mobile education units (like Bookmobiles) to bring portable seminars or materials from local colleges to work sites.

•   *Enhance support (financing, role models) for internal corporate entrepreneurship – new ventures, buyouts, and spin-off businesses* around promising ideas that the established company cannot use. These are important sources of innovation in technology, and they smooth employment fluctuations. Such programs also have a cascade effect inside the established company: they encourage entrepreneurial behaviour, which can come back into the core business in the form of innovation that helps the company compete, saving jobs.

•   *Expand entrepreneurship skills at both ends of the life cycle.* Beginning in early grades, give children opportunities to practice running small businesses, using programs such as Junior Achievement or An Income of Her Own, which helps teenage girls start businesses. Create 'Senior Achievement' programs to help older workers get education and encouragement to start new businesses after hours, while still employed, as a transition to retirement or an alternative to large-company employment.

•   *Develop school-to-work apprenticeship programs that offer students the chance to gain job skills.* Extend apprenticeship programs to displaced workers from declining industries – such as the Massachusetts Software Council's internship program to help people move from hardware to software jobs.

Actions for civic engagement and leadership development

•   *Mount tours of other cities for current and future community leaders, to stretch their horizons and give them a shared view of what is possible.* Use cosmopolitans with experience in other places as critics, idea sources, or sounding boards.

•   *Mount a civic leadership development programme in which the next generation of leaders forms project teams to tackle community problems.* Younger and rising leaders can be sponsored by companies from downtown, urban neighborhoods, the suburbs, and the exurbs, where emerging companies reside, inclusive by race, gender, and foreign experience, and fellowships can be offered to not-for-profit and government organizations. This would address critical issues while enlarging the connections across ethnic groups and organizations, building social capital and future networks.

•   *Link community service to employee development and training, and help volunteers get credit or credentials as part of their career development.* Companies that send managers on Outward Bound expeditions for team building might send them to 'Inward Bound' community service programmes that are equally physically challenging and team oriented, the way Timberland uses its City Year involvement. Time for volunteering is limited, but time for training is increasing. Smaller companies that do not have their own staff development programs could be included, and emerging new technology companies that see little value in traditional cultural or civic organizations would see value in leadership development for their staff.

•   *Use the agenda of newer, technology-based industries as the basis for new forms of civic contribution.* Offer schools and not-for-profit institutions as sites for demonstrating the value of new technologies. Take advantage of the professional resources and products emerging companies offer to make the community a laboratory for testing and using the latest technologies – which also helps spread technology know-how, includes locals, and reinforces the city's status as a world class centre.

•   *Create short-term private sector–public sector job exchanges to foster greater understanding and develop multifaceted leaders.* Teacher apprenticeships in industry would be one example; another, businesspeople on loan to the public sector. Develop databases of available people and opportunities.

• *Help not-for-profit organizations understand the implications of strategic philanthropy and become better business partners by teaming with others to multiply their impact.* Find ways to measure the quantity and quality of community service, so that not-for-profits can value the time they receive (not just the dollars) and those engaged in community service can reach a high performance standard and get 'résumé credit' for their activities.

Actions like these are important elements of a community's strategy for joining the world class – its domestic economic development plan and its foreign policy. Such actions should not be isolated initiatives; they should be linked to a vision for excellence based on the city's distinctive global capabilities.

Source: Kanter, 1995, pp.371–8

READING 7B
Michael E. Porter: 'The competitive advantage of the inner city'

The economic distress of America's inner cities may be the most pressing issue facing the nation. The lack of businesses and jobs in disadvantaged urban areas fuels not only a crushing cycle of poverty but also crippling social problems, such as drug abuse and crime. And, as the inner cities continue to deteriorate, the debate on how to aid them grows increasingly divisive.

The sad reality is that the efforts of the past few decades to revitalize the inner cities have failed. The establishment of a sustainable economic base – and with it employment opportunities, wealth creation, role models, and improved local infrastructure – still eludes us despite the investment of substantial resources.

Past efforts have been guided by a social model built around meeting the needs of individuals. Aid to inner cities, then, has largely taken the form of relief programs such as income assistance, housing subsidies, and food stamps, all of which address highly visible – and real – social needs.

Programs aimed more directly at economic development have been fragmented and ineffective. These piecemeal approaches have usually taken the form of subsidies, preference programs, or expensive efforts to stimulate economic activity in tangential fields such as housing, real estate, and neighborhood development. Lacking an overall strategy, such programs have treated the inner city as an island isolated from the surrounding economy and subject to its own unique laws of competition. They have encouraged and supported small, sub-scale businesses designed to serve the local community but ill equipped to attract the community's own spending power, much less *export* outside it. In short, the social model has inadvertently undermined the creation of economically viable companies. Without such companies and the jobs they create, the social problems will only worsen.

The time has come to recognize that revitalizing the inner city will require a radically different approach. While social programs will continue to play a critical role in meeting human needs and improving education, they must support – and not undermine – a coherent economic strategy. The question we should be asking is how inner-city-based businesses and nearby employment opportunities for inner city residents can proliferate and grow. A sustainable economic base *can* be created in the inner city, but only as it has been created elsewhere: through private, for-profit initiatives and investment based on economic self-interest and genuine competitive advantage – not through artificial inducements, charity, or government mandates.

We must stop trying to cure the inner city's problems by perpetually increasing social investment and hoping for economic activity to follow. Instead, an economic model must begin with the premise that inner city businesses should be profitable and positioned to compete on a regional, national, and even international scale. These businesses should be capable not only of serving the local community but also of exporting goods and services to the surrounding economy. The cornerstone of such a model is to identify and exploit the competitive advantages of inner cities that will translate into truly profitable businesses.

Our policies and programs have fallen into the trap of redistributing wealth. The real need – and the real opportunity – is to create wealth.

Toward a new model: location and business development

Economic activity in and around inner cities will take root if it enjoys a competitive advantage and occupies a niche that is hard to replicate elsewhere. If companies are to prosper, they must find a compelling competitive reason for locating in the inner city. A coherent strategy for development starts with that fundamental economic principle, as the contrasting experiences of the following companies illustrate.

Alpha Electronics (the company's name has been disguised), a 28-person company that designed and manufactured multimedia computer peripherals, was initially based in lower Manhattan. In 1987, the New York City Office of Economic Development set out to orchestrate an economic 'renaissance' in the South Bronx by inducing companies to relocate there. Alpha, a small but growing company, was sincerely interested in contributing to the community and eager to take advantage of the city's willingness to subsidize its operations. The city, in turn, was happy that a high-tech company would begin to stabilize a distressed neighborhood and create jobs. In exchange for relocating, the city provided Alpha with numerous incentives that would lower costs and boost profits. It appeared to be an ideal strategy.

By 1994, however, the relocation effort had proved a failure for all concerned. Despite the rapid growth of its industry, Alpha was left with only 8 of its original 28 employees. Unable to attract high-quality employees to the South Bronx or to train local residents, the company was forced to outsource its manufacturing and some of its design work. Potential suppliers and customers refused to visit Alpha's offices. Without the city's attention to security, the company was plagued by theft.

What went wrong? Good intentions notwithstanding, the arrangement failed the test of business logic. Before undertaking the move, Alpha and the city would have been wise to ask themselves why none of the South Bronx's thriving businesses was in electronics. The South Bronx as a location offered no specific advantages to support Alpha's business, and it had several disadvantages that would prove fatal. Isolated from the lower Manhattan hub of computer-design and software companies, Alpha was cut off from vital connections with customers, suppliers, and electronic designers.

In contrast, Matrix Exhibits, a $2 million supplier of trade-show exhibits that has 30 employees, is thriving in Atlanta's inner city. When Tennessee-based Matrix decided to enter the Atlanta market in 1985, it could have chosen a variety of locations. All the other companies that create and rent trade-show exhibits are based in Atlanta's suburbs. But the Atlanta World Congress Center, the city's major exhibition space, is just a six-minute drive from the inner city, and Matrix chose the location because it provided a real competitive advantage. Today Matrix offers customers superior response time, delivering trade-show exhibits faster than its suburban competitors. Matrix benefits from low rental rates for warehouse space – about half the rate its competitors pay for similar space in the suburbs – and draws half its employees from the local community. The commitment of local police has helped the company avoid any serious security problems. Today Matrix is one of the top five exhibition houses in Georgia.

Alpha and Matrix demonstrate how location can be critical to the success or failure of a business. Every location – whether it be a nation, a region, or a city – has a set of unique local conditions that underpin the ability of companies based there to compete in a particular field. The competitive advantage of a location does not usually arise in isolated companies but in clusters of companies – in other words, in companies that are in the same industry or otherwise linked together through customer, supplier, or similar relationships. Clusters represent critical masses of skill, information, relationships, and infrastructure in a given field. Unusual or sophisticated local demand gives companies insight into customers' needs. Take Massachusetts's highly competitive cluster of information-technology industries: it includes companies specializing in semiconductors, workstations, supercomputers, software, networking equipment, databases, market research, and computer magazines.

Clusters arise in a particular location for specific historical or geographic reasons – reasons that may cease to matter over time as the cluster itself becomes powerful and competitively self-sustaining. In successful clusters such as Hollywood, Silicon Valley, Wall Street, and Detroit, several competitors often push one another to improve products and processes. The presence of a group of competing companies contributes to the formation of new suppliers, the growth of companies in related fields, the formation of specialized training programs, and the emergence of technological centres of excellence in colleges and universities. The clusters also provide newcomers with access to expertise, connections, and infrastructure that they in turn can learn and exploit to their own economic advantage.

If locations (and the events of history) give rise to clusters, it is clusters that drive economic development. They create new capabilities, new companies, and new industries. I initially described this theory of location in *The Competitive Advantage of Nations* (Free Press, 1990), applying it to the relatively large geographic areas of nations and states. But it is just as relevant to smaller areas such as the inner city. To bring the theory to bear on the inner city, we must first identify the inner city's competitive advantages and the ways inner city businesses can forge connections with the surrounding urban and regional economies.

The true advantages of the inner city

The first step toward developing an economic model is identifying the inner city's true competitive advantages. There is a common misperception that the inner city enjoys two main advantages: low-cost real estate and labor. These so-called advantages are more illusory than real. Real estate and labor costs are often higher in the inner city than in suburban and rural areas. And even if inner cities were able to offer lower-cost labor and real estate compared with other locations in the United States, basic input costs can no longer give companies from relatively prosperous nations a competitive edge in the global economy. Inner cities would inevitably lose jobs to countries like Mexico or China, where labor and real estate are far cheaper.

Only attributes that are unique to inner cities will support viable businesses. My ongoing research of urban areas across the Untied States identifies four main advantages of the inner city: strategic location, local market demand, integration with regional clusters, and human resources. Various companies and programs have identified and exploited each of those advantages from time to time. To date, however, no systematic effort has been mounted to harness them.

Strategic location. Inner cities are located in what *should* be economically valuable areas. They sit near congested high-rent areas, major business centers, and transportation and communications nodes. As a result, inner cities can offer a competitive edge to companies that benefit from proximity to downtown business districts, logistical infrastructure, entertainment or tourist centers, and concentrations of companies …

There is significant potential, then, for expanding the inner-city business base by building on the advantage of strategic location. Among the initial prospects are location-sensitive industries now situated elsewhere, nearby companies and industries that face space constraints, and back-office or support functions amenable to relocation or outsourcing. Consider Boston's Longwood medical area, a huge concentration of world-class health care facilities. Longwood is located near the inner city neighborhoods of Roxbury and Jamaica Plain. Today such activities as laundry services, building maintenance, and just-in-time delivery of supplies are performed in-house or by suburban vendors. But, because of Longwood's proximity to the inner city, activities like these could be shifted to businesses based in Roxbury or Jamaica Plain – especially if basic infrastructure such as roads could be improved.

Local market demand. The inner city market itself represents the most immediate opportunity for inner-city-based entrepreneurs and businesses. At a time when most other markets are saturated, inner city markets remain poorly served – especially in retailing, financial services, and personal services. In Los Angeles, for example, retail penetration per resident in the inner city compared with the rest of the city is 35% in supermarkets, 40% in department stores, and 50% in hobby, toy, and game stores.

The first notable quality of the inner city market is its size. Even though average inner city incomes are relatively low, high population density translates into an immense market with substantial purchasing power. Boston's inner city, for example, has an estimated total family income of $3.4 billion. Spending power per acre is comparable with the rest of the city despite a 21% lower average household income level than in the rest of Boston, and, more significantly, higher than in the surrounding suburbs. In addition, the market is young and growing rapidly, owing in part to immigration and relatively high birth rates …

Ultimately, what will attract the inner city consumer more than anything else is a new breed of company that is not small and high-cost but a professionally managed major business employing the latest in technology, marketing, and management techniques. This kind of company, much more than exhortation, will attract spending power and recycle capital within the inner city community.

Integration with regional clusters. The most exciting prospects for the future of inner city economic development lie in capitalizing on nearby regional clusters: those unique-to-a-region collections of related companies that are competitive nationally and even globally. For example, Boston's inner city is next door to world-class financial-services and health-care clusters. South Central Los Angeles is close to an enormous entertainment cluster and a large logistical-services and wholesaling complex.

The ability to access competitive clusters is a very different attribute – and one much more far reaching in economic implication – than the more generic advantage of proximity to a large downtown area with concentrated activity. Competitive clusters create two types of potential advantages. The first is for business formation. Companies providing supplies, components, and support services could be created to take advantage of the inner city's proximity to multiple nearby customers in the cluster. For example, Detroit-based Mexican Industries has emerged as one of the most respected suppliers of head rests, arm rests, air bags, and other auto parts by forging close relationships with General Motors, Ford, Chrysler, and Volkswagen of America. Last year, the company had more than 1,000 employees, most of whom live in the inner city, and revenues of more than $100 million. Bing Steel, a 54-person company with $57 million in sales, has made similar connections, supplying flat roll steel and coils to the auto industry.

The second advantage of these clusters is the potential they offer inner city companies to compete in downstream products and services. For example, an inner city company could draw on Boston's strength in financial services to provide services tailored to inner city needs – such as secured credit cards, factoring, and mutual funds – both within and outside the inner city in Boston and elsewhere in the country. Boston Bank of Commerce (BBOC) is a trusted local institution in the inner city with strong ties to the community. It has many small non-profit customers, such as the Dimock Community Health Center in Roxbury, which has a $1 million endowment. There are many non-profit organizations like Dimock whose funds are sitting idle in low-interest savings accounts because they lack the investment savvy and size to attract sophisticated money managers …

Few of these opportunities are currently being pursued. Most of today's inner city businesses either have not been export oriented, selling only within the local community rather than outside it, or have seen their opportunities principally in terms defined by government preference programs. Consequently, networks and relationships with surrounding companies are woefully underdeveloped. New private sector initiatives will be needed to make these connections and to increase inner city entrepreneurs' awareness of their value. Integration with regional clusters is potentially the inner city's most powerful and sustainable competitive advantage over the long term. It also provides tremendous leverage for development efforts: by focusing on upgrading existing and nascent clusters, rather than on supporting isolated companies or industries, public and private investments in training, infrastructure, and technology can benefit multiple companies simultaneously.

Human resources. The inner city's fourth advantage takes on a number of deeply entrenched myths about the nature of its residents. The first myth is that inner city residents do not want to work and opt for welfare over gainful employment. Although there is a pressing need to deal with inner city residents who are unprepared for work, most inner city residents are industrious and eager to work. For moderate-wage jobs ($6 to $10 per hour) that require little formal education (for instance, warehouse workers, production-line workers, and truck drivers), employers report that they find hardworking, dedicated employees in the inner city …

Admittedly, many of the jobs currently available to inner city residents provide limited opportunities for advancement. But the fact is that they are jobs; and the inner city and its residents need many more of them close to home. Proposals that workers commute to jobs in distant suburbs – or move to be near those jobs – underestimate the barriers that travel time and relative skill level represent for inner city residents. Moreover, in deciding what types of businesses are appropriate to locate in the inner city, it is critical to be realistic about the pool of potential employees. Attracting high-tech companies might make for better press, but it is of little benefit to inner city residents. Recall the contrasting experiences of Alpha Electronics and Matrix Exhibits. In the case of Alpha, there was a compete mismatch between the company's need for highly skilled professionals and the available labor pool in the local community. In contrast, Matrix carefully considered the available workforce when it established its Atlanta office. Unlike the Tennessee headquarters, which custom-designs and creates exhibits for each client, the Atlanta office specializes in rentals made from prefabricated components – work requiring less-skilled labor, which can be drawn from the inner city. Given the workforce, low-skill jobs are realistic and economically viable: they represent the first rung on the economic ladder for many individuals who otherwise would be unemployed. Over time, successful job creation will trigger a self-reinforcing process that raises skill and wage levels.

The second myth is that the inner city's only entrepreneurs are drug dealers. In fact, there is a real capacity for legitimate entrepreneurship among inner city residents, most of which has been channelled into the provision of social services. For instance, Boston's inner city has numerous social service providers as well as social, fraternal, and religious organizations. Behind the creation and building of those organizations is a whole cadre of local entrepreneurs who have responded to intense local demand for social services and to funding opportunities provided by government, foundations, and private sector sponsors. The challenge is to redirect some of that talent and energy toward building for-profit businesses and creating wealth.

The third myth is that skilled minorities, many of whom grew up in or near inner cities, have abandoned their roots. Today's large and growing pool of talented minority managers represents a new generation of potential inner city entrepreneurs. Many have been trained at the nation's leading business schools and have gained experience in the nation's leading companies. Approximately 2,800 African Americans and 1,400 Hispanics graduate from MBA programs every year compared with only a handful 20 yeas ago. Thousands of highly trained minorities are working at leading companies such as Morgan Stanley, Citibank, Ford, Hewlett-Packard, and McKinsey & Company. Many of these managers have developed the skills, network, capital base, and confidence to begin thinking about joining or starting entrepreneurial companies in the inner city …

Changing roles and responsibilities for inner city development

… **The new role of the private sector.** The economic model challenges the private sector to assume the leading role. First, however, it must adopt new attitudes towards the inner city. Most private sector initiatives today are driven by preference programs or charity. Such activities would never stand on their own merits in the marketplace. It is inevitable, then, that they contribute to growing cynicism. The private sector will be most effective if it focuses on what it does best: creating and supporting economically viable businesses built on true competitive advantage. It should pursue four immediate opportunities as it assumes its new role.

1. Create and expand business activity in the inner city. The most important contribution companies can make to inner cities is simply to do business there. Inner cities hold untapped potential for the profitable businesses. Companies and entrepreneurs must seek out and seize those opportunities that build on the true advantages of the inner city. In particular,

retailers, franchisers, and financial services companies have immediate opportunities. Franchises represent an especially attractive model for inner city entrepreneurship because they provide not only a business concept but also training and support.

Businesses can learn from the mistakes that many outside companies have made in the inner city. One error is the failure of retail and service businesses to tailor their goods and services to the local market. The needs and preferences of the inner city market can vary greatly – something that companies like Goldblatt Brothers have recognized. The Chicago retailer understands that its inner city customers buy to meet immediate needs, and it has tailored its retail merchandise and purchasing planning to its customers' buying habits. For example, unlike most stores, which stock winter coats in the fall, Goldblatt Brothers stocks its coats in the winter.

Another common mistake is the failure to build relationships within the community and to hire locally. Hiring local residents builds loyalty from neighborhood customers, and local employees of retail and service businesses can help stores customize their products. Evidence suggests that companies that were perceived to be in touch with the community had far fewer security problems, whether or not the owners lived in the community. For example, Americas' Food Basket hires locally and is widely viewed as a good citizen of the community. As a result, management reports that it has not had to hire a security guard and that neighbors often call if they witness anything amiss …

2. Establish business relationships with inner city companies. By entering into joint ventures or customer-supplier relationships, outside companies will help inner city companies by encouraging them to export and by forcing them to be competitive. In the long run, both sides will benefit. For example, AB&W Engineering, a Dorchester-based metal fabricator, has built a close working relationship with General Motors. GM has given AB&W management assistance and a computerized ordering system and has referred a lot of new business to AB&W. In turn, AB&W has become a high-performing and reliable supplier. Such relationships, based not on charity but on mutual self-interest, are sustainable ones; every major company should develop them.

3. Redirect corporate philanthropy from social services to business-to-business efforts. Countless companies give many millions of dollars each year to worthy inner-city social-service agencies. But philanthropic efforts will be more effective if they also focus on building business-to-business relationships that, in the long run, will reduce the need for social services.

First, corporations could have a tremendous impact on training. The existing system for job training in the United States is ineffective. Training programs are fragmented, overhead intensive, and disconnected from the needs of industry. Many programs train people for nonexistent jobs in industries with no projected growth. Although reforming training will require the help of government, the private sector must determine how and where resources should be allocated to ensure that the specific employment needs of local and regional businesses are met. Ultimately, employers, not government, should certify all training programs based on relevant criteria and likely job availability.

Training programs led by the private sector could be built around industry clusters located in both the inner city (for example, restaurants, food service, and food processing in Boston) and the nearby regional economy (for example, financial services and health care in Boston). Industry associations and trade groups, supported by government incentives, could sponsor their own training programs in collaboration with local training institutions …

Second, the private sector could make an equally substantial impact by providing management assistance to inner city companies. As with training, current programs financed or operated by the government are inadequate. Outside companies have much to offer companies in the inner city: talent, know-how, and contacts. One approach to upgrading management skills is to emphasize networking with companies in the regional economy that either are part of the same cluster (customers, suppliers, and related businesses) or have expertise in needed areas. An inner city company could team up with a partner in the region who provides management assistance; or a consortium of companies with a required expertise, such as information technology, could provide assistance to inner city businesses in need of upgrading their systems …

4. Adopt the right model for equity capital investments. The investment community – especially venture capitalists – must be convinced of the viability of investing in the inner city. There is a small but growing number of minority-oriented equity providers (although none specifically focus on inner cities). A successful model for inner city investing will probably not look like the familiar venture-capital model created primarily for technology companies. Instead, it may resemble the equity funds operating in the emerging economies of Russia or Hungary – investing in such mundane but potentially profitable projects as supermarkets and laundries. Ultimately, inner-city-based businesses that follow the principles of

competitive advantage will generate appropriate returns to investors – particularly if aided by appropriate incentives, such as tax exclusions for capital gains and dividends for qualifying inner city businesses.

The new role of government. … Undeniably, inner cities suffer from a long history of discrimination. However, the way for government to move forward is not by looking behind. Government can assume a more effective role by supporting the private sector in new economic initiatives. It must shift its focus from direct involvement and intervention to creating a favourable environment for business. This is not to say that public funds will not be necessary. But subsidies must be spent in ways that do not distort business incentives, focusing instead on providing the infrastructure to support genuinely profitable businesses. Government at all levels should focus on four goals as it takes on its new role.

1. Direct resources to the areas of greatest economic need. The crisis in our inner cities demands that they be first in line for government assistance. This may seem an obvious assertion. But the fact is that many programs in areas such as infrastructure, crime prevention, environmental cleanup, land development, and purchasing preference spread funds across constituencies for political reasons. For example, most transportation infrastructure spending goes to creating still more attractive suburban areas. In addition, a majority of preference-program assistance does not go to companies located in low-income neighborhoods.

Investments that boost the economic potential of inner cities must receive priority. For example, Superfund cleanup dollars should go to sites in high-unemployment inner city areas before they go to low-unemployment suburban sites. Infrastructure improvements should go to making inner city areas more attractive business locations. And crime prevention resources should go to high-crime inner city areas. Spending federal, state, and local money in that way will have the added benefit of easing critical social problems, thus reducing social service spending …

2. Increase the economic value of the inner city as a business location. In order to stimulate economic development, government must recognize that it is a part of the problem. Today its priorities often run counter to business needs. Artificial and outdated government-induced costs must be stripped away in the effort to make the inner city a profitable location for business. Doing so will require rethinking policies and programs in a wide range of areas. There is early evidence that self-inflicted regulatory costs can be overcome. Consider the success of the Indianapolis

Regulatory Study Commission in Indiana. In two short years, Indianapolis ended its taxi monopoly, streamlined its building permitting process, and eliminated a wide range of needless regulations.

Indeed, there are numerous possibilities for reform. Imagine, for example, policy aimed at eliminating the substantial land and building cost penalties that businesses face in the inner city. Ongoing rent subsidies run the risk of attracting companies for which an inner city location offers no other economic value. Instead, the goal should be to provide building-ready sites at market prices. A single government entity could be charged with assembling parcels of land and with subsidizing demolition, environmental cleanup, and other costs. The same entity could also streamline all aspects of building – including zoning, permitting, inspections, and other approvals …

3. Deliver economic development programs and services through mainstream, private sector institutions. There has been a tendency to rely on small community-based nonprofits, quasi-governmental organizations, and special-purpose entities, such as community development banks and specialized small-business investment corporations, to provide capital and business-related services. Social service institutions have a role, but it is not this. With few exceptions, nonprofit and government organizations cannot provide the quality of training, advice, and support to substantial companies that mainstream, private sector organizations can. Compared with private sector entities such as commercial banks and venture capital companies, special-purpose institutions and nonprofits are plagued by high overhead costs; they have difficulty attracting and retaining high-quality personnel, providing competitive compensation, or offering a breadth of experience in dealing with companies of scale …

The most important way to bring debt and equity investment to the inner city is by engaging the private sector. Resources currently going to government or quasi-public financing would be better channelled through other private sector investors. Minority-owned banks that have superior knowledge of the inner city market could gain a competitive advantage by developing business-lending expertise in inner city areas …

4. Align incentives built into government programs with true economic performance. Aligning incentives with business principles should be the goal of every government program. Most programs today would fail such a test. For example, preference programs in effect guarantee companies a market. Like other forms of protectionism, they dull motivation and retard cost and quality improvement …

Direct subsidies to businesses do not work. Instead, government funds should be used for site assembly, extra security, environmental cleanup, and other investments designed to improve the business environment. Companies then will be left to make decisions based on true profit.

The new role of community-based organizations. Recently, there has been renewed activity among community-based organizations (CBOs) to become directly involved in business development. CBOs can, and must, play an important supporting role in the process. But choosing the proper strategy is critical, and many CBOs will have to change fundamentally the way they operate. While it is difficult to make a general set of recommendations to such a diverse group of organizations, four principles should guide community-based organizations in developing their new role.

1. Identify and build on strengths. Like every other player, CBOs must identify their unique competitive advantages and participate in economic development based on a realistic assessment of their capabilities, resources, and limitations. Community-based organizations have played a much-needed role in developing low-income housing, social programs, and civic infrastructure. However, while there have been a few notable successes, the vast majority of businesses owned or managed by CBOs have been failures. Most CBOs lack the skills, attitudes, and incentives to advise, lend to, or operate substantial businesses. They were able to master low-income housing development, in which there were major public subsidies and a vacuum of institutional capabilities. But, when it comes to financing and assisting for-profit business development, CBOs simply can't compete with existing private sector institutions.

Moreover, CBOs naturally tend to focus on community entrepreneurship: small retail and service businesses that are often owned by neighborhood residents. The relatively limited resources of CBOs, as well as their focus on relatively small neighborhoods, is not well-suited to developing the more substantial companies that are necessary for economic vitality …

2. Work to change workforce and community attitudes. Community-based organizations have a unique advantage in their intimate knowledge of and influence within inner city communities, and they can use that advantage to help promote business development. CBOs can help create a hospitable environment for business by working to change community and workforce attitudes and acting as a liaison with residents to quell unfounded opposition to new businesses. When BayBank wanted to open a new branch in Dorchester, for example, a local community development corporation was instrumental in smoothing relations with a few vocal critics who could have delayed the project or even driven the bank away.

3. Create work-readiness and job-referral systems. Community-based organizations can play an active role in repairing, screening, and referring employees to local businesses. A pressing need among many inner city residents is work-readiness training, which includes communication, self-development, and workplace practices. CBOs, with their intimate knowledge of the local community, are well equipped to provide this service in close collaboration with industry …

CBOs can also help inner city residents by actively developing screening and referral systems. Admittedly, some inner-city-based businesses do not hire many local residents. The reasons are varied and complex but seem to revolve around a few bad experiences that owners have had with individual employees and their work attitudes, absenteeism, false injury claims, or drug use. A study of the impoverished Red Hook neighborhood in Brooklyn points to the importance of social networks – networks that are often lacking in inner cities – as informal job referral systems. The study found that a local development corporation, the South Brooklyn LDC, played an important role in helping local residents get jobs by developing relationships with nearby businesses and screening and referring employees to them.

4. Facilitate commercial site improvement and development. Community-based organizations (especially community development corporations) can also leverage their expertise in real estate and act as a catalyst to facilitate environmental cleanup and the development of commercial and industrial property. For example, the Codman Square Neighborhood Development Corporation in Boston was part of a group including the Boston Public Facilities Department, local merchants, and the local health center that encouraged 36 businesses to move into a depressed neighborhood. The group used its considerable community organizing talent to help merchants form an association to identify the neighborhood's needs as well as barriers to meeting them. It negotiated with the police to increase patrols in the area and pushed the mayor's office to board up abandoned buildings and to rid the area of trash and abandoned cars. After bringing together many different constituencies, it led a campaign to encourage businesses to locate in the neighborhood.

Overcoming impediments to progress

This economic model provides a new and comprehensive approach to reviving our nation's distressed urban communities. However, agreeing on and implementing it will not be without its challenges. The private sector, government, inner city residents, and the public at large all hold entrenched attitudes and prejudices about the inner city and its problems. These will be slow to change. Rethinking the inner city in economic rather than social terms will be uncomfortable for many who have devoted years to social causes and who view profit and business in general with suspicion. Activists accustomed to lobbying for more government resources will find it difficult to embrace a strategy for fostering wealth creation. Elected officials used to framing urban problems in social terms will be resistant to changing legislation, redirecting resources, and taking on recalcitrant bureaucracies. Government entities may find it hard to cede power and control accumulated through past programs. Local leaders who have built social service organizations and merchants who have run mom-and-pop stores could feel threatened by the creation of new initiatives and centers of power. Local politicians schooled in old-style community organizing and confrontational politics will have to tread unfamiliar ground in facilitating cooperation between business and residents.

These changes will be difficult ones for both individuals and institutions. Nonetheless, they must be made. The private sector, government, and community-based organizations all have vital new parts to play in revitalizing the economy of the inner city. Business people, entrepreneurs, and investors must assume a lead role, and community activists, social service providers, and government bureaucrats must support them. The time has come to embrace a rational economic strategy and to stem the intolerable costs of outdated approaches.

Source: Porter, 1995, pp.55–71

CHAPTER 8

On the openness of cities

by Michael Pryke

1 *The future is open*

As you have progressed through the chapters in this book you may well have reached the view that cities are open, mobile, mixed places. In many senses cities are collections of processes that are formed and reformed through a host of interconnections, which is, after all, a direct reflection of their openness. This approach is helpful in thinking about the future of cities and what the nature of today's openness might hold for the lives of those within them. As you can imagine, there are many aspects to this question that are worth consideration. Let's consider three of them here. A city's *dynamism,* the degree of *inequality* within a city and the nature of *power* within cities in a world of flows, all hinge around cities' openness. Before moving on to the activity below, you might like to dwell upon what openness – as you have understood its use in this book – might mean for city futures. Recalling the characteristics of today's global connections, does the term suggest, for instance, opportunities, the chance for new ideas to arrive and mix in cities? Or does openness almost inevitably mean the production of cities as spaces of uncertainty, vulnerability and indeterminacy?

ACTIVITY 8.1   Bearing in mind your thoughts on the meaning of openness, turn to Reading 8A by Patralekha Chatterjee, which discusses the impact on city life of the collapse of the financial markets in South-East Asia in the late 1990s. While reading the article, jot down what is suggestive of the *increasing openness* or permeability of cities. How has the South-East Asian 'meltdown', as the article terms it, fuelled *urban tensions* and possibly altered the futures of people within these cities? Where does the power behind this type of urban upheaval lie – within or beyond the cities of this region? ◆

What is striking about this account is how the dynamism of these cities suddenly froze. Following the collapse of the financial markets and the contagion that spread through previously empowering interconnections – connections which closely knitted together this region's cities and the global economy – a cityscape emerged in which 'cranes stand idle' and 'half finished buildings dot the landscape'. The bustle and movement of economic prosperity was suddenly replaced by relative stillness among the idle cranes and vacant office blocks. And while the changing yet 'prosperous times' hid inequality, the financial meltdown exposed and aggravated this growing divide, creating new tensions as inflation and lay-offs began to wreak havoc in the social and ethnic fabric of cities in this part of the world. The straw-hatted women – recent migrants from Thailand's poorer northern regions – protesting outside their employer's factory in Bangkok for the restoration of overtime, and the report of attacks on ethnic Chinese shopkeepers in Indonesian cities are good illustrations of this point.

The broader issue to come from the reading is that

•   hardly anything that has taken place in these cities as a consequence of the financial meltdown can be understood completely, or in some cases even partially, by limiting analysis solely to what goes on *within* each city.

It was these cities' openness to financial flows that played a significant part in creating the 'Asian miracle' and which led to the collapse that impacted 'so pervasively to change the very character of some cities'. It was the flow of migrants from countries and countrysides to cities that helped to build the miracle and which produced the tensions that have now surfaced within cities. Equally, it was the near collapse of financial markets that produced new tensions between the 'countryside' and 'cities' in Indonesia and Thailand, as established flows between the two were abruptly reversed. This reaffirms the difficulty of stating confidently where a city begins and ends. The coalescing of these and other uncertainties created a 'mixture of unrest and optimism on the streets'. How to stabilize them into a more certain future is a task which itself appears more open to the influence and stipulations of international organizations such as the International Monetary Fund (IMF) and private sector finance, than it is to the preferences of the inhabitants of these cities. Thus we read of the 'bail-out plan' backed by the IMF, we see the observation of a western investment bank, Jardine Fleming, and we are told of the ever present financial monitoring – here in the form of a US investment bank's 'International Indices' – to assess the relative performance of countries/markets in the region.

1.1   ... OR MAYBE NOT THAT OPEN?

In this book we have attempted to highlight the manner in which the city is both in constant movement, yet settled in its formation. This apparent paradox has served to help us understand city life, as has the tension between community and difference. More than that, however, the focus on mobility, movement and change has enabled us to produce a whole set of understandings, not simply about what goes on in and between cities, but also, as we have explored in Reading 8A, about the openness of cities and their futures. A reasonable question might now be not simply 'to what exactly are cities open?' – a question that we have already begun to answer – but rather 'are cities always and everywhere open to the same flows and influences?'. The type of openness written about in Reading 8A relates predominantly to things economic. This serves as a useful reminder about how we understand and employ the idea of openness and begins to address the second question:

●   'Openness' does not mean that cities always and everywhere are open to all types of potential futures, but rather that certain cities, as a consequence of the sort of networks into which they are woven, are open to *certain* possible futures.

For example, as we saw in Chapter 5, cities such as Paris (and also Milan) have *made* themselves open to networks of fashion. No matter how hard other cities might try, they are unlikely instantly to become centres for western fashion.

Though this may seem an obvious point, and one far removed from the financial meltdown in South-East Asia, it serves as a reminder that cities, and the social relations that make and remake them, are constrained by

- the history of networks of which they are a part – one consequence of which is that certain networks may simply bypass them; and by

- the feasibility of fabricating themselves socially, as well as technically, culturally, and so on, into any one of a number of new futures.

As we have seen from the later chapters especially, this juggling of a few (rather than many) possible futures is made a more slippery task because of the flows and influences circulating through the workings and interconnections of today's global economy.

It is worth remembering too that cities can become *disconnected* from a set of networks through, for example, *active* political choice – or at least attempts can be made to regulate the nature of interconnection. Again, something of this active disconnection emerges from Chatterjee's article. While reform-minded governments might be tempted to 'look inside' for solutions, the decision actively to disconnect is constrained by the influence of working within the potentially unsettling rhythms of the global financial system and its chief international institutions. Cities may well choose to try, as it were, to keep the door ajar, rather than either closed or fully open. Part of this chapter's broader purpose is to review the ways in which the openness of cities is constructed through interconnections – both within and between cities. Indeed, this has been our aim throughout the book.

1.2 OPENNESS TO MULTIPLE SOCIAL RELATIONS?

We have been trying to portray cities as sites of multiple social relations in which people are affected by, and coexist with, not just other people from a range of ethnic backgrounds, but also non-humans – 'nature' in all its forms. The varieties of animal, plant and bird life which move and settle among the rhythms and buildings of any city and the multiplicity of the social relations they engender cannot but affect the city. The city, as we have come to understand it, is a *place of intensity*, a place of dense interaction. Yet the different movements within and around the city – given their variety and the ways in which they sometimes settle uneasily alongside one another – contain within them latent *tensions*. These tensions are born out of the heterogeneity described in earlier chapters. Such tensions come from the presence within cities of a whole variety of social groups, with their differing mobilities, incomes, employment chances, and so on. Along neighbouring streets or even along the same street you may well find people from a range of ethnic backgrounds – highly paid employees of multinational media corporations, old-age pensioners, gay couples, poorly paid cleaners, young unemployed – all of whom will be experiencing 'cityness' according to very different rhythms, with their movements and juxtapositions potentially harbouring latent tension.

The overall impression gained from the early chapters is that this multiplicity of social relations is not self-contained within the city, but in many cases forms part of wider links to other parts of other cities. For these chapters stressed that the formation, transformation and fortunes of cities should be understood in the wider context of city relationships and viewed in terms of the *networks* of which they may form a part. We would stress 'may form a part' simply to reaffirm that such networks are not predetermined, and should not be seen as a guarantee of a city's future.

Another point that began to emerge in the discussion of networks is that they expose cities to processes and influences that are not just outside of that city, but increasingly outside of that city's nation-state. You may well have picked this out from your reading of Chapters 6 and 7 – and indeed of Reading 8A – and their discussion of the ways in which global economic change has more recently come to influence day-to-day city life within a range of cities in developing countries.

2 *Cities as places of multiple connections*

So how can we formulate an approach to understanding today's cities? It may be helpful to reiterate some of the main points that were raised in Chapter 1 and develop them further.

You will recall from the book introduction that at the heart of this text is a concern to think spatially about cities. Much of the discussion so far has been in keeping with that aim. For example, we have noted the heterogeneity of spaces within cities, and the flows through and between cities that arise from networks of stretched social relations that are all part of the formation of cities. But how can we make sense of all that is going on within and between contemporary cities? The problem becomes more complicated when we recall that today's cities are increasingly exposed to new and substantial global influences. The stretching of social relations and institutions *and* the growing intensity of contact between cities that is made possible by transport and communication technologies, not only shape today's cities but greatly multiply the types of tension experienced within cities, some of which we have already touched on.

As Amin and Graham comment in Chapter 1, travel, migration, global consumerism and television, to take but a few examples, all disturb local traditions, in even the more 'provincial' cities today. These new experiences offer continual contrasts which bring tensions to life within any city that add to those brought about by sources outside of that city. The point here is to reflect on the *permeability* of cities. This is another way of talking about cities as 'places of multiple connections'. One consequence of this is that we cannot be satisfied with thinking about cities in isolation or as having a 'fixed coherence' – something we have already remarked upon in relation to Reading 8A and the openness of cities. This notion of permeability is useful in helping us to understand cities. It is exemplified in Chapter 1, where the authors emphasize 'the idea of cities as places of intersection between many webs of social, cultural and technological flow, and the superimposition of these relational webs on the physical spaces of cities'. Such a *relational perspective* is very much about *thinking through cities spatially*, and it is a perspective that has influenced all the chapters in this book.

Building on this perspective allows us to piece together our thoughts about cities in ways that equip us 'to explore the interplay between cities as places and cities as nodes within diverse webs of social, economic and cultural relations' (see Chapter 1, section 2.4). To reiterate, cities should not just be seen as familiar places or place names on the map, but in terms also of their geographies of economic, social, political and cultural interconnections. Glasgow, Cardiff and Nairobi are 'the places they are' because of their various and different connections. The relational perspective enables us to appreciate cities in this way, but it also prevents us from slipping into what might be thought of as unproductive ways of understanding cities. Given that we have already noted the need to be aware of the impact of global processes in shaping cities, perhaps the most fruitful outcome of viewing cities relationally is that the complexity of what 'pours' into today's cities can be understood for its cross-cutting nature as opposed to a simple prioritization of the global over the local. This is illustrated well in Reading 8A and its description of the interlacing of ethnicities, expectations, experiences and presences within the cities of South-East Asia. The Mercedes sedans, the ethnic Chinese, the straw-hatted women from the countryside, the western designer clothes, the private hospitals and investment bankers are part of the global-to-local interconnections that coexist within cities.

You may recall the discussion in Chapter 1 of the urban landscape of the Lower East Side and Manhattan, New York. This example of two quite different places in one city, though only a few blocks apart, illustrates a further aspect of the relational perspective: the *social outcomes* of the mix of diverse connections that occur within cities. Amin and Graham refer to this meeting and missing of connections as 'fundamentally ambivalent tensions' (section 2.4). What do they mean by this term? It is an important though perhaps unfamiliar phrase, but it helps us get a better grasp of the heterogeneity of cities. The openness of cities to a whole range of people from diverse social, cultural, economic and ethnic backgrounds raises issues about how such heterogeneity is experienced and negotiated within cities. It is a theme that was followed through in various ways in Chapters 2, 3 and 4. It also raises the question of how this meeting of difference – whether viewed as cultural, sexual, economic or whatever – is reflected in the public/private division of cities. In certain cities, such as São Paulo or Los Angeles, what might be thought of as a utopian dream of cultural mixing has produced instead a retreat into privatized, high-security spaces in the form of gated communities.

This raises a question about whether the increasing openness of cities, and the flows and connections through and between them, will lead to increasingly fragmented cities, as anxieties and fears translate into material boundaries across cityscapes. How is such a mix to be managed? Furthermore, again in tune with the relational perspective, should such a mixing, crossing and mingling of various cultures and social flows always be viewed in negative terms? Surely, as Reading 1A by Hannerz suggested, such mixing can lead to positive transformations, demonstrating how the openness of cities means that their multiple overlapping spaces can spawn not only new experiences, economic innovations, tensions and so on, but also new identities as relational worlds fuse in contemporary cities.

It is worth staying with this theme a little longer to consider the implications that arise from Chapter 1's interpretation of the impact of new communications and information technologies in city spaces. One of the key points raised by the authors was the uneven 'weaving' of these technologies into the social fabric of cities, so that parts of cities may be 'physically close but relationally distant'. This is a process which seems likely to exacerbate unevenness both between and within cities and also to heighten tensions as highly unequal worlds are juxtaposed within them: 'the relational connections between the social worlds of urban places and their subtle links with electronically mediated experience' need to be thought through simultaneously.

Our discussion of Chapter 1 would seem to suggest that the authors aim to unsettle traditional thinking on the city, but in section 4 of the chapter Amin and Graham provide three related ways in which our approaches to cities can be resettled. In doing this, they allow us to keep in mind points about the increasing density of cities, the interconnectedness of cities, and the multiplicity of times and spaces that these interconnections represent within cities, as well as allowing us to view cities as places of diverse social encounter. These are all themes that have run through the later chapters and which help us to think through the city spatially.

2.1   NETWORKS AND CO-PRESENCE

The theme of co-presence signals the shear density of city life that is created by all the various connections and stretched social relations that simultaneously fan out from cities and connect into them. All of these flows *intensify* social, economic and cultural life within cities. Such density and heightened intensity can give cities a certain buzz which can be turned to the advantage of some parts of cities, while other parts of the same city can be locked into a less advantageous density. In the City of London, for example, there is a crowding, a dense agglomeration of the offices of firms involved in international finance which employ some extremely well-paid people. This is an agglomeration of 'economic' advantage reproduced through an intensity of international connections into global finance which spawns cultural, social as well as economic advantage for many of those working there. It is an agglomeration that is juxtaposed with council housing which accommodates, among others, a large Bangladeshi community, whose members, although part of an international network, find themselves less advantaged as they go about their day-to-day lives in this part of London.

A further point which extends the analysis of **Massey (1999)** elsewhere in this series, is the multiplicity of time-spaces within contemporary cities. Such multiplicity would seem to emerge from the permeable nature of today's cities: cities are exposed, as we have seen, to a whole variety of connections, a wide range of stretched social relations, some empowered by new communications technology. The point is that these connections do not then simply bring together different ethnicities, for example, to produce a variety of ethnic spaces; they import at the same moment the social times indigenous to these ethnic groups. While such a mix may represent a fairly lively mélange of space-times which produces a cityscape that moves to a host of different rhythms, the same point – that there is no single or dominant time-space dimension to cities – alerts us to the idea of the city as a continual *superimposition* of times and spaces. This need not be a case of thinking in terms of the colossal superimposition of dominant rhythms as talked about in, say, Chapter 6; it can just as easily be understood as an encouragement to 'listen geographically', as it were, to the practices of different ethnic groups arriving in a city and the social spaces that result. These practices may range from 'different' eating times, to the observation of 'non-traditional' holy days, to the cultural innovation around, say, music, clubs and nightlife that is sparked particularly by the young within ethnic groups.

- The idea of superimposition is a reminder not only of how dominant rhythms may pass through interconnections and disrupt a city's time-spaces, as the examples of life in Harare and Dar es Salaam in Chapter 6 illustrate, but also an encouragement to be aware of the range of time-spaces within cities and the nature of the connections that generate them.

The nature of connections recalls the theme of mobility and interconnectedness and the rhythmic metaphor that has run through many of the chapters in this book. As we have seen, particularly in the later chapters, what goes on within certain cities, or more accurately within particular parts of cities such as New York, Tokyo and London, can serve to *influence* what happens within the many spaces of other cities. The interconnectedness of cities can be viewed as serving to transmit potentially dominant times which can disrupt and influence the experience of city life. So, while it may be seen as beneficial, if not the ultimate goal, for cities to open themselves to as many connections as possible, this has to be reconciled with the temporal imperatives, the dominant rhythms, that may work through these connections. One issue this raises is the political tension that lies at the heart of deciding a city's future. As we remarked earlier, should the goal be to open a city to a set of supposedly advantageous connections in order to achieve a better future, when the control of what flows through those interconnections increasingly lies outside of cities?

This point applies equally to the cultural diversity of cities. The way in which cities stretch across and between a range of cultural, economic and social influences should not be read as a celebration of diversity, for it equally represents social inequality. This reaffirms the importance of the question posed in the conclusion to Chapter 1: with such an intense mixture, change and ambivalence at the heart of contemporary urban life, is it possible to generalize about cities? Is it pointless to search for a wholeness to a city? One might conclude all too easily that cities are simply collections of *fragments*. Yet, as the conclusion to Chapter 1 argues, it would be a mistake to see cities only as loosely connected bits and pieces. The point of unsettling our thoughts about cities was to make us see and appreciate the connections that run through the 'fragments' – that is, to recognize the sheer physicality of cities and the array of social processes and spaces that are juxtaposed and superimposed through the collection of buildings we call cities.

Yet viewing cities in this way is not to go back to viewing them in isolation. What the built form of cities represents, and what the different social groups based within these buildings are part of, are *global flows*.

2.2   LIVING WITH DIFFERENCE

If city buildings serve to stabilize particular relational webs and connections, the people within them can then be understood to move to the rhythms of wider national and international connections. In the same way, the 'cityness' that results from the meeting or juxtaposition of all this movement and settlement (whether this be culturally positive or socially antagonistic) and its co-presence, can be understood actively to generate something new, something different. You only have to think about how 'new' musical and dance styles, such as Salsa or Zouk, emerge within particular cities as a result of the fusion of distant and sometimes disparate influences, to understand something of how cities, as arenas, can generate difference from the mixing of their connections.

This may be familiar to you from Chapter 3, with its focus on urban migration and movement, on the social differences between people moving in and between cities, on how these differences settle, and on how they manifest themselves in cities. An important part of this chapter was its exploration of how it is that movement and flows into cities come to settle in the ways they do. While the main aim was on the settling of these flows within cities – where the flows began, that is – the wider spatial networks of which they were a part were also brought into the argument.

It is worth considering this point further, particularly in the light of what was said in Chapter 1 about how cities create new identities through the mixing of relational webs. Recalling the account in Chapter 3 of how the nineteenth-century West European cities grew from greater freedom and mobility within cities, and through greater numbers of people selling their labour power, think about how this 'reordering of people's lives' provoked changes in the use of the city. As migration to cities increases, how might this reorder the use of city space? What will this lead to in terms of the use of, and access to, public space? How do people cope with such a range of difference? How do they deal with the tensions of city life? Finally – channelling answers to these questions in a particular direction – how can cities (be made to) function on the basis of such a range of movement, diversity and difference?

Chapter 2's discussion of movement to and through the city as part of transport networks raised a number of issues around the theme of community and difference. For example, transport is an important feature in shaping the city, but access to it is unequal. The unevenness of access to public and private forms of transport and the different qualities of each, exacerbate feelings and experiences of tensions. The ability to travel a certain distance across a city can be highly unequal for different people, depending on their income, race, gender, able-bodiedness, and so on. For example, a low-paid single mother will face a wholly different set of constraints in moving around a city, to a single, childless, well-paid professional. Thinking spatially here produces different geographies which in their own way describe a form of segregation within cities.

These points echo our earlier stress on the increasing openness of cities and the questioning of what it means. For what is suggested here is that the presumed openness of cities quickly closes down once we begin to ask 'for whom?' and 'by whom?' It might be worth reflecting for a moment on how you would, and indeed if you could, imagine a city of differences – of income, gender, ethnicity, age, sexuality – which did not lead to a city of segregation and, at the extreme, of gated communities. As cities become a more intense mix of ethnicities and sexualities, the desire by many to be among familiar groups, rather than in the midst of difference, needs somehow to be balanced in ways that ensure such unities of *similarity* do not breed intolerance of *difference*. You might like to spend some time reconsidering the questions raised towards the end of section 10 of Chapter 3. They are not altogether easy to answer, but they do seem pertinent to the imagination and creation of cities as better places to live.

This quest for better cities is much more than a question of how we learn to live with human strangers; it should also, as Chapter 4 reminds us, be extended to our non-human neighbours. You may recollect how this chapter played with the idea of the stranger in urban theory – an idea discussed at length in Chapter 3 – to begin to think about 'ways of living with a mix of city-natures'. You might also remember that this quest followed from the recognition that cities of difference and diversity are formed not just through movements and encounters of culturally and socially diverse people, but also through flows and settling of non-humans – of materials, plants, energies, animals, and so on.

The issues raised in Chapter 4 stimulate thoughts about the 'living together of city natures'. This all has to be thought through alongside a search for social justice and new forms of living together by humans in cities – not an easy but a necessary task. For as Chapter 4, section 5 notes, 'city-nature movements, embodied in plant and animal species, are vital resources for many different kinds of urban living and settlement'. In an increasingly urban world in which the issue of 'sustainability' seems to be being voiced by a growing number of people, such an unsettling of how we think about 'cities' and 'nature' may be just what is needed: 'Redefinitions of what it means to be urban that include recognition of city-natures may help to produce more sustainable forms of urban living' (section 5.1).

Trying to resettle our thoughts about city-nature formations brings us back to the interconnectedness of cities and what these connections 'contain'.

3 Cities as sites of multiple influences

To remain with city-nature formations for a moment, Chapter 4 offers us an important reminder about movement and settlement: 'formations don't simply exist in contained spaces within cities; they also move across and between cities, and incorporate places and processes that we normally associate with the non-urban' (section 5.2). Chapter 4 reworks the idea of the ecological footprint in terms of relational webs or networks focused on cities. Making use of a quote from David Harvey (Activity 4.9), the chapter reminds us of how and why our thinking about city-nature formations – about the neat boxes labelled 'society' and 'environment', to use Harvey's language – as well as our understanding of cities more generally, might be boosted by applying the metaphor of networks. 'These networks', we are reminded, 'which are forever forming, collapsing and reforming, are important devices for understanding the ordering of ecosystems both within and outside the conventional city boundaries … Harvey's point is that even the remotest of landscapes … are touched by the threads of these networks' (section 5.2). This is an important point and echoes the emphasis of Chapter 1 of this book, although the language used there spoke of 'relational webs' rather than networks.

3.1   NETWORKS AND RHYTHMS

The idea of cities as parts of relational webs or networks, or as points of intersection, formed the conceptual backbone to Chapters 5, 6 and 7. It is useful to remind ourselves that these chapters, in addition to noting how cities are parts of networks, raised a number of further issues and questions (some of which we have already touched on here) which gave us an understanding of the *influences* running through these networks, how these can *work* to change the day-to-day rhythms of city life, and how these networks are *organized,* institutionally or otherwise. Our reading of these later chapters helps us to understand some of the points made in the earlier chapters. For instance, we are better able to appreciate the flows of migration into cities at various moments, the social and cultural histories carried by these movements, and how they come together in cities, through the idea of city networks (or relational webs). This reaffirms the need to be attentive to the rhythms of the networks, to the times they carry and to the city spaces they help produce. In addition, noting the messages of Chapters 5 and 6, we are urged to attune our senses to what these connections transmit from city to city. The example we were given highlighted flows of economic influence: about how to establish the 'correct' macroeconomic policy and the supposedly enabling power of private finance. Contemporary cities, then, and what goes on within them – the pace, rhythms, tensions, intensities of cities – are seen to be inextricably bound up with how successful they are in making themselves part of networks and on what terms they can do this. Viewing cities in this way should make us aware that city futures and the *variety* of futures that may confront any one city are not predetermined. Such futures are increasingly dependent upon, and actively made through, networks. As we saw from Reading 8A, cities may also be just as easily and swiftly undone by the same networks.

But which cities and which groups within them start the ball rolling, as it were? Do ideas generated in Lima and New York about how to organize city life and run cities in a particular way carry the same potential to affect urbanization processes in Chicago or Glasgow? How is it that certain ideas are transmitted to greater effect than others? If a city such as New York or London is said to be part of an influential and 'successful network', does this mean that everyone in that city benefits? And once a city has worked its way into a successful network, does that mean its future is more or less guaranteed? These sorts of questions seem to be particularly relevant to those cities of the South that are set to be the location of huge increases in urbanization and which are actively trying to locate themselves in successful networks. The world's first 'smart cities', Putrajaya and Cyberjaya in Malaysia, which were discussed in section 4.2 of Chapter 5, are good illustrations of this point. Malaysia has set about 'disentangling itself from the declining networks of British Rule', according to the government's official brochure, and establishing itself at the centre of the 'Information Age'. It is a country 'where information, ideas, people, goods and services move across borders in the most cost-effective ways', as in an effort to be at the forefront of the information revolution (Chapter 5, section 2). But as we are reminded in the chapter, the success of such a strategy is more than a question of simply listing the various 'factors' that come together in cities through various networks; it is rather how a city's different strengths and connections come together in such a way as to produce an *intense* set of interactions.

It is this intensity, and the creativity that comes from it, that has the potential to make a city powerful and influential. This notion of 'creativity', it seems, is crucial to understanding what makes certain cities powerful, for it is linked, particularly in Chapter 5 and Reading 5A by Sassen, to the insistence on the *active production* of economic power. The greater the concentration of flows, the more intense becomes the power of a city. You might like to ask yourself how feasible it is for all cities to try to put together the right mix or concentration of flows that are supposedly needed to make a city powerful. Can they all be successful? Or must there also be 'losers' in the competition between cities for a 'successful' future? These questions relate closely to what is meant by the 'openness' of cities.

3.2   NETWORKS AND INFLUENCE

In thinking about such questions, the temptation to equate power with cities should be avoided. This, you may recall, is an important point about networks in relation to cities, and is also an important part of thinking about cities spatially. 'City powers', we are told by Castells, 'are mobilized *through* networks; it is what flows *through* the networks which empowers particular groups and generates certain cities as sites of power' (Chapter 5, section 3.2). As a result, we are told to avoid viewing the global city (or any city, for that matter) as a place, but to see it rather, Castells argues, 'as a process'. Whether you agreed with Sassen or Castells in Chapter 5, the manner in which the two writers think about cities helps in thinking about cities as active centres, as nodes or hubs of dynamic connections.

Before going any further, you might like to re-read quickly section 4 of Chapter 5, as it raises a number of issues that run through Chapters 6 and 7. One, which we noted above, is the dynamic nature of connections which may serve to transform cities in any number of ways. Another, equally relevant issue – one that emerges from earlier points regarding the active production of networks and city powers – is that just as a city's place in a network is actively made, a city may just as easily fall out of such a network. Equally, a city may find that a particular network is no longer of significant advantage. Moscow is a case in point.

The emphasis on networks, whether these be 'city networks' or 'networks of cities', focuses attention on what it is that is passing through these networks, what is active within the connections, and how all this activity, this dynamism, can influence processes of urbanization.

For example, as we saw in Chapter 3, networks of migration affect the cultural and social mix of cities, which in turn produces new social geographies and new times and spaces within cities. Similarly, Chapters 6 and 7 drew attention to the 'big ideas' and influences – 'conventions' as they are termed in these chapters – that are actively binding cities into particular urban futures, or at least attempting to do so. In Chapter 7, for instance, it was argued that amidst the rise of neo-liberalism, cities have played two important roles as the 'hunter and the hunted': 'Cities are both the spaces *upon* which neo-liberal practices of governmentality and industry operate and the spaces *through* which they operate' (section 2). We saw too that these ideas and practices about how supposedly to run efficient economies, and how to think about economic organization, are not simply passive; as they filter into cities, they actively disrupt daily routines and rhythms. We explored this further by recognizing how today's 'economic' connections in many senses actively set and change the rhythm of time and can thus alter the day-to-day experience of city life.

ACTIVITY 8.2 Briefly reconsider how you would now use the term 'openness' to explain the varied processes that form and reform contemporary cities. Consider too the sorts of qualification you might suggest to the claim that 'the future of cities is open'. ◆

The open nature of cities, as places through which the activities and influences of the global economy pass and reverberate, can be seen to make them spaces of instability and contingency, just as easily as offering them the potential to become spaces of opportunity and empowerment. Again, these are points which came through in Reading 8A.

It would seem that in today's global economy, immediate city futures are bound into a mix of uncertain opportunities, chiefly because they are permeable places recreated through the constant exposure of often unequal cultural, social and economic pasts to global influences. The idea of the openness of city futures needs, then, to be qualified by recognizing that every city will be the site of very particular encounters between specific webs of social, cultural and economic histories and the tendencies of globalization. Such futures, therefore, will not take a linear form, for just as there was no single Asian miracle, so there is no single city future.

REFERENCES

Chatterjee, P. (1998) 'A new economic reality on Asian city streets', *Urban Age*, vol.5, no.4, pp.6–9.

Massey, D. (1999) 'On space and the city', in Massey, D., Allen, J. and Pile, S. (eds) *City Worlds*, London, Routledge/The Open University (Book 1 in this series).

READING 8A
Patralekha Chatterjee: 'A new economic reality on Asian city streets'

Across Southeast Asia, cities are reeling from the impact of the financial collapse that struck much of the region last year. Half-finished buildings dot the landscapes, and closed businesses line many streets. Residents who had enjoyed prosperous times only a few months ago now face a much darker future. Many have seen their savings disappear and their carefully built financial portfolios turn to dust. Some have lost their jobs – even their homes – and are scrambling to build new lives. Although local governments and international organizations are trying to control the situation and some of the economic free-fall appears to be under control, city dwellers across the region say that their lives have changed forever.

The impact is so pervasive that the very character of some cities has changed. There are fewer Mercedes sedans on the streets of Bangkok because the owners could not pay their loan installments and banks repossessed the cars. A visitor to this city, once infamous for its traffic congestion, can now get from the airport to the central district in about an hour. Cranes still dominate the Bangkok skyline, but they are idle. The construction boom is over – one of many unavoidable reminders that Thailand's economy and its people are suffering …

The sharp dip in purchasing power has meant radical lifestyle changes for many Bangkok residents. Thailand witnessed phenomenal growth in the 1980s and early 1990s. In 1988, the country's gross domestic product growth hit a record 13.2 per cent. The boom years also saw a growing separation between rich and poor, with the rich clustered around thriving Bangkok. Foreign money poured in. Hotels, luxury condos and gleaming office complexes shot up by the dozens into the city's polluted skies. The streets were choked with new automobiles, and the new middle class developed a taste for imported goods.

If the great Asian meltdown was only about giving up gold watches, designer clothes and imported chocolates, the region's politicians and business magnates would sleep more soundly. But, as firms go bust or downsize, labour unrest is rearing its head. Demonstrators are a familiar sight in front of Thai prime minister Chuan Leekpai's office. A posse of policemen stand guard. One afternoon, there is a mammoth gathering of women. Many wear straw hats typical of migrants from Thailand's poorer northeastern regions. They are workers from a toy factory. They chant protest songs to the beat of a drum, calling for restoration of overtime and annual bonuses.

So far, these demonstrations have been peaceful, in contrast to the food riots of Indonesia. But as lay-offs mount and pay packets shrink, it is hard to tell what lies ahead. Last year, the Japanese company Sanyo decided to give one month's bonus instead of the usual three. Shortly thereafter, one of its factories was burned down.

In Indonesia, a rise of almost 24 per cent in food prices over the past ten months has sparked social unrest at levels not seen for 30 years. This year alone, violence broke out in at least 12 Indonesian towns because of increases in prices of essential goods. In mid-February, there were reports of at least three people killed and 154 detained for looting in provincial towns.

The worst damage was in the coastal town of Pamanukan, in Java, where shops, religious places and doctors' surgeries were set on fire or damaged, according to *Suara Pembaruan*, a local newspaper. Recently, groups of young men reportedly attacked stores owned by ethnic Chinese shopkeepers in Eastern and Central Java along a highway that leads to Surabaya, the nation's second largest city.

Early in February, hundreds were reported to be chanting 'hungry, hungry' at a rally in Jakarta. Protesters have marched to the office of the state agency that distributes rice and other basic commodities. But displays of force by truncheon-wielding riot police have protected the Indonesian capital from violence – so far.

'We are still in the early stages. Food price rises hurt because, beyond a point, you cannot cut down on food consumption. The food inflation has been the worst in Indonesia', said Singapore-based Rajiv Malik, regional economist for the investment firm of Jardine Fleming.

Despite the bail-out plan backed by the International Monetary Fund (IMF), Jakarta faces a triple-digit percentage price explosion for imported food staples in the near future. This could further fuel anger from Indonesians who are already struggling with a 73 per cent decline in the value of their currency – the rupiah – against the US dollar since July.

On the face of it, it looks like a no-win situation. When there's a drought and crops fail, people traditionally move to the cities in search of jobs. When inflation and layoffs hit cities, people return to the countryside. Now, thousands of Indonesians are victims of both a drought and the debt crisis. 'To stem urban tension, the government is creating jobs in the countryside, heavily subsidizing rice and offering one-way tickets out of Jakarta', one recent news report noted.

The Indonesian cityscape may deteriorate sharply after April. Increases in fuel and electricity tariffs, stipulated by IMF, will begin. At about the same time, state-held monopolies will be eliminated, which will erase subsidies that have kept the prices of key imported commodities low. Rupiah prices may jump 250 to 500 per cent for items such as sugar, soybeans and wheat flour, unless the Indonesian government steps in to cushion the blow. Currently, Indonesia imports 100 per cent of its wheat, one-third of its sugar and soybeans, and as much as 10 per cent of its rice. Rice production in Indonesia has been steadily declining as thousands of hectares of agricultural land are converted to industrial and residential uses every year.

Many workers in Indonesian cities found locked gates and no jobs upon returning from their annual holiday in February. In Jakarta, a spokesman of the Indonesian Manpower Ministry said manufacturers could not afford a much-needed minimum wage increase this year, leaving the minimum monthly salary at 172,000 rupiah, or $16.70 at current exchange rates.

Many infrastructure projects are also being kept on hold. In Bangkok, one of the biggest casualties may be the elevated railway that was supposed to ease traffic congestion. Another may be the construction of a second airport.

Private hospitals, once a thriving business in Bangkok, are also in dire straits, as patients seek cheaper care. 'People are more careful about spending the money, even in matters of health. Instead of visiting hospitals for minor illnesses, they would purchase medicines from drugstores', says Dr Viroj Tangcharoensathien of the Health Systems Research Institute.

Medical benefits of many middle-class, white-collar workers are likely to be reduced, forcing them to shift to government hospitals. Many private hospitals face closure. The loss is evident among hospitals registered under the Stock Exchange of Thailand (SET). Only two hospitals registered on the SET made any profit last year. The other nine have all registered losses of more than five million baht. They had sought foreign loans to open new branches. Now they cannot repay their debts.

The economic downturn is spawning novel ideas in urban lifestyles across Southeast Asia. 'With living costs rising, cultivating your own veggies is a great way to save cash', says a report in the Bangkok-based

Nation newspaper. In Kuala Lumpur, Malaysian deputy prime minister Anwar Ibrahim feels the same. Ibrahim has asked people to start vegetable patches at home. This would, in part, help reduce Malaysia's food import bill, estimated at ten billion ringgit, or $3.85 billion. Malaysia currently imports 40 per cent of its vegetables, according to the *New Straits Times*.

Despite the gloomy stories one hears in the streets of Bangkok, in South Korea – the cradle of the economic firestorm ravaging Southeast Asia since last July – the economy is off to a good start in 1998. Morgan Stanley Capital International Indices noted that South Korea ranked first in terms of performance in the emerging markets, with its stock indices in January rising by 70.6 per cent. Thailand came second with a growth of 34.2 per cent in the month.

Political change

In both countries, economic troubles have ushered in political changes. Economic turmoil brought down one government in Bangkok, and a reform-minded president has taken office in South Korea. Popular acclaim has greeted the new Chuan Leekpai government in Thailand and the Kim Dae Jung administration in South Korea. Thailand has adopted a new constitution, and analysts say financial and legislative reforms in both countries will augur well for future stability. But even the most dogged optimists concede that it is still a long haul before shiny indices translate into good times for ordinary people.

In many ways, South Korea seems to have turned the corner. The value of the national currency, the won, has rebounded with the political reforms. An agreement has been reached with a group of international banks to roll over $24 billion in short-term debt.

But Seoul's decision to facilitate retrenchment as part of an economic restructuring package is likely to face stiff opposition. As businesses step up the pace of layoffs, feminist groups in the South Korean capital say women are often the first in the firing line. Choi Myong-Sook, senior executive at Women Link – a group dedicated to women's welfare in South Korea – says that since last November she has received up to 20 calls a day from depressed or angry women who lost their jobs only because they were women. Until the crisis struck, South Korea had a labour shortage. At the end of 1997, its unemployment rate had risen to 3 per cent.

The cities, more than the countryside, reflected the Asian economic miracle. The cities now bear the brunt of the downturn. Bangkok residents 'are more affected than others because the credit system collapsed, the banks and financial systems collapsed. But the middle class still has to pay back car loans and housing instalments. Many among the salaried class have lost their jobs; they have to live on their savings', says Thanong Khantong, assistant editor of the *Nation*.

The recession is likely to stem the tide of people moving from villages to cities. During Thailand's boom years, thousands from the country's impoverished northeast regions flocked to Bangkok in search of work. Now many find themselves jobless and are packing their bags to go home. But the question is whether the agricultural sector can absorb them. A lot of people who left their farms to work in cities have forgotten how to grow rice. Will they be forced to return to the cities, thus fuelling more urban tension? Time alone will tell.

As residents in Bangkok and cities across the region get used to new realities and less affluent lifestyles, one can sense a mixture of unrest and optimism on the streets. Offices and apartments blocks may be half-empty, but traffic is flowing more smoothly in the Thai capital. The urge to splurge may, like bank balances, be low, but middle-class families are discovering the virtues of staying home with the family. In countries where there has been a change of guard politically, the mood is upbeat. As new leaders, backed by popular mandate, try to push through long-needed reforms, one gets the feeling that it is too soon to write off the East Asian miracle. The next phase may not be as dazzling as the first, but it will be more sustainable.

Source: Chatterjee, 1998, pp.6–9

Acknowledgements

Grateful acknowledgement is made to the following sources for permission to reproduce material in this book:

CHAPTER 1

Text

Cosgrove, D. (1996) 'Windows on the city', *Urban Studies*, 33 (8), Carfax Publishing Limited, PO Box 25, Abingdon, Oxfordshire, OX14 3UE; Hannerz, U. (1996) *Transnational Connections, Culture, People, Places*, Routledge; Madon, S. (1997) 'Information-based global economy and socioeconomic development: the case of Bangalore', *The Information Society*, 13 (3), Taylor and Francis Ltd.

Figures

Figure 1.2: from *Intelligent Environments: Spatial Aspects of the Information Revolution*, edited by Peter Droege, Elsevier, Amsterdam, 1997, page 31; photo: Dan Wiley; *Figure 1.3:* Getty Images; *Figure 1.4:* from *Fortress America – Gated Communities in the United States* by Edward J. Blakely and Mary Gail Snyder, Brookings Institution Press, Cambridge, Mass., 1997; photo: Edward J. Blakely; *Figure 1.5:* photo: Sue Cunningham Photographic; *Figure 1.6:* Heldur Netocny/ Panos Pictures; *Figure 1.7:* © Mick Brownfield.

CHAPTER 2

Text

UNCHS (1996) *An Urbanizing World: Global Report on Human Settlements 1996: Executive Summary*, United Nations Centre for Human Settlements; Sit, V.F.S. (1996) 'Beijing: urban transport issues in a socialist Third World setting (1949–1992)', *Journal of Transport Geography*, 4 (4), December 1996, Elsevier Science Ltd. All rights reserved; Klein, J. and Olson, M. (1997) *Taken for a Ride*, Community Media Productions; Hau, T. D. (1995) 'Transport for urban development in Hong Kong', *Transport and Communications for Urban Development, Report of the Habitat II Global Workshop*, United Nations Centre for Human Settlement; Graham, S. (1997) 'Urban planning in the "information society" ', *Town and Country Planning*, 66 (11), Town and Country Planning Association.

Figures

Figure 2.1: Venezia, 1572, from *The City Maps of Europe*, Braun and Hogenburg, 1991, vol.1, no.43, pp.116–17. © Random House, London; *Figure 2.2:* Venetian 'toppo'; photo: Osvaldo Böhm Fotografo, Venice; *Figure 2.3:* photo: J. Allan Cash Ltd; *Figure 2.4:* photo: Julio Etchart/Panos Pictures; *Figure 2.5:* © The

Advertising Archive Ltd; *Figure 2.6:* UNCHS (1995) *Economic Instruments and Regulatory Measures for the Demand Management of Urban Transport*, United Nations Centre for Human Settlement; *Figure 2.7:* Faiz, A. and Gautam, S. P. (1994) *Motorization, Urbanization and Air Pollution – Discussion Paper*, The World Bank; *Figure 2.8:* Canal and Broadway by Alan Wolfson, 1982, 6.5 x 11.5 x 11 inches, mixed media diorama; photograph: courtesy Louis K. Meisel Gallery, New York; *Figure 2.9:* photo: Susan Hoyle.

Table

Table 2.2: from 'Transport and the taxi mafia in South Africa', by Khosa and Mill in *Urban Edge*, 1.

CHAPTER 3

Text

Maspero, F. (1994) in Jones, P. (trans.) *Roissy Express: A Journey Through the Paris Suburbs,* Verso.

Figures

Figure 3.1: © Editions du Seuil, 1990, in François Maspero, *Les Passagers du Roissy-Express; Figures 3.2, 3.3, 3.4 and 3.5:* © Anaïk Frantz.

CHAPTER 4

Text

Simon, J. (1996) 'Mexico City trashed', in Dion, M. and Rockman, A. (eds) *Concrete Jungle*, Juno Books; Wilson, E. (1995) 'The rhetoric of urban space', *New Left Review*, 209, January/February 1995, New Left Review Ltd, London; Wolch, J.R., West, K. and Gaines, T.E. (1995) 'Transspecies urban theory', *Environment and Planning D*, 13 (6), pp.746–9, Pion Ltd, London.

Figures

Figure 4.1: photo: Mary Evans Picture Library; *Figure 4.3:* Pysek, P. (1993) 'Factors affecting the diversity of flora and vegetation in central European settlements', *Vegetatio*, 106, Fig.2, p.92, © 1993 Kluwer Academic Publishers. With kind permission from Kluwer Academic Publishers; *Figures 4.5 and 4.6:* adapted from 'Social constructions of nature: a case study of conflicts over the development of Rainham Marshes' by Carolyn M. Harrison and Jacqueline Burgess, in *Transactions of the Institute of British Geographers*, 19 (3), 1994; *Figure 4.7:* courtesy of Stoke-on-Trent City Council.

CHAPTER 5

Text

Reprinted by permission of the publisher from Worlds of Production by Michael Storper and Robert Salais, Cambridge, Mass.: Harvard University Press, Copyright © 1997 by the President and Fellows of Harvard College; Richardson, R. (1997) *Towards the Information Society in Southeast Asia? Experiences in Singapore and Malaysia*, Centre for Urban and Regional Development Studies, University of Newcastle; *Unlocking the Full Potential of the Information Age* (1998) Multimedia Super Corridor, Multimedia Development Corporation; Sassen, S. (1995) 'On concentration and centrality in the global city', in Knox, P.L. and Taylor, P.J. (eds) *World Cities in a World-System*, Cambridge University Press; Skeldon, R. (1997) 'Hong Kong: colonial city to global city to provincial city?', *Cities*, 14 (5), with permission from Elsevier Science.

Figures

Figure 5.1: reproduced from: M.F. Pochna, *Dior*, 1996, Editions Assouline, Paris; photo: Laziz Hamani; *Figure 5.2:* photo: Spectrum Colour Library; *Figures 5.3 and 5.5:* Friedmann, J. (1986) 'The world city hypothesis (editor's introduction)', *Development and Change*, 17 (1), Sage Publications Ltd. Copyright © 1986 Institute of Social Studies; *Figure 5.4:* Llewelyn-Davies, UCL Bartlett School of Planning and Comedia (1996) *Four World Cities: A Comparative Study of London, Paris, New York and Tokyo*, © Crown Copyright is reproduced with the permission of the Controller of Her Majesty's Stationery Office. With special thanks for help from Bally Meeda at Llewelyn-Davies; *Figure 5.6:* photo: Michael Pryke; *Figure 5.7:* photo: Tayacan/Panos Pictures; *Figure 5.8 (right): Cyberjaya: The Model Intelligent City in the Making,* Multimedia Super Corridor, Multimedia Development Corporation, Cyberjaya; *Figure 5.9 (top):* photo: David Read/Panos Pictures; *Figure 5.9 (bottom):* photo: Bruce Paton/Panos Pictures; *Figure 5.10 (right, top and bottom):* Spectrum Colour Library; (*left, top*) G.R. Richardson/Spectrum Colour Library; (*left, bottom*) D & J Heaton/Spectrum Colour Library.

CHAPTER 6

Text

Kanji, N. (1995) 'Gender, poverty and economic adjustment in Harare, Zimbabwe', *Environment and Urbanization*, 7 (1), International Institute for Environment and Development; Tripp, A.M. (1994) 'Deindustrialization and the growth of women's economic associations and networks in urban Tanzania', in Rowbotham, S. and Mitter, S. (eds) *Dignity and Daily Bread*, Routledge; Gilbert, A. (1997) 'Work and poverty during economic restructuring: the experience of Bogotá, Colombia', *IDS Bulletin*, 28 (2), The Institute of Development Studies, University of Sussex.

Figures

Figure 6.1: photo: Phillip Edwards/Panos Pictures; *Figure 6.2:* New York Stock Exchange, photo: Spectrum Colour Library. Wall Street view, Washington, DC and London, photos: Michael Pryke. West Shinjuku Business District, Tokyo, photo: Chris Stowers/Panos Pictures. São Paulo Stock Exchange traders, Associated Press/Dario Lopez-Mills; *Figure 6.3:* from *The State in a Changing World, World Development Report, 1997*, Oxford University Press, Inc; *Figure 6.4:* photo: Ron Giling/Panos Pictures; *Figure 6.5:* photo: A.A. Johnson/Spectrum Colour Library; *Figure 6.6a:* photo: Jeremy Horner/Panos Pictures; *Figure 6.6b:* photo: Paul Mark Smith/Panos Pictures.

Tables

Tables 6.1 and 6.2: The World Bank (1997) *Global Development Finance 1997*, The International Bank for Reconstruction and Development/The World Bank.

CHAPTER 7

Text

Mitchell, K. (1997) 'Transnational subjects', in Ong, A. and Nonini, D.M. (eds) *Ungrounded Empires, The Cultural Politics of Modern Chinese Transnationalism*, Routledge; Reprinted with the permission of Simon and Schuster from *World Class* by Rosabeth Moss Kanter. Copyright © 1995 by Rosabeth Moss Kanter; Reprinted by permission of Harvard Business Review. Excerpted from 'The competitive advantage of the inner city', by Michael E. Porter, May/June 1995. Copyright © 1995 by the President and Fellows of Harvard College, all rights reserved.

Figures

Figure 7.1: advert for World Economic Forum, courtesy of Thomas Scherer Human Resources, Switzerland; *Figure 7.2:* Reproduced from *The Economist*, 19 April 1997; *Figure 7.3:* Reuters/Popperfoto; *Figure 7.5:* photo: David Loh/ Reuters/Popperfoto; *Figure 7.6:* © South China Morning Post; *Figure 7.7:* photo: Tourism British Columbia; *Figure 7.8:* MK Let Net, City Discovery Centre, Milton Keynes.

Table

Table 7.1: Clarke, J. and Newman, J. (1997) *The Managerial State: Power, Politics and Ideology in the Remaking of Social Welfare*, Sage Publications Ltd. © John Clarke and Janet Newman 1997.

CHAPTER 8

Text

Chatterjee, P. (1998) 'A new economic reality on Asian city streets', *Urban Age*, 5 (4), reprinted with permission, Urban Age magazine, 1998.

COVER

Photo: Oldrich Karasek/Tony Stone Images

Index